T0332122

Managing Resources for Futuristic Wireless Networks

Mamata Rath
Birla School of Management, Birla Global University, India

A volume in the Advances in
Wireless Technologies and
Telecommunication (AWTT) Book
Series

Published in the United States of America by
    IGI Global
    Information Science Reference (an imprint of IGI Global)
    701 E. Chocolate Avenue
    Hershey PA, USA 17033
    Tel: 717-533-8845
    Fax: 717-533-8661
    E-mail: cust@igi-global.com
    Web site: http://www.igi-global.com

Library of Congress Cataloging-in-Publication Data

Names: Rath, Mamata, editor.
Title: Managing resources for futuristic wireless networks / Mamata Rath,
    editor.
Description: Hershey, PA : Information Science Reference(an imprint of IGI
    Global), [2020] | Summary: "This book examines the issues, algorithms,
    and solutions for achieving best resource utilization in vehicular ad
    hoc networks"-- Provided by publisher.
Identifiers: LCCN 2019007937 | ISBN 9781522594932 (h/c) | ISBN
    9781522594949 (s/c) | ISBN 9781522594956 (eISBN)
Subjects: LCSH: Vehicular ad hoc networks (Computer networks)--Data
    processing. | Wireless communication systems--Data processing.
Classification: LCC TE228.37 .M36 2020 | DDC 004.6--dc23
LC record available at https://lccn.loc.gov/2019007937

This book is published in the IGI Global book series Advances in Wireless Technologies and
Telecommunication (AWTT) (ISSN: 2327-3305; eISSN: 2327-3313)

British Cataloguing in Publication Data
A Cataloguing in Publication record for this book is available from the British Library.

All work contributed to this book is new, previously-unpublished material.
The views expressed in this book are those of the authors, but not necessarily of the publisher.

For electronic access to this publication, please contact: eresources@igi-global.com.

Advances in Wireless Technologies and Telecommunication (AWTT) Book Series

ISSN:2327-3305
EISSN:2327-3313

Editor-in-Chief: Xiaoge Xu, University of Nottingham Ningbo China, China

MISSION

The wireless computing industry is constantly evolving, redesigning the ways in which individuals share information. Wireless technology and telecommunication remain one of the most important technologies in business organizations. The utilization of these technologies has enhanced business efficiency by enabling dynamic resources in all aspects of society.

The **Advances in Wireless Technologies and Telecommunication Book Series** aims to provide researchers and academic communities with quality research on the concepts and developments in the wireless technology fields. Developers, engineers, students, research strategists, and IT managers will find this series useful to gain insight into next generation wireless technologies and telecommunication.

COVERAGE

- Mobile Communications
- Broadcasting
- Cellular Networks
- Global Telecommunications
- Wireless Broadband
- Mobile Technology
- Radio Communication
- Virtual Network Operations
- Network Management
- Mobile Web Services

IGI Global is currently accepting manuscripts for publication within this series. To submit a proposal for a volume in this series, please contact our Acquisition Editors at Acquisitions@igi-global.com or visit: http://www.igi-global.com/publish/.

Titles in this Series

For a list of additional titles in this series, please visit:
https://www.igi-global.com/book-series/advances-wireless-technologies-telecommunication/73684

Wireless Sensor Network-Based Approaches to Digital Image Processing in the IoT
Manoj Diwakar (DIT University, India) and Kaushik Ghosh (DIT University, India)
Information Science Reference • © 2020 • 300pp • H/C (ISBN: 9781799832812) • US
$195.00

Innovative Perspectives on Interactive Communication Systems and Technologies
Muhammad Sarfraz (Kuwait University, Kuwait)
Information Science Reference • © 2020 • 330pp • H/C (ISBN: 9781799833550) • US
$195.00

Handbook of Research on the Political Economy of Communications and Media
Serpil Karlidag (Baskent University, Turkey) and Selda Bulut (Ankara Haci Bayram Veli
University, Turkey)
Information Science Reference • © 2020 • 404pp • H/C (ISBN: 9781799832706) • US
$195.00

Forensic Investigations and Risk Management in Mobile and Wireless Communications
Kavita Sharma (National Institute of Technology Kurukshetra, India) Mitsunori Makino
(Chuo University, Japan) Gulshan Shrivastava (National Institute of Technology Patna, India)
and Basant Agarwal (Indian Institute of Information Technology Kota, India)
Information Science Reference • © 2020 • 314pp • H/C (ISBN: 9781522595540) • US
$195.00

*IoT and WSN Applications for Modern Agricultural Advancements Emerging Research
and Opportunities*
Proshikshya Mukherjee (KIIT University (Deemed), India) Prasant Kumar Pattnaik (KIIT
University (Deemed), India) and Surya Narayan Panda (Chitkara University, India)
Engineering Science Reference • © 2020 • 145pp • H/C (ISBN: 9781522590040) • US
$175.00

701 East Chocolate Avenue, Hershey, PA 17033, USA
Tel: 717-533-8845 x100 • Fax: 717-533-8661
E-Mail: cust@igi-global.com • www.igi-global.com

Editorial Advisory Board

Table of Contents

Detailed Table of Contents

Chapter 1

Vaishali Vilas Sarbhukan, Terna Engineering College, University of
Mumbai, India

Lata Ragha, Father Conceicao Rodrigues Institute of Technology,
University of Mumbai, India

The mobility and scalability brought by wireless networks made it possible in many applications. In MANET, discovering the stable, secure, and reliable routes is a challenging research problem due to the open nature of wireless communications. In Zhexiong Wei Routing Trust scheme (ZWRT), Bayesian approach and Dempster Shafer theory is used to evaluate more realistic trust value. Mohamed et al proposed Establishing STAble and Reliable Routes (ESTAR) system in heterogeneous multihop wireless networks. The above existing methods failed to solve the complete problem of information loss in MANETs. There are several other reasons due to which the information may loss in MANETS, such as mobility and congestion of mobile nodes. Therefore, the novel Enhanced Trust-based Secure Routing (ETSR) scheme is presented to deliver stable, reliable, and secure data communication in MANETs. The simulation results demonstrate that ETSR routing protocol improve QoS performance as compared to state-of-the-art techniques.

Chapter 2

Utkarsha Sumedh Pacharaney, Datta Meghe College of Engineering, India
Ranjan Bala Jain, Vivekanand Education Society's Institute of
*   Technology, India*
Rajiv Kumar Gupta, Terna Engineering College, University of Mumbai,
*   India*

The chapter focuses on minimizing the amount of wireless transmission in sensory data gathering for correlated data field monitoring in wireless sensor networks (WSN), which is a major source of power consumption. Compressive sensing (CS) is a new in-node compression technique that is economically used for data gathering in an energy-constrained WSN. Among existing CS-based routing, cluster-based methods offer the most transmission-efficient architecture. Most CS-based clustering methods randomly choose nodes to form clusters, neglecting the topology structure. A novel base station (BS)-assisted cluster, spatially correlated cluster using compressive sensing (SCC_CS), is proposed to reduce number of transmissions in and form the cluster by exploiting spatial correlation based on geographical proximity. The proposed BS-assisted clustering scheme follows hexagonal deployment strategy. In SCC_CS, cluster heads are solely involved in data gathering and transmitting CS measurements to BS, saving intra-cluster communication cost, and thus, network life increases as proved by simulation.

Chapter 3

Meena T., VIT-AP University, India
Ravi Sankar Sangam, VIT-AP University, India

In recent days, the usage of drones was increased and extended to various domains such as surveillance, photography, military, rescue, etc. Drones are small flying computers with on-board sensors and camera with a limited battery and coverage area. Due to the limited coverage area, usage of standalone drones in the above-mentioned domains such as rescue and military is restrictive. Multi-drones with self-organizing network can help to solve the above discussed issues. Hence, this chapter presents an extensive review on drone networks in which the core areas such as coverage, connectivity, link establishment, etc. are discussed. Finally, this chapter concludes by leveraging the challenges in state-of-the-art technologies in drone networking.

Chapter 4

Hakima Khelifi, Beijing Institute of Technology, China
Senlin Luo, Beijing Institute of Technology, China
Boubakr Nour, Beijing Institute of Technology, China
Hassine Moungla, Paris Descartes University, France
Syed Hassan Ahmed, Georgia Southern University, USA

The challenging characteristics of the vehicular environment such as high mobility, diversity of applications, dynamic topologies, unreliable broadcast channels, and short-lived connectivity call into the need to extend the IP-based network to fulfill the user and VANETs requirements. Researchers are developing new network communication models to transfer the future internet. The information-centric networking (ICN) paradigm is a promising solution that may overcome the issues mentioned above. ICN involves a named content, name-based routing, in-network caching, and content-based security, which make it a suitable architecture for VANET applications. In this chapter, the authors present recent advances in VANET solutions that rely on named-data networking (NDN), which is the most active ICN implementation. The issues of the current host-centric model, mapping between NDN and VANET, is also discussed along with future research directions.

Chapter 5

Daniel Minoli, DVI Communications, USA
Benedict Occhiogrosso, DVI Communications, USA

Cyber physical systems (CPSs) are software-intensive smart distributed systems that support physical components endowed with integrated computational capabilities. Tiered, often wireless, networks are typically used to collect or push the data generated or required by a distributed set of CPS-based devices. The edge-to-core traffic flows on the tiered networks can become overwhelming. Thus, appropriate traffic engineering (TE) algorithms are required to manage the flows, while at the same time meeting the delivery requirements in terms of latency, jitter, and packet loss. This chapter provides a basic overview of CPSs followed by a discussion of a newly developed TE method called 'constrained average', where traffic is by design allowed to be delayed up to a specified, but small value epsilon, but with zero packet loss.

Chapter 6

Muhammad Irfan, COMSATS University Islamabad, Wah Campus, Pakistan
Ayaz Ahmad, COMSATS University Islamabad, Wah Campus, Pakistan
Raheel Ahmed, COMSATS University Islamabad, Wah Campus, Pakistan

Single carrier frequency division multiple access (SC-FDMA) is a promising uplink transmission technique that has the characteristic of low peak to average power ratio. The mobile terminal uplink transmission depends on the batteries with limited power budget. Moreover, the increasing number of mobile users needs to be accommodated in the limited available radio spectrum. Therefore, efficient resource allocation schemes are essential for optimizing the energy consumption and improving the spectrum efficiency. This chapter presents a comprehensive and systematic survey of resource allocation in SC-FDMA networks. The survey is carried out under two major categories that include centralized and distributed approaches. The schemes are also classified under various rubrics including optimization objectives and constraints considered, single-cell and multi-cell scenarios, solution types, and perfect/imperfect channel knowledge-based schemes. The advantages and limitations pertaining to these categories/rubrics have been highlighted, and directions for future research are identified.

Chapter 7

Puspanjali Mallik, Shailabala Women's Autonomous College, India

The internet of things (IoT) fulfils abundant demands of present society by facilitating the services of cutting-edge technology in terms of smart home, smart healthcare, smart city, smart vehicles, and many more, which enables present day objects in our environment to have network communication and the capability to exchange data. These wide range of applications are collected, computed, and provided by thousands of IoT elements placed in open spaces. The highly interconnected heterogeneous structure faces new types of challenges from a security and privacy concern. Previously, security platforms were not so capable of handling these complex platforms due to different communication stacks and protocols. It seems to be of the utmost importance to keep concern about security issues relating to several attacks and vulnerabilities. The main motive of this chapter is to analyze the broad overview of security vulnerabilities and its counteractions. Generally, it discusses the major security techniques and protocols adopted by the IoT and analyzes the attacks against IoT devices.

Chapter 8

Mirza Waseem Hussain, Baba Ghulam Shah Badshah University, India
Sanjay Jamwal, Baba Ghulam Shah Badshah University, India
Tabasum Mirza, Baba Ghulam Shah Badshah University, India
Malik Mubasher Hassan, Baba Ghulam Shah Badshah University, India

The communication platform in the computing field is increasing at a rapid pace. Technology is constantly budding with the materialization of new technological devices, specifically in the communication industry. The internet is expanding exponentially. Internet-enabled devices are becoming part and parcel of our daily lives. It has turned out to be almost impossible to think about the world without the internet. The internet structures might be reinforced to meet coming prerequisites in mobile communication. Congestion plays a vital role in regulating the flow of data to accelerate the exchange of data in between the wired and wireless devices. In this chapter, the authors try to highlight various network congestion techniques with their limitations proposed from time to time by various researchers. This chapter plays a vital role in highlighting the history of networking congestion detection/avoidance techniques starting from the early days of networking.

Chapter 9

Archana Sharma, Institute of Management Studies Noida, India

Truthful authentication with secure communication is necessary in location-based services to protect from various risks. The purpose of this research is to identify security risks in mobile transactions especially in location-based services like mobile banking. The factors need to be identified the reasons of customer distrust in mobile banking. In addition, the security issues with mobile banking systems and mobile devices are highlighted. The chapter finds which approach is more suitable and secure for mobile banking transaction between customer and bank. The research predominantly focuses upon customer trust, security issues, and transaction costs owing to different technology standards of mobile commerce. The first phase highlights the various location-based services in m-commerce, various technology standards, customer trust, and perceived risk, and further, at next level, it highlights the various problems associated mobile database and a comparative study of various replication protocols, transaction security issues, and LBS security challenges.

Chapter 10

L. Naga Durgaprasad Reddy, Sri Yerramilli Narayanamurthy College, India

This chapter researches in the area of software-defined networking. Software-defined networking was developed in an attempt to simplify networking and make it more secure. By separating the control plane (the controller)—which decides where packets are sent—from the data plane (the physical network)—which forwards traffic to its destination—the creators of SDN hoped to achieve scalability and agility in network management. The application layer (virtual services) is also separate. SDN increasingly uses elastic cloud architectures and dynamic resource allocation to achieve its infrastructure goals.

Chapter 11
  Jhum Swain, Institute of Technical Education and Research, Bhubaneswar, India

A mobile ad hoc network (MANETs) is an assortment of a variety of portable nodes that are linked collectively in a greater number in a wireless medium that has no permanent infrastructure. Here, all the nodes in the node partake in acting as both router and host and is in charge for accelerating packets to other nodes. This chapter discusses the various attacks on different layers and on various security protocols. So, designing a secure routing protocol is a main challenge in MANET. As we all know, this is a mobile ad hoc network so nodes in the network dynamically establish paths among each other so it is vulnerable to different kinds of threats. So, in this case, we need secured communication among the nodes present in the network.

Chapter 12
  Mamata Rath, Birla School of Management, Birla Global University, India
  Sushruta Mishra, Kalinga Institute of Industrial Technology, India

Vehicular ad hoc networks (VANETs) have evolved as an invigorating network system and application domain in current communication technology. In smart city applications context, there are smart vehicles embedded with sensors and dynamically programmed IoT devices, which are to be managed and controlled energetically. Progressively, vehicles are being furnished with surrounded actuators, handling signals, and wireless communication abilities. This chapter focuses on the fact that this special network has opened various possible outcomes for intense and potential extraordinary applications on security, effectiveness, comfort, confidentiality effort, and interest while they are significantly vibrant. Irrespective of many challenges such as high frequency of topology change and link failure possibility, routing management in VANET has been successful in traffic scenario during vehicle-to-vehicle communication.

Preface

OVERVIEW

While planning about the management of resources in futuristic wireless networks, the key parameter which needs to be considered is balanced approach of resource distribution. Balanced approach is necessary to provide an unbiased working environment for the distribution, sharing, allocation, and supply of resources among the devices of the wireless network. From traditional to emerging wireless networks, equality of distribution plays an important role at every layer of communication. Equal resource distribution maintains balance and stability between the operations of communication systems and thus improves the performance of wireless networks. Proper resource management is a key factor to achieve the optimal results for the parameters like throughput, Quality of Service (QoS), Spectrum utilization, Power, and Resource Management in wireless network.

The recent era of communication systems has introduced promising wireless technologies and concepts. These emerging technologies have evolved the concept of futuristic wireless networks. For the smooth operation and performance of these networks, a key factor is to establish fairness among the devices of these networks. Right management of resource influences t the capabilities of these wireless network. The objective of this book is to present research articles and results related to control and management of key parameters of bandwidth, spectrum sensing, channel selection, resource sharing, and task scheduling.

TARGET AUDIENCE

The target audience of this book will be composed of professionals and researchers working in the field of networking, information and knowledge management in various disciplines, e.g. library, information and communication sciences, administrative sciences and management, education, adult education, sociology, computer science, and information technology. Moreover, the book will provide insights and support executives concerned with the management of expertise, knowledge, information and organizational development in different types of work communities and environments.

IMPORTANCE OF EACH OF THE CHAPTER SUBMISSIONS

Chapter 1 describes about mobile ad hoc networks security enhancement with trust management. The mobility and scalability brought by wireless network made it possible in many applications. In MANET, discovering the stable, secure and reliable routes is challenging research problem due to the open nature of wireless communications. In Zhexiong Wei Routing Trust scheme (ZWRT), Bayesian approach and Dempster Shafer Theory is used to evaluate more realistic trust value. Mohamed et al proposed Establishing STAble and Reliable Routes (ESTAR) system in heterogeneous multihop wireless networks. Above existing methods failed to solve the complete problem of information loss in MANETs. There are several other reasons due to which the information may loss in MANETS such as mobility and congestion of mobile nodes. Therefore the novel Enhanced Trust based Secure Routing (ETSR) scheme is presented to deliver stable, reliable and secure data communication in MANETs. The simulation results demonstrate that ETSR routing protocol improve QoS performance as compared to state of art techniques.

Chapter 2 is about clustering and compressive data gathering for transmission efficient wireless sensor networks. The chapter focuses on minimizing amount of wireless transmission in sensory data gathering for correlated data field monitoring in Wireless Sensor Network (WSN) which is major source of power consumption. Compressive Sensing (CS) is a new in-node compression technique which is economically used for data gathering in an energy constrained WSN. Among existing CS based routing, cluster-based methods offer most transmission efficient architecture. Most CS based clustering methods randomly choose nodes to form clusters, neglecting the topology structure. A novel Base Station (BS) assisted cluster, Spatially Correlated Cluster using Compressive Sensing (SCC_CS), is proposed to reduce number of transmissions in and form the cluster by exploiting spatial correlation based on geographical proximity. The proposed BS assisted clustering scheme follows hexagonal deployment strategy. In SCC_CS, cluster heads are solely involve in data gathering and transmitting CS measurements to BS, saving intra-cluster communication cost and thus network life increases as proved by simulation.

Chapter 3 is about study of self-organizing coordination for multi-UAV systems. In recent days, the usage of drones was increased and extended to various domains such as surveillance, photography, military, rescue etc. Drones are small flying computers with on-board sensors and camera with a limited battery and coverage area. Due to the limited coverage area, usage of standalone drones in the above mentioned domains such as rescue, military is restrictive. Multi drones with self-organizing network can help to solve the above discussed issues. Hence, this chapter presents an extensive review on drone network in which the core areas such as coverage, connectivity, link establishment etc., are discussed. Finally, this chapter concludes by leveraging the challenges in state of the art technologies in drone networking.

Chapter 4 is about vehicular networks in the eyes of future internet architectures. The challenging characteristics of the vehicular environment such as high mobility, diversity of applications, dynamic topologies, unreliable broadcast channels, and short-lived connectivity, call into the need to extend the IP-based network to fulfill the user and VANETs requirements. Researchers are developing new network communication models to transfer the future Internet. The Information-Centric Networking (ICN) paradigm is a promising solution that may overcome the issues mentioned above. ICN involves a named content, name-based routing, in-network caching, and content-based security which make it a suitable architecture for VANET applications. In this chapter, we present recent advances in VANET solutions that rely on Named-Data Networking (NDN), which is the most active ICN implementation. The issues of the current host-centric model, mapping between NDN and VANET is also discussed along with future research directions.

Chapter 5 is about constrained average design method for QoS-based traffic engineering at the edge/gateway boundary in VANETs and cyber-physical environments. Cyber physical systems (CPSs) are software-intensive smart distributed systems that support physical components endowed with integrated computational capabilities. Tiered, often wireless, networks are typically used to collect or push the data generated or required by a distributed set of CPS-based devices. The edge-to-core traffic flows on the tiered networks can become overwhelming. Thus, appropriate Traffic Engineering (TE) algorithms are required to manage the flows, while at the same time meeting the delivery requirements in terms of latency, jitter, and packet loss. This chapter provides a basic overview of CPSs followed by a discussion of a newly developed TE method called 'constrained average', where traffic is by design allowed to be delayed up to a specified, but small value epsilon, but with zero packet loss.

Chapter 6 is about resource allocation techniques for SC-FDMA networks. Single Carrier Frequency Division Multiple Access (SC-FDMA) is a promising uplink transmission technique which has the characteristic of low Peak to Average Power Ratio. The mobile terminals' uplink transmission depends on the batteries with limited power budget. Moreover, the increasing number of mobile users needs to be accommodated in the limited available radio spectrum. Therefore, efficient resource allocation schemes are essential for optimizing the energy consumption and improving the spectrum efficiency. This chapter presents, a comprehensive and systematic survey of resource allocation in SC-FDMA networks. The survey is carried out under two major categories that includes centralized, and distributed approaches. The schemes are also classified under various rubrics including optimization objectives and constraints considered, single-cell and multi-cell scenarios, solution types, and perfect/ imperfect channel knowledge based schemes. The advantages and limitations pertaining to these categories/rubrics have been highlighted and directions for future research are identified.

Chapter 7 is about analysis of vulnerabilities in IoT and its solutions. The Internet of Things (IoT) fulfils abundant demands of present society by facilitating services of cutting edge technology in terms of smart home, smart healthcare, smart city, smart vehicles and many more which enables present day objects in our environment to have network communication and capability to exchange data. These wide range of applications are collected, computed and provided by thousands of IoT elements placed in open spaces. The highly interconnected heterogeneous structure faces new types of challenges from a security and privacy concern. Previously, security platforms were not so capable to handle these complex platforms due to different communication stacks and protocols. It seems to be the utmost important to keep concern about security issues relating to several attacks and vulnerabilities. The main motive of this review chapter is to analyze the broad overview of security vulnerabilities and it's counteractions. Generally, it discusses the major security techniques and protocols adopted by the IoT and analyzes the attacks against IoT devices.

Chapter 8 is about taxonomy of computer network congestion control/avoidance methods. The communication platform in the computing field is increasing at a rapid pace. Technology is constantly budding with the materialization of new technological devices, specifically in the communication industry. The Internet is expanding exponentially. Internet enabled devices are becoming part and parcel of our daily. It has turned out to be almost impossible to think about the world without the Internet. The Internet structures might be reinforced to meet coming prerequisites in mobile communication. Congestion plays a vital role in regulating the flow of data to accelerate the exchange of data in between the wired and wireless devices. In this chapter the author's try to highlight various network congestion techniques with their limitations proposed from time to time by various researchers. This chapter plays a vital role in highlighting the history of networking congestion detection/ avoidance techniques starting from early days of networking till date.

Chapter 9 is about m-commerce location based services. Truthful authentication with secure communication are necessary in Location Based Services to protect from various risks. The purpose of this research is to identify security risks in mobile transactions especially in location based services like mobile banking. The factors need to be identified the reasons of customer distrust in mobile banking. In addition, the security issues with mobile banking systems and mobile devices to be highlighted. Finding which approach is more suitable and secure for mobile banking transaction between customer and bank. The research predominantly focuses upon customer trust, security issues and transaction costs owing to different technology standards of Mobile Commerce in twofold, the first phase highlights the various location based services in M-Commerce, various technology standards, customer trust and perceived risk and further at next level it highlights the various problems associated mobile database and, comparative study of various replication protocols, transaction security Issues and LBS security challenges.

Chapter 10 is about software-defined networking (SDN) emerging technology. Software Defined Networking was developed in an attempt to simplify networking and make it more secure. By separating the control plane (the controller)—which decides where packets are sent—from the data plane (the physical network)—which forwards traffic to its destination. The creators of SDN hoped to achieve scalability and agility in network management. SDN increasingly uses elastic cloud architectures and dynamic resource allocation to achieve its infrastructure goals.

Chapter 11 is about security management in mobile ad hoc network. A Mobile Ad Hoc Networks (MANETs) is an assortment of a variety of portable nodes that are linked collectively in a greater number in a wireless medium that has no permanent infrastructure. Here, all the nodes in the node partake in acting as both router and host and is in charge for accelerating packets to other nodes. This chapter discusses about the various attacks on different layers and on various security protocols. So, designing a secure routing protocol is a main challenge in MANET. As we all know this is a mobile ad hoc network so nodes in the network dynamically establish paths among each other so its vulnerable to different kinds of threats. So in this case we need secured communication among the nodes present in the network.

Chapter 12 describes enhancement of network performance in VANET using dynamic routing strategies. Vehicular Ad hoc Networks (VANETs) have been evolved as an invigorating network system and application domain in current communication technology. In smart city applications context, there are smart vehicles embedded with sensors and dynamically programmed IoT devices which are to be managed and controlled yet energetically. Progressively vehicles are being furnished with surrounded actuators, handling signals and wireless communication abilities. This chapter focuses on the fact that this special network has opened a various possible outcomes for intense and potential extraordinary applications on security, effectiveness, comfort, confidentiality effort and interest while they are significantly vibrant. Irrespective of many challenges such as high frequency of topology change and link failure possibility, routing management in VANET has been successful in traffic scenario during vehicle-to-vehicle communication.

CONCLUSION

In today's digital world, communication and networking is the heart of the technology. Any new advancement is impossible without a strong networking in the background. Significance of computer networking and its resources can be realized while using the web continuously without interruption. It's what makes the internet possible. It's how businesses support sprawling multinational footprints. It's even stitching together the appliances in our homes into one, smart, convenient fabric. For all of its importance, though, networking technology as a whole doesn't tend to change much. For example, the internet itself runs on an iteration of a protocol which was built long back. Most of the developments in networking over that time have revolved around capacity and speed – not capability. As we move into the 2020 there are a number of trends and technologies that are poised to create real change in the world of networking. In the book Managing Resources for Futuristic Wireless Networks different flavors of networking concepts are included starting from Wireless networks, Mobile ad-hoc networks, Resource Management, Clustering, Spectrum Management, Sharing of Networks, SDN(Software Defined Network), Security issues in networks, UAV (Unmanned Aerial Vehicles (UAV), resource allocation techniques, etc.

Mamata Rath
Birla School of Management, Birla Global University, India

Acknowledgment

The editor is grateful for the assistance of the Reviewers and Editorial Advisory Board members.

The editor is also thankful to all the contributors and IGI Global staff.

Introduction

Resource management is concerned about the organization and properly handling entire wireless mobile network to provide uninterrupted network connectivity to many mobile devices moving together in the mobile network. This task is mostly important for ubiquitous computing, which commonly means anytime, anywhere computing and communication. Most of the 3G and entire 4G and beyond wireless communication technology is all-IP. This growing use of IP devices in portable applications has created the demand for mobility support for entire networks of IP devices. Futuristic wireless networks are developing rapidly to fulfill demand of new generation devices equipped with artificial intelligence and machine learning. The proposed book focuses on various challenges faced by different networking systems. Topics included in this book are related to emergence of advanced networks, security issues, SDN, UAV, VANET, resource allocation, congestion control, etc.

Chapter 1

Mobile Ad Hoc Networks Security Enhancement With Trust Management

Vaishali Vilas Sarbhukan
Terna Engineering College, University of Mumbai, India

Lata Ragha
Father Conceicao Rodrigues Institute of Technology, University of Mumbai, India

ABSTRACT

The mobility and scalability brought by wireless networks made it possible in many applications. In MANET, discovering the stable, secure, and reliable routes is a challenging research problem due to the open nature of wireless communications. In Zhexiong Wei Routing Trust scheme (ZWRT), Bayesian approach and Dempster Shafer theory is used to evaluate more realistic trust value. Mohamed et al proposed Establishing STAble and Reliable Routes (ESTAR) system in heterogeneous multihop wireless networks. The above existing methods failed to solve the complete problem of information loss in MANETs. There are several other reasons due to which the information may loss in MANETS, such as mobility and congestion of mobile nodes. Therefore, the novel Enhanced Trust-based Secure Routing (ETSR) scheme is presented to deliver stable, reliable, and secure data communication in MANETs. The simulation results demonstrate that ETSR routing protocol improve QoS performance as compared to state-of-the-art techniques.

DOI: 10.4018/978-1-5225-9493-2.ch001

INTRODUCTION

Due to self-setup and self-upkeep abilities of MANETs, nowadays mobile ad hoc networks (MANETs) have become a mainstream inquire about subject. A dynamic network can be designed with the help of wireless hubs which do not require a settled infrastructure. Because of portability and flexibility brought by remote framework made it conceivable in number of applications. Among all the modern remote systems, Mobile Ad hoc Network (MANET) is a champion amongst the foremost essential and distinctive applications. Tragically, the open medium, distributed nature and dynamic topology of MANET make it helpless against different sorts of assaults (Yu et al., 2013), (Wang et al., 2014). These assaults include black hole assault, grey hole assault, sybil assault, packet dropping assault and sleep deprivation assault etc. Hence, security is principle snag in strategic MANETs (Chapin & Chan, 2011). There are primarily two methodologies that can give security in MANETs. These methodologies are prevention based and detection based methodologies (Bu et al., 2011a), (Bu et al., 2011 b). Prevention based methodologies are based on cryptography and detection based methodologies focus on trust threshold. Prevention based methodologies are contemplated completely in MANETs (Fang, 2009), (Yu et al., 2010). One disadvantage of prevention based methodologies is that they require a centralized key administration foundation. But it is not possible practically in conveyed systems, for example, MANETs. Likewise, a centralized infrastructure will be the primary focus of opponents in war zones. On the off chance that the framework is annihilated, the entire system might be incapacitated. Moreover, in spite of the fact that prevention based methodologies can avert bad conduct, but still there are some possibilities stayed for noxious hubs. Noxious hubs can take an interest in the routing strategy and exasperate appropriate routing foundation. Albeit some fantastic work has been done on detection based approaches like reputation based schemes, payment schemes, cryptographic schemes and trust schemes, these schemes cannot guarantee route stability, reliability and security at the same time. Basically most of authors focus only on malicious users in network but while working in MANET authors have to consider parameters like mobility, energy level, density of nodes etc. Also most existing trust based approaches in MANETs, do not consider first-hand information and second-hand information at the same time to evaluate the trust of watched hub. Therefore, it results in inaccurate trust value estimation. Existing techniques failed to solve the complete problem of information loss in MANETs. There are several other reasons due to which the information may loss in MANETS such as mobility and congestion of mobile nodes. By considering above loopholes of existing methods a novel enhanced trust based secure routing scheme (ETSR) is designed to establish stable, reliable and secure routing path for data transmission.

LITERATURE REVIEW

Here detailed survey is given on detection based security mechanisms used in MANET which are based on reputation schemes, trust schemes, cryptographic and payment schemes.

1. Bu et al. (2011a) developed a scheme to provide security in MANET .Here authors used two concepts with data fusion. These concepts are authentication and intrusion detection. According to authors, the biometric systems perform positive verification process in authentication mode. Positive verification is the process of comparing coordinating score. In this process coordinating score between the info test and the enlisted layout with a decision threshold is compared and yields a binary decision. Decision may be acknowledge or dismiss. Authors used Markov model. L sensors are chosen for authentication and intrusion detection at each time slot to observe the security state of the network. To solve the fusion problem Dempster–Shafer theory (DST) used.

2. Buchegger & Le Boudec (2004) developed a robust reputation framework for misbehaviour discovery in MANETs. In this methodology, two types of ratings are utilized about everyone else who is of intrigue. These ratings are reputation rating and trust rating. This methodology is completely dispersed and none of the agreement is important. To exploit scattered reputation information, i.e. to gain from others' perceptions before learning by claim understanding, some technique is required to fuse the reputation ratings into the perspectives of others. Major objective of the whole reputation framework is that to make decisions about different hubs depending upon ratings. In reputation based framework each hub utilizes its trust rating and reputation rating to periodically classify other hubs. The hubs are classified on the basis of two criteria. First is normal/misbehaving and second is trustworthy/untrustworthy.

3. Shakshuki et al. (2013) developed Enhanced Adaptive Acknowledgement Scheme (EAACK) for MANETs. EAACK is intrusion detection framework which is specially designed for MANETs. EAACK has higher malicious-behaviour-detection rates in specific conditions as compared to existing traditional methodologies. EAACK methodology does not incredibly influence the system performances. EAACK has better exhibitions as compared to Watchdog, TWOACK, and AACK in terms of beneficiary crash, compelled transmission control, and false misconduct report.

4. Mean Field Game Theoretic Approach provides an amazing mathematical tool to issues with an extensive number of players to give security in Mobile Ad hoc Networks (Wang et al., 2014). This plan helps to empower an individual hub in MANETs to settle on vital security protection choices without united

association. In addition, since security resistance components devour valuable framework assets (e.g. vitality), the Mean Field Game Theoretic Approach considers the security essential of MANETs and also the structure resources.

5. Das & Islam (2012) created security and protection issues which have turned out to be basically imperative with the quick extension of multiagent frameworks. To adapt to the deliberately altering behaviour of malignant operators to circulate outstanding burden as equally as conceivable among service providers; they introduced a dynamic trust calculation demonstrate called as "Secured Trust". In this paper, they originally broke down the various parts identified with assessing the trust of an operator and afterward proposed a thorough quantitative model for assessing such trust. Authors also proposed a novel load-balancing algorithm based on the different factors which are defined in

Model given in this chapter. Authors provide a dynamic trust computation model for effectively assessing the trust of agents even in the presence of very swaying malevolent conduct.

6. Wei et al. (2014a) built up a framework for Trust Establishment with Data Fusion for Secure Routing in MANETs. This plan improves routing security subject to trust. Here indirect perceptions are used to improve the precision of trust esteems with indirect perceptions. For data fusion that is to intertwine onlookers' opinions Dempster-Shafer theory is utilized. Consequently, an increasingly correct trust esteem can be determined from indirect perceptions. At that point the trust esteem is utilized to overhaul the MANET routing protocol security.

7. Zheng et al. (2012) built up a Game Theoretic Approach for Security and Quality of Service (QoS) Co-Design in MANETs with Cooperative Communications. Here authors proposed a game theoretic way to deal with quantitatively investigate the assault methodologies of the aggressor in order to settle on rational decision on transfer determination and the validation parameter adjustment to achieve the exchange off among security and Quality of Service (QoS) in CO-MANETs. Game theoretic approaches have been proposed to upgrade organize security. Game theory tends to issues in which numerous players with clashing impetuses or objectives rival one another. Thus it can give a mathematical structure to showing and examining choice issues. Authors utilized a dynamic Bayesian game theoretic approach to deal with empower a hub to settle on strategic decisions on transfer determination and validation parameter adjustment.

8. Kraounakis et al. (2015) developed distributed systems built in open competitive and highly dynamic pervasive environments. These distributed systems are

made up of autonomous elements which demonstrate and communicate in an intelligent and flexible manner. It helps to achieve their own objectives and purposes. Framework entities may be characterized into two primary classifications which are named as, in principle, in conflict. Here major purpose is to exhibit a computational model for trust foundation dependent on a reputation system. This computational model consolidates direct service resource requestors (SRRs') experiences and information disseminated from observer SRRs on the basis of their past encounters with the service resource providers (SRPs). The planned instrument segregates between unreasonable feedback ratings purposefully and unexpectedly. It also consider potential changes to suppliers' conduct, and weighs more recent events in the assessment of the overall reputation ratings.

9.   According to Tan et al. (2016), there is vast development in the field of ad hoc network technology. Such development covers areas like vehicular ad hoc network (VANET), wireless sensor network (WSN), emergency and military communications etc. In order to secure the data plane of ad hoc networks, authors proposed a novel trust management framework based on fuzzy logic to secure the data plane of ad hoc networks. Here fuzzy logic is utilized to define imprecise empirical knowledge which helps to assess path trust esteem. Along with fuzzy logic they used graph theory. Graph theory is used to fabricate a novel trust model for evaluating hub trust esteem.

10.   Mahmoud et al. (2015) proposed hybrid approach dependent on trust model and payment frameworks to choose more reliable and energy aware paths for data transmission. The E-STAR routing method proposed for setting up steady and dependable routes in heterogeneous multihop wireless networks. In E-STAR scheme, payment and trust frameworks are combined using a trust-based routing protocol and energy-aware routing protocol. Incorporating the payment and trust systems with the routing protocol results in many favorable circumstances which can be condensed as pursues. At first, trust among the hubs is empowered by making information about the hubs' past conduct available. Relaying packets by obscure hubs involves an explicit segment of danger, therefore a source hub requires to confide the hubs that hand-off its packets. Second, this coordination can convey messages through dependable routes and empower the source hubs to recommend their required dimension of trust. Third, the hubs that break courses by giving more tendency to the profoundly believed hubs in course choice are punished and thus in acquiring credits, are punished. Fourth, the reconciliation of the payment and trust frameworks with the directing convention can punish the hubs that report erroneous vitality ability. This is on the grounds that the courses will be broken at these hubs and their trust esteems will corrupt. Finally, a hub may utilize an avaricious

philosophy: never earn excessively unneeded credits and quit transferring others' packets after procuring adequate credits. The coordination of the payment and trust frameworks stimulates the hubs to collaborate in transferring packets to earn credits as well as stimulates the wealthy hubs to cooperate to maintain good trust esteems. This is because the hubs lose trust over time if they do not coordinate. By this way, in addition to payment, trust is another motivating force for participation. However, consideration of both payment framework and trust framework along with cryptographic authentication can increment routing overhead and end to end delay. This chapter uses concepts of E-STAR in order to achieve the reliable and secure communication for MANET (Mahmoud et al., 2015). To minimize the routing overhead and delay, the onion routing based secure data transmission is used from source node to destination node via the selected stable and reliable paths (Liu & Yu, 2014). The trust esteems for each hub are computed using the different key parameters such as energy level, mobility, load parameter etc (Khan et al., 2017).

11. Wei et al. (2014b) used uncertain reasoning which is the part of artificial intelligence. This model comprises of first–hand information module, second–hand information module, trust evaluation and update module, trust repository, networking component and application component. Trust evaluation and update module obtained evidences from first–hand information module and second–hand information module. Trust evaluation and update module used Bayesian inference approach to evaluate and update trust values based on first–hand information and also used Dempster Shafer Theory-Belief function approach to evaluate and update trust values based on second–hand information. Then the calculated trust esteems are stored in the module of trust repository. Networking component is related with routing mechanism. On the basis of trust repository module secure routing path is established between sources and destination using routing mechanism provided by network component. Finally application part can send information through secure routing path.

12. In (Ou et al., 2009), authors presented Trusted Computing concept. Trusted Computing is a basic research field in information security. Major Key issue to be resolved is trust assessment for trust model. It is incredible essentialness for guaranteeing security of trust model for trusted computing to dissect regularly and confirm in configuration procedure of utilization model for trusted computing and examine its trust on a basic level. Here the issues of security and trust for TPM (Trust Platform Module) were considered. Also a trust assessment model dependent on TPM was developed. Validity of trust demonstrate for TPM was dissected in principle. In this trust assessment model believability of trust consolidates direct trust, recommendation trust and indirect trust. Direct trust is also known as coordinate trust. Recommendation trust is also called as

suggestion trust and indirect trust is called as roundabout trust. Coordinate trust is communicated by fulfillment of subject and it principally originates from advancement or dissatisfaction of two subject's correspondences. Suggestion trust is a subject trusts or embraces encounter information recommended by the other subject. Roundabout trust is composed by trust chain made up of a series of recommendations or suggestions.

13. Kiefhaber et al. (2010) demonstrated that trust is a vital factor for independent dispersed frameworks. In order to manage complex distributed frameworks self-organizing strategy is used. This paper acquaints two strategies to measure and figure trust esteems with respect to the reliability of other hubs by direct perception. These strategies have least overhead. Authors exhibit this methodology utilizing a simulator and exhibit that both algorithms meet towards the settled genuine trust esteem and remain inside close limits of it. Assessments additionally demonstrate the measure of estimations expected to compute a significant trust an incentive and additionally the significance of the hubs' own reliability esteem when estimating another hub. Also, authors present an enhancement of one strategy to ascertain a significant trust even by an unreliable observer.

14. Theodorakopoulos & Baras (2006) studied the issue of surveying the trust level as speculation of the shortest path problem in an oriented graph. Here the edges correspond to the opinion that a hub has about other hub. The fundamental objective is to empower the hubs to roundabout way assemble trust relationships utilizing solely observed data.

15. Velloso et al (2010) presented human-based model. This model is used to produce a trust connection between hubs in ad hoc network. Without the prerequisite for global trust knowledge, they have shown a convention that scales productively for expansive systems.

ISSUES IN MANET

Challenges in MANET Protocols

- Major goals of a distributed Mobile Ad Hoc Network (MANET) are reliability, availability, scalability, and reconfigurability. But when collaboration or cooperation becomes critical in order to achieve the mission and above mentioned framework goals then in that case managing trust in a distributed Mobile Ad Hoc Network becomes a challenging task.
- Security protocol architects for MANETs face technical difficulties because of serious asset imperatives in transmission limit, memory measure,

battery life, computational power, and exceptional remote attributes such as openness to eavesdropping, absence of explicit entrance and exit points, high security threats, vulnerability, unreliable correspondence, and fast changes in topologies or memberships because of user portability or hub failure.

- An additional alert is required to design security protocols for military MANETs because battlefield communication networks must adapt with certain components. These components include hostile environments, hub heterogeneity, frequently stringent performance limitations, hub disruption, and high tempo activities prompting to fast changes in system topology and service necessities, and dynamically shaped networks of intrigue wherein members may not have predefined trust relationships. To adjust to these components, systems must have the capacity to reconfigure flawlessly, by means of low-multifaceted nature distributed network management schemes.

Issues With Mobile Ad Hoc Network Routing

- Uneven joins: Maximum of the wired structures depend upon the isosceles associations which are constantly stable. Be that as it may, this isn't applicable to ad hoc systems as the hubs are versatile and continually switching their scenario inner device. For instance think about a MANET wherever first hub transmits a flag to second hub but this doesn't enlighten something concerning the nature of an association within the turnaround heading.

- Steering Overhead: Steering in Advert -Hoc networks has been a testing assignment in case of remote systems. The good sized cause at the rear of this can be the regular amendment in prepare topography in mild of abnormal state of hub versatility. Along these lines, some stale courses are created in the steering table which prompts superfluous steering overhead.

- Interference: It is the real issue with Advert-Hoc networks as connections go back and forth relying upon the transmission qualities, one transmission may meddle with another and hub may catch transmissions of different hubs and the aggregate transmission can decline.

- Dynamic Topology: It is the real issue in MANET because of inconsistent topology feature. The versatile hub may move or medium attributes may additionally trade. In advert-Hoc networks, routing tables need to with the aid of one approach or some other reñect these adjustments in topology and routing calculations need to be adjusted. As an instance in a ðxed arrange steering desk clean occurs for each 30sec. This refreshing recurrence may be low for Advert-Hoc networks.

PROBLEM STATEMENT

Albeit some fantastic research has been done on detection based approaches like reputation based schemes, payment schemes, cryptographic schemes and trust schemes, these schemes can't ensure route stability, reliability and security at the same time. Reliable path generation is threat not only by vindictive users but also by mobility of hubs. As mobility of hubs increases performance of MANET diminishes so mobility is another threat. Second factor is energy level which also influences performance of MANET. After that density of hubs in same area has significant impact to congestion in network which can drop packets extensively. Actually these are reasons while designing security systems like trust based methods and cryptography based methods. Basically most of authors focus only on malicious users in network but while working in MANET authors have to consider above mentioned parameters like mobility, energy level, density of nodes etc. In Zhexiong Wei Routing Trust model (ZWRT), DST (Dempster Shafer Theory) technique is utilized to observe each hub in network and task is performed without secure trusted party (TP) (Wei et al., 2014b). This is one of significant issue with ZWRT. Another challenge of this method was that it enhances throughput and packet delivery ratio (PDR) but average end to end delay and overhead is too high. Reason behind this is that algorithm used by authors in this paper. In this algorithm each node in network cross verify each and every neighbours, this process takes a lot of time and obviously overhead and delay will increase. In short ZWRT concentrates only on evaluation of realistic and accurate trust value but does not consider energy level and secure routing path. Also most existing trust based approaches in MANETs, do not exploit direct and indirect perceptions (also called second-hand information that is obtained from third-party nodes) at the same time to evaluate the trust of watched hub. Moreover, indirect perception in most approaches is only used to evaluate the reliability of hubs, which are not in the range of the observer hub. Therefore, inaccurate trust values may be inferred. In addition, most techniques for trust assessment from direct perception do not separate data packets and control packets. However, in MANETs, control packets usually are more important than data packets. In ESTAR (Establishing STAble and Reliable route) paper, Mahmoud et al., (2015) considered energy level along with trust value to generate stable route. Here payment system is incorporated with trust system to set up stable, reliable and secure routing path. Security is provided utilizing Secure Hash Algorithm (SHA) algorithm. SHA is cryptographic strategy utilized for secure data exchange from source to goal. But issue was that time taken by this algorithm is more for routing communication. Other issue that to set up stable route they consider only energy level but there are other factors like mobility and load on any particular intermediate hubs. Mobility and load on any particular intermediate hubs influences performance of framework. In short several solutions presented based on cryptography techniques, trust based methods, and hybrid solutions to protect MANET's communications, however most of the recent methods focused on identifying the malicious nodes and prevent them from data communication process only. Such methods failed to solve the complete problem of information loss in MANETs. There are several other reasons due to which the information may loss in MANETS such as mobility and congestion of mobile nodes. By considering above loopholes of existing techniques a novel enhanced trust based secure routing scheme (ETSR) is proposed to set up stable, reliable and secure routing path for data transmission. ETSR is based on hybrid terminologies of both prevention based approach and detection based approach. In ETSR prevention based approach is used for secure data transmission and detection based approach is used for reliability purpose. ETSR method consider energy level, mobility, density of nodes, realistic trust value and load on intermediate nodes. Therefore stable, reliable and secure unified trust management scheme is proposed that enhances the security in MANETs using onion routing cryptographic strategy.

OBJECTIVES AND SCOPE

In MANET trust is translated as degrees of the conviction that a hub in a system (or an operator in a distributed framework) will complete tasks that it ought to be. Because of the particular attributes of MANETs, trust in MANETs has five fundamental properties as subjectivity, dynamicity, asymmetry, setting versatility and fragmented transitivity.

Objectives of research are as follows.

- To design trust model for the nodes evaluation based on the factors such as node energy, node mobility and node load.
- To evaluate more accurate and realistic trust value.
- Finally the stable and shorter route selection for relying the data from source node to destination node in network.
- To propose the onion routing based cryptographic algorithm for secure data transmission which works in all phases of routing such as route discovery, route establishment, and route maintenance.
- To separate between data packets and control packets.
- To develop scheme for detection of malicious behaviours of nodes such as packet dropping and modifying the packets.
- To improve performance of system in terms of throughput, packet delivery ratio (PDR), average end-to-end delay, overhead of messages, packet loss and jitter.

TRUST MODEL IN MANET

In MANET trust is translated as degrees of the conviction that a hub in a system (or an agent in a distributed framework) will carry out tasks that it should. In view of the express characteristics of MANETs, trust in MANETs has five crucial properties as showed up in figure 1. Properties of trust in MANET as subjectivity, dynamicity, asymmetry, setting resilience and fragmented transitivity.

Subjectivity-It implies spectator hub has the privilege to decide put stock in estimation of eyewitness hub.

Dynamicity-It infers trust of hub ought to be changed relying upon its conduct.

Asymmetry-It implies that if first hub trusts second hub then second hub does not really trust first hub.

Setting resilience -It implies that trust appraisal is ordinarily in light of conduct of a hub.

Fragmented transitivity-It implies that if first hub trusts second hub and second hub trusts third hub then first hub does not really put trust in third hub.

Figure 1. Trust Properties in MANET (Cho et al., 2011)

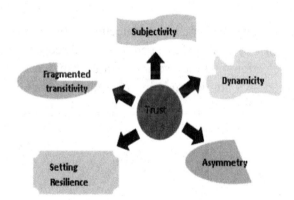

PROPOSED METHODOLOGY

In this section the novel ETSR (Enhanced Trust based Secure Routing) methodology is elaborated. The main intention of designing novel approach for MANETs is to overcome the challenges and loopholes of existing systems. All these operations are performed using the onion routing. The technique onion routing is proven for minimum security overhead already in (Liu &Yu 2014). The ETSR trust model is used to select most stable and reliable path based on intermediate nodes trust

Figure 2. Architecture of ETSR scheme

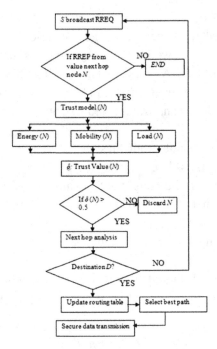

evaluation. Figure 2 shows the complete architecture of ETSR scheme. The ETSR scheme mainly consist of four phases, such as route discovery, trust based node evaluation, route selection, and then secure data transmission from the selected route. All these processes are designed by using the cryptographic functions of onion routing.

Algorithm1: *ETSR*
Inputs
S: Source node
D: Destination node
N: Current next hop node
δ: Maximum route discovery limit
$\Delta=0$: Route discovery timer
$\alpha1$: Energy level probability of N
$\alpha2$: Mobility level probability of N
$\alpha3$: Load level probability of N
$\Phi=0.5$: Trust threshold value

  1. S initiate route discovery
  2. S broadcasting *RREQ* using onion function

3. Start timer Δ
4. IF (*S* Received *RREP* from valid *N*)
5. $\alpha1$: Estimate energy level of *N*

$$(\alpha1)^N = \frac{IE^N - CE^N}{IE^N}$$

6. $\alpha2$: Estimate mobility of *N*

$$\alpha2 = 1 - LN$$

7. $\alpha3$: Estimate load of *N*

$$\alpha3 = 1 - (T)N$$

8. END IF
9. Calculate Trust (*N*) using

$$\text{Trust (N)} = (\alpha1 + \alpha2 + \alpha3)/3$$

10. IF (Trust (*N*) > Φ)
11. Select *N*
12. Update Routing table entries
13. Resending *RREQ* packets from *N* onwards
14. Repeat process from 4-14 till *D* node
15. Δ++
16. ELSE
17. Select next *N*
18. Δ++
19. END IF
20. IF ($\Delta > \delta$)
21. STOP
22. ELSE
23. Broadcasting *RREQ* packets till *D*
24. END IF
25. Trace the reverse paths and store into routing table.
26. Sort all selected paths in increasing hop count order.
27. Select best path and perform data transmission using onion routing.

In above algorithm 1, the source node S initiates the process of route discovery from source node to destination node. Before initiating the route discovery, current public key, session key, and private keys are generated and updated in *RREQ* packet as per the onion routing methodology. The detailed steps of onion routing based data transmission is discussed in reminder of this section. The algorithm 1 is mainly demonstrating the overall secure and reliable route discovery process using the trust model which is applied on each next hop node N. The RREP received from the one hop neighbors of current node S or previous intermediate node N. Once the RREP received from number of next hop nodes, on each node the trust model based evaluation is applied in which the three parameters are computed such energy level, mobility, and load. The combined probability of each node N is computed in range of 0 to 1 and this probability is called as proposed trust model value. Further the current trust value is compared with the set threshold value 0.5. The threshold 0.5 is enough to justify the behavior of current node in network as the network conditions are frequently changing. The process of trust evaluation is repeated until all paths discovered from S to D. Authors believe that all discovered paths are stable, energetic, and reliable for initiating the data transfer, therefore authors sort all selected paths in increasing order. The path with minimum hop counts is selected for further data transmission process. In case, if the designed ETSR algorithm failed to discover any satisfying path, then to prevent the data loss authors update the threshold value by decreasing to 0.3 and select the path for data transmission. If that selected path breaks, then threshold value is set to 0.5 again for next route discovery. The dynamic invention of threshold value is out of scope of our discussion. The trust evaluation parameters discussed below:

1. **Energy:** The energy drainage of any intermediate node in network leads to route break and hence dropping the QoS performance. Energy level of node is also important parameter in deciding the trust behavior of node as explained in (Khan et al.,2017). The current energy level $\alpha1$ of node N is computed as:

$$\left(\alpha1\right)^{N} = \frac{IE^{N} - CE^{N}}{IE^{N}} \tag{1}$$

Where, IE^{N} is initial energy of node N and CE^{N} consumed energy of node N.

2. **Mobility:** Mobility is another threat for more unreliable communications in network. Higher the mobility, frequent the route breaks. The mobility of nodes in its neighborhood is determined by computing the neighborhoods link changes rate L. The L at node N is computed as following:

$$(L)^N = \frac{LR^N + LB^N}{\max(LR^N) + max(LB^N)} \qquad (2)$$

Where, LR^N is the link arrival rate and LB^N is the link breakage rate of node N. Further mobility based probability $\alpha2$ of node N is estimated as:

$$\alpha2 = 1 - LN \qquad (3)$$

3. **Load:** The load is another factor that causes the QoS drop and frequent route breaks in communications. Thus in ETSR system, load parameter is considered for the trust evaluation of current node N. This can be done by using the Traffic load intensity (T) at the neighboring nodes of S. The T estimation of node N computed as following:

$$(T)^N = \frac{ML^N}{q_{max}^N} \qquad (4)$$

Where q_{max}^N the length of interference in queue of node N and ML^N is mean traffic load at the node N (Khan et al., 2017). Finally the load level probability $\alpha3$ of node N is estimated as:

$$\alpha3 = 1 - (T)N \qquad (5)$$

Thus the probabilities computed above shows that higher the values higher the reliability and stability in routes. The trust value of node is computed by using all three probabilities as:

$$Trust\ (N) = (\alpha1 + \alpha2 + \alpha3)/3 \qquad (6)$$

The onion routing concept used for the secure data transmission are summarized in below steps demonstrated in algorithm 2:

Algorithm 2: *Onion Routing*
Inputs
S: Source node
D: Destination node
IN: Current intermediate node
PKI: Public Key Infrastructure

K^{pu}: Public key

K^{gpr}: Group private key

K^{pr}: Private key

K^{ss}: Session key

P: Current packet

1. PKI Generation
   1.1 PKI generation via broadcasting S ID
   1.2 Extract the K^{pu}, K^{pr}, K^{ss}
   1.3 Update routing table
2. At S Node
   2.1 Fetch routing information
   2.2 Get current K^{ss}
   2.3 Generate new K^{ss}
   2.4 Updated routing table with new key
   2.5 Discover the route using algorithm 1
   2.6 Apply key encryption onion at IN and D using K^{pu}, K^{pr}
   2.7 Signing by S using K^{gpr}
   2.8 Transmit current P towards next hop IN
3. At IN Node
   3.1 Success = verify (P) using K^{gpr}
   3.2 IF (Success == true)
   3.3 IF ($IN == D$)
   3.4 Decrypt current packet using K^{gpr}
   3.5 ELSE
   3.6 Perform onion routing operations on received packet and forward to next IN in selected path using set of keys
   3.7 ELSE
   3.8 Packet is received from malicious node
   3.9 Drop (P)
   3.10 END IF
4. STOP

The overhead of onion routing based security method is less as compared to other cryptographic methods. Authors considered that the shared key algorithm uses a keyed hashed MAC such as MD5. The approximate time required for various cryptographic operations such as RREQ (49.68ms), RREP (1.4ms), and Data (1.12ms) on PIII processor (Mahmoud et al., 2015).If authors compare time taken by onion routing in (Liu & Yu et al., 2014) AASR (Authenticated Anonymous Secure Routing) with time taken by SHA algorithm then it clearly depicts that there is vast difference between these values 1.12ms and 8.5ms respectively. Therefore onion routing is preferred in ETSR proposed system rather than using SHA. Proposed methodology uses Enhanced Trust based Secure Routing Protocol to establish stable reliable and secure routing path. The routing protocols have three processes: i) Route Request Packet (RREQ) delivery; ii) Route selection; and iii) Route Reply Packet (RREP) delivery.

IMPLEMENTATION PLATFORM USED AND PERFORMANCE MEASURES

There are five performance metrics considered in studying performance of ETSR conspire, These metrics are as follows: 1) Throughput, which is the total size of data packets correctly received by a destination node every second; 2) Packet Delivery Ratio (PDR), which is the ratio of the number of data packets received by a destination node and the number of data packets generated by a source node; 3) average end-to-end delay, which is the mean of end-to-end delay between a source node and a destination node with Constant Bit Rate (CBR) traffic; 4) message overhead, which is the size of TLV(Type Length Value) blocks in total messages that are used to carry trust values; 5) Packet loss which is the difference between the number of packets generated by node and the number of packets received successfully by destinations during the simulation; and 6)Jitter which is the variation/fluctuation of end to end delay between the two packets.

Authors of this chapter presented evaluation of ETSR scheme using the NS2 simulator. Authors compared the performance of ETSR method with the non-security protocols such as AODV, DSDV and benchmark security methods such ZWRT and ESTAR. The NS2 version 2.34 used to implement and evaluate the performance of ETSR method. The performance evaluation is conducted by designing the mobility speed variations and density variations in presence of malicious users in network. The key performance metrics such as average throughput, average delay, packet delivery ratio (PDR), routing overhead and packet loss are measured for the comparative study as explained above. The attack model is designed for each routing protocol in which the malicious nodes are introduced in network. The malicious nodes perform the selfish behavior and dropping the packets. In all the networks the malicious nodes are 10% of total mobile nodes estimate the effectiveness of ETSR routing protocol. Table 1 shows the list of network parameters used in simulation.

Table 1. Network Parameters of ETSR

Simulator	Network Simulator 2.34
Number of Nodes	10-200
Traffic Patterns	CBR (Constant Bit Rate)
Network Size (X * Y)	1000 x 1000
Mobility	10 – 35 m/s
Simulation Time	80 seconds
Number of Malicious Nodes	10%
Transmission range	250m
Non-secure routing protocols	AODV, DSDV
Secured routing schemes	ZWRT, ESTAR, ETSR
MAC Protocol	802.11b
Initial Energy	0.5J
Mobility model	Random waypoint
Topology	Random
Interface Type	Phy/WirelessPhy
Queue Type	Droptail/Priority Queue
Queue Length	50 Packets
Transport Agent	UDP
Antenna Type	Omni Antenna
Propagation Type	Two Ray Ground

EXPERIMENTAL ANALYSIS

For Scenario1 results are as follows: Scenario 1 is related with mobility that is varying velocity of nodes. Figure 3 to Figure 7 are results for Scenario 1: mobility for basic AODV, basic DSDV, ZWRT, ESTAR and ETSR. Figures 3 and 4 show that ETSR has best performance in terms of throughput and PDR respectively. Figures 4 and 5 show node velocity versus delay and overhead respectively.

In figure 5, as mobility of nodes increases from 10 m/s to 35 m/s delay increases slightly which is negligible if consider other parameters like throughput, PDR, overhead and packet loss. Next to DSDV, ETSR has less delay than remaining methods with considering factors like throughput, PDR. Figure 6 shows that ETSR has best performance among all existing methods in terms of overhead.

Figure 7 shows mobility versus packet loss. Packet loss is less for ETSR than AODV, DSDV, ZWRT and ESTAR as secure data transmission is used.

Figure 3. Simulation results for scenario 1- Throughput

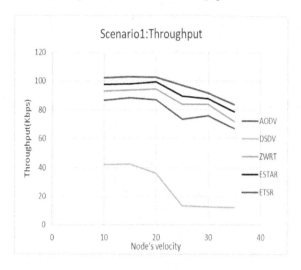

Figure 4. Simulation results for scenario 1- PDR

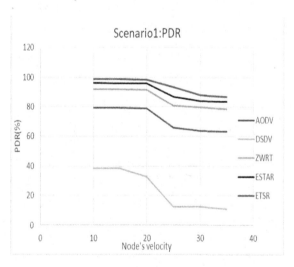

For Scenario 2 results are as follows: Scenario 2 is related with density that is varying number of nodes. Figure 8 to Figure 12 are results for Scenario 2 to evaluate performance of ETSR in terms of throughput, PDR, delay, overhead and packet loss respectively for AODV, DSDV, ZWRT, ESTAR and ETSR. Here for different number of nodes velocity is constant. Here speed of node is 10 m/s.

Figure 5. Simulation results for scenario 1- Delay

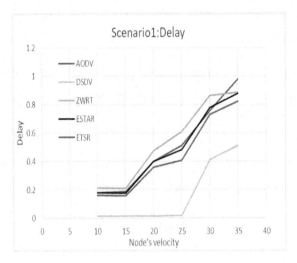

Figure 6. Simulation results for scenario 1- Overhead

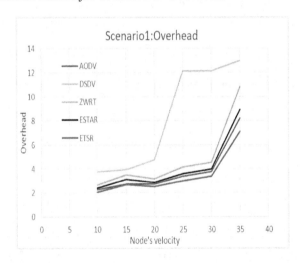

Here in figure 8, as number of nodes increases from 10 to 200 throughput decreases slightly as there are more chances of collision between nodes as number of nodes increases. Comparing AODV, DSDV, ZWRT and ESTAR approaches ETSR works better in terms of throughput. In figure 9, it is shown that ETSR based system has more PDR than remaining methods as throughput is more in ETSR than other approaches.

Figure 7. Simulation results for scenario 1- Packet Loss

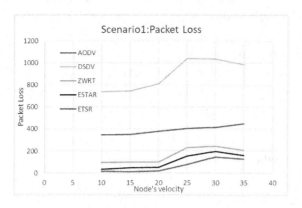

Figure 8. Simulation results for scenario 2- Throughput

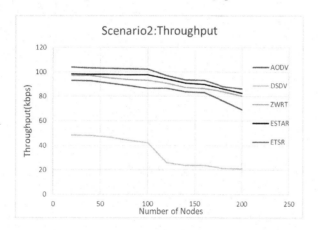

Figure 9. Simulation results for scenario 2- PDR

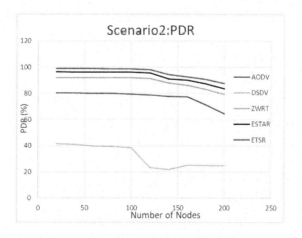

Figure 10 shows that as number of nodes increases delay increases in all protocols. DSDV and ETSR have less delay than AODV, ZWRT and ESTAR but throughput and PDR factors are very poor in DSDV so overall ETSR is the best in terms of delay. Figure 11 depicts that as number of nodes increases overhead is more. As compared to AODV, DSDV, ZWRT and ESTAR based approaches ETSR has better performance in terms of overhead. From figure 12 it is observed that packet loss is lowest for ETSR scheme than remaining methods.

Figure 10. Simulation results for scenario 2- Delay

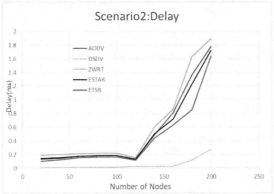

Figure 11. Simulation results for scenario 2- Overhead

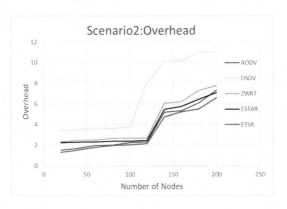

Figure 12. Simulation results for scenario 2- Packet Loss

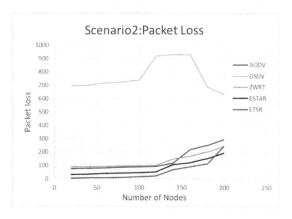

Impact of Packet Size on Average Jitter in MANET

Jitter is the variation/fluctuation of end to end delay between the two packets. Packet arrival time is expected to be very low while calculation of jitter parameter. From figures 13 and 14 authors conclude that the most important factor which plays most crucial role for Average Jitter network performance metric is packet size, followed by routing protocol as effect of packet size alone is more than routing protocol. After considering routing protocol authors have to consider DSSS (Direct Sequence Spread Spectrum) rate.

Figure 13. Average Jitter versus Number of Packets Transmitted per Second (Packet Size 256 bytes)

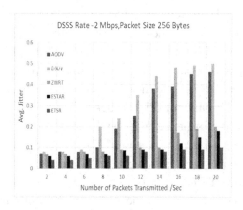

Figure 14. Average Jitter versus Number of Packets Transmitted per Second (Packet Size 512 bytes)

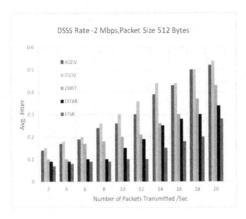

CONCLUSION AND FUTURE SCOPE

The objective of ETSR scheme is to design novel hybrid security solution for mobile ad hoc networks. Enhanced trust based model along with the onion based cryptography approach is designed to improve the QoS performance of MANETs under the presence of malicious users in network. The process of trust based nodes evaluation is performed as the time of route discovery only, hence it leads the more stable and reliable communication performed using the onion based approach. For the trust evaluation authors exploited the nodes mobility, the load of node, and energy level of node. In an onion routing, messages are embodied in layers of encryption, similar to layers of an onion. The encoded information is transmitted through a progression of system hubs called onion switches, every one of which "strips" away a solitary layer, revealing the information's next goal. At the point when the last layer is decoded, the message arrives at its goal. The simulation results shows the effectiveness of ETSR method in which the overall QoS performance of ETSR is significantly improved as compared to state-of-art security solutions like ZWRT and ESTAR considering two different scenarios. Scenario 1 is mobility and Scenario 2 is density. Throughput, PDR (Packet Delivery Ratio), delay, overhead and packet loss, these performance parameters are used in comparative analysis. Packet size has impact on average jitter in MANET. As packet size increases jitter increases. For future work, various aspects of malicious nodes and attacker models are considered using the ETSR protocol. Additionally current strategic activities have complex correspondence and figuring necessities, frequently including distinctive alliance groups that can't be bolstered by the present mobile ad hoc networks. So Software Defined Networking (SDN) – empowered mobile ad hoc networks can be planned in the strategic field in which ETSR can be utilized.

REFERENCES

Bu, S., Yu, F. R., Liu, X. P., Mason, P., & Tang, H. (2011a). Distributed combined authentication and intrusion detection with data fusion in high-security mobile ad hoc networks. *IEEE Transactions on Vehicular Technology, 60*(3), 1025–1036. doi:10.1109/TVT.2010.2103098

Bu, S., Yu, F. R., Liu, X. P., & Tang, H. (2011b). Structural results for combined continuous user authentication and intrusion detection in high security mobile ad-hoc networks. *IEEE Transactions on Wireless Communications, 10*(9), 3064–3073. doi:10.1109/TWC.2011.071411.102123

Buchegger, S., & Le Boudec, J. Y. (2004). A robust reputation system for peer-to-peer and mobile ad-hoc networks. In *P2PEcon 2004* (pp. 2004–2009). No. LCA-CONF.

Chapin, J. M., & Chan, V. W. (2011, November). The next 10 years of DoD wireless networking research. In *Military Communications Conference, 2011-MILCOM 2011* (pp. 2238–2245). IEEE. doi:10.1109/MILCOM.2011.6127653

Cho, J. H., Swami, A., & Chen, R. (2011). A survey on trust management for mobile ad hoc networks. *IEEE Communications Surveys and Tutorials, 13*(4), 562–583. doi:10.1109/SURV.2011.092110.00088

Das, A., & Islam, M. M. (2012). Secured Trust: A dynamic trust computation model for secured communication in multiagent systems. *IEEE Transactions on Dependable and Secure Computing, 9*(2), 261–274. doi:10.1109/TDSC.2011.57

Fang, Y., Zhu, X., & Zhang, Y. (2009). Securing resource-constrained wireless ad hoc networks. *IEEE Wireless Communications, 16*(2).

Khan, M. S., Midi, D., Khan, M. I., & Bertino, E. (2017). Fine-Grained Analysis of Packet Loss in MANETs. *IEEE Access: Practical Innovations, Open Solutions, 5*, 7798–7807. doi:10.1109/ACCESS.2017.2694467

Kiefhaber, R., Satzger, B., Schmitt, J., Roth, M., & Ungerer, T. (2010, December). Trust measurement methods in organic computing systems by direct observation. In *2010 IEEE/IFIP International Conference on Embedded and Ubiquitous Computing* (pp. 105-111). IEEE. 10.1109/EUC.2010.25

Kraounakis, S., Demetropoulos, I. N., Michalas, A., Obaidat, M. S., Sarigiannidis, P. G., & Louta, M. D. (2015). A robust reputation-based computational model for trust establishment in pervasive systems. *IEEE Systems Journal, 9*(3), 878–891. doi:10.1109/JSYST.2014.2345912

Liu, W., & Yu, M. (2014). AASR: Authenticated anonymous secure routing for MANETs in adversarial environments. *IEEE Transactions on Vehicular Technology*, *63*(9), 4585–4593. doi:10.1109/TVT.2014.2313180

Mahmoud, M. M., Lin, X., & Shen, X. S. (2015). Secure and reliable routing protocols for heterogeneous multihop wireless networks. *IEEE Transactions on Parallel and Distributed Systems*, *26*(4), 1140–1153. doi:10.1109/TPDS.2013.138

Ou, W., Wang, X., Han, W., & Wang, Y. (2009, December). Research on Trust Evaluation Model Based on TPM. In *Frontier of Computer Science and Technology, 2009. FCST'09. Fourth International Conference on* (pp. 593-597). IEEE. 10.1109/FCST.2009.10

Shakshuki, E. M., Kang, N., & Sheltami, T. R. (2013). EAACK—A secure intrusion-detection system for MANETs. *IEEE Transactions on Industrial Electronics*, *60*(3), 1089–1098. doi:10.1109/TIE.2012.2196010

Tan, S., Li, X., & Dong, Q. (2016). A Trust Management System for Securing Data Plane of Ad-Hoc Networks. *IEEE Transactions on Vehicular Technology*, *65*(9), 7579–7592. doi:10.1109/TVT.2015.2495325

Theodorakopoulos, G., & Baras, J. S. (2006). On trust models and trust evaluation metrics for ad-hoc networks. *IEEE Journal on Selected Areas in Communications*, *24*, 318-328.

Velloso, P. B., Laufer, R. P., Cunha, D. D. O., Duarte, O. C. M., & Pujolle, G. (2010). Trust management in mobile ad hoc networks using a scalable maturity-based model. *IEEE eTransactions on Network and Service Management*, *7*(3), 172–185. doi:10.1109/TNSM.2010.1009.I9P0339

Wang, Y., Yu, F. R., Tang, H., & Huang, M. (2014). A mean field game theoretic approach for security enhancements in mobile ad hoc networks. *IEEE Transactions on Wireless Communications*, *13*(3), 1616–1627. doi:10.1109/TWC.2013.122313.131118

Wei, Z., Tang, H., Yu, F. R., Wang, M., & Mason, P. (2014a, June). Trust establishment with data fusion for secure routing in MANETs. In *Communications (ICC), 2014 IEEE International Conference on* (pp. 671-676). IEEE 10.1109/ICC.2014.6883396

Wei, Z., Tang, H., Yu, F. R., Wang, M., & Mason, P. C. (2014b). Security Enhancements for Mobile Ad Hoc Networks with Trust Management Using Uncertain Reasoning. *IEEE Transactions on Vehicular Technology*, *63*(9), 4647–4658. doi:10.1109/TVT.2014.2313865

Yu, F. R., Tang, H., Bu, S., & Zheng, D. (2013). Security and quality of service (QoS) co-design in cooperative mobile ad hoc networks. *EURASIP Journal on Wireless Communications and Networking*, *2013*(1), 188. doi:10.1186/1687-1499-2013-188

Yu, F. R., Tang, H., Mason, P. C., & Wang, F. (2010). A hierarchical identity based key management scheme in tactical mobile ad hoc networks. *IEEE eTransactions on Network and Service Management*, *7*(4), 258–267. doi:10.1109/TNSM.2010.1012.0362

Zheng, D., Tang, H., & Yu, F. R. (2012, October). A game theoretic approach for security and quality of service (QoS) co-design in MANETs with cooperative communications. In *Military Communications Conference, 2012-MILCOM 2012* (pp. 1–6). IEEE. doi:10.1109/MILCOM.2012.6415562

Chapter 2

Clustering and Compressive Data Gathering for Transmission Efficient Wireless Sensor Networks

Utkarsha Sumedh Pacharaney
Datta Meghe College of Engineering, India

Ranjan Bala Jain
Vivekanand Education Society's Institute of Technology, India

Rajiv Kumar Gupta
Terna Engineering College, University of Mumbai, India

ABSTRACT

The chapter focuses on minimizing the amount of wireless transmission in sensory data gathering for correlated data field monitoring in wireless sensor networks (WSN), which is a major source of power consumption. Compressive sensing (CS) is a new in-node compression technique that is economically used for data gathering in an energy-constrained WSN. Among existing CS-based routing, cluster-based methods offer the most transmission-efficient architecture. Most CS-based clustering methods randomly choose nodes to form clusters, neglecting the topology structure. A novel base station (BS)-assisted cluster, spatially correlated cluster using compressive sensing (SCC_CS), is proposed to reduce number of transmissions in and form the cluster by exploiting spatial correlation based on geographical proximity. The proposed BS-assisted clustering scheme follows hexagonal deployment strategy. In SCC_CS, cluster heads are solely involved in data gathering and transmitting CS measurements to BS, saving intra-cluster communication cost, and thus, network life increases as proved by simulation.

DOI: 10.4018/978-1-5225-9493-2.ch002

INTRODUCTION

Wireless Sensor Network (WSN) is an agglomeration of randomly scattered tiny sensor nodes, whose primary objective is to gather data for the specific application they have been deployed in an Adhoc fashion. This gathered data is wirelessly transmitted to the Base Station (BS)/Sink. Wireless Communication is the main contributor to a sensor's energy consumption. Hence, even though sensory data gathering is the fundamental task in WSN, it is a major source of power consumption. To reduce the number of data packet transmission required for data gathering usually compression techniques are employed. However, conventional compression techniques introduce excessive in-node computations and control overheads. Compressive Sensing (CS) is a new in-node compression technique that compresses sensory data and accurately recovers it at the BS. It can be very economically used for data gathering in energy constrained WSN. A brief overview of CS is as follows:

Compressive sensing is a new framework developed for single-signal sensing and compression. It exploits the fact that many natural occurring signals are sparse or compressible if represented on a proper basis and represented concisely, then recovery from a small number of projections is guaranteed or traceable (Donoho David L.,2006). Compressive sensing data compression is accomplished in the following three steps.

1.   Sparse representation of the signal
2.   Sampling the signal
3.   Recovery of the original signal.

- **Sparse representation of the signal**
  Consider a signal f^l to be a real-valued discrete-time signal with finite length N. Vectorally represented as

$$f^d = \left[f_1, f_2, \ldots\ldots\ldots\ldots f_N \right] \in \mathbb{R}^N \tag{1.1}$$

It is defined as k-sparse if it has a sparse representation in a proper basis

$$\psi = \left[\psi_{ij} \right] \in \mathbb{R}^{NXN} \tag{1.2}$$

Where $f^l = \psi x$ and x has only k non-zero elements

- **Sampling the signal**

  The k -sparse signal can be under-sampled and be recovered from $M \ll N$ random measurements.

$$Y = \left[Y_1, Y_2, \ldots \ldots \ldots Y_M \right]^T \in \mathbb{R}^M \tag{1.3}$$

The random measurements are generated by

$$Y = \varnothing f^d \tag{1.4}$$

Where $\varnothing = \left[\varnothing_{ij} \right] \in \mathbb{R}^{MXN}$ is called the measurement matrix.

The measurement vector Y for N element is formed by

$$Y_i = \sum_{i=1}^{M} \sum_{j=1}^{N} \varnothing_{ij} f_j^d \tag{1.5}$$

- **Recovery of the original signal**

  It has been shown that reconstruction of a k-sparse signal with high probability from only $M = O\left(k \log \dfrac{N}{k} \right)$ CS measurements employing l_1 Optimization problem is possible. l_1-norm minimization is given by

$$\min x_{l_1} \; subject \; to \; Y = \varnothing f^d \tag{1.6}$$

CS obeys the rule of Restricted Isometric Property (RIP) or Uncertainty Principle (UUP) i.e. sensing and measurement matrix be incoherent with each other. Fig. 1 shows the CS framework.

In CS, for n data length signal, $m \ll n$, data samples are transmitted, where m corresponds to the number of CS measurements required for signal recovery. The transmission cost in CS is decided by two factors, the number of CS measurements and each CS measurement transmission cost (Wu, X. et.al., 2014). To reduce CS measurements hybrid CS is proposed and to reduce CS measurement transmission cost routing is integrated with CS. Clustering-based routing protocols significantly reduces energy consumption as compared to non-clustering routing methods. Hence, clusters utilizing CS are extensively used in WSN (Xie, R. et.al., 2014)(Chen, J.et. al., 2016)(Zhang, C. et.al., 2017). When employing this method, the clustering algorithm can affect the number of data transmissions. Most CS-based clustering

Figure 1. Compressive Sensing Framework

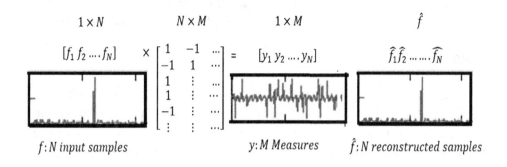

methods randomly choose nodes to form clusters neglecting the topology structure or property of sensor node readings such as spatial characteristics. WSNs are densely deployed, in this scenario one could use the information from one node to predict information in another, thus achieving overall lower communication cost. In (Leinonen, M. . et.al., 2015), the capabilities of various CS methods to reduce communication of correlated data gathering applications in a multihop WSN were studied. In (Nguyen, M. T. et.al., 2016) integration of CS and clustering to reduce power consumption related to data collection was proposed with the CCS algorithm. But the clusters are randomly formed. However, to learn correlation this requires communication between sensors or prior knowledge, adding to a new communication cost. Also, spatially correlated clusters in literature are elastic with no uniform measure of distance and error tolerance. This motivates us to design a novel cluster that promotes energy-efficient data collection in CS-based WSN.

In this work, we have designed a novel Spatially Correlated Cluster with Compressive Sensing (SCC_CS), which promotes energy-efficient data gathering in WSN to prolong lifetime. Different from other spatially correlated clusters, these clusters follow Tobler's Law "Everything is related to everything else, but near things are more related than distant things" (Miller, H. J.,2004). Hence, correlation is not based on sensor readings or distance but on spatial proximity. The proposed novel cluster is BS assisted while spatial correlation in the cluster is estimated based on geographical proximity within the sensing range r_s. Herein, a relation between spatial proximity, impact area and correlation is derived.

Motivation

Sensors in WSNs are often densely deployed in a harsh environment with very little or no maintenance. Therefore, the operation of these networks relies on these small and inexpensive devices under severe energy constraints. Saving energy in such a network is always a critical problem that has a direct impact on the lifetime of the network. Also, the spatial correlation of the sensor readings in WSN results in an inherent sparsity of data hence facilitates the application of CS technique in WSN.

Objective

A primary objective of data gathering is to obtain an approximation of the signal field with as little energy expenditure as possible. Sensor nodes are deployed in high density as the individual competency of these nodes really does not matter but the collaboration between them makes a difference. The collaboration is energy expensive in wireless media. In this media, the task of transmitting information from one point to another is a common and well-understood exercise, but the problem of efficiently transmitting or sharing information from and among a vast number of distributed nodes remains a great challenge especially in an energy-constrained network. Thus, the objective is to design energy-efficient data collection for energy-restricted wireless sensor networks by exploiting spatial correlation in a dense WSN. The fundamental aim of the work is to exploit spatial correlation in the cluster to reduce the number of transmissions and apply CS at the cluster head to save energy consumption to prolong the life-span of the wireless sensor network.

Significant Contribution

The main contribution of this chapter is-

- Energy-efficient data gathering scheme is proposed to improve network lifetime with the impact of spatial correlation on the routing method which is further integrated with compressive sensing.
- The sensing range parameter is adopted to overcome the challenge of grouping nodes randomly while forming clusters in the existing methods of compressive sensing integrated with clustering.
- Deriving a relation between spatial proximity, impact area, and correlation and prove that impact area for spatial proximity is equal to the sensing range parameter of a sensor node.
- Designing a novel Spatially Correlated Cluster based on the concept of spatial proximity within the impact area with Compressive Sensing (SCC_CS). Two approaches are used for cluster formation viz: Cluster First and Leader First approach. We also prove that for hexagonal topology both these approaches couple with each other.

- The cluster head is only the active member in the entire cluster while the rest of the members in the cluster are put to sleep. Thus, solely CH senses and transmits compressed data to the base station saving on the intra-cluster communication cost.

Background

In general, for minimizing high data transmission and increasing the lifetime of WSN, Clustering and data compression is one of the best techniques. Duarte et al., 2012 have overviewed and detailed an array of proposed compression methods. In particular, CS is a new approach to simultaneously sense and compress, and that promises to reduce the sampling and computational cost and is used very economically in an energy-constrained sensor network. The First practical implementation of CS called Compressive Data Gathering (CDG) was done by Luo, C et.al., 2009, which uses a tree-based aggregation. In CDG instead of receiving individual sensor readings, the sink will be sent a few weighted sums of all the readings, from which it will restore the original data. But this burdens the leaf nodes with unnecessary transmissions. To alleviate this problem, a hybrid version of CS and raw data collection was proposed by Jun Luo, et. al., 2010 and applied by Xiang, L., et. al., 2011, in a tree-based aggregation. Since clustering, have many advantages over tree-based method, integration of clustering and CS is proposed in the literature. Xie, R. et.al., 2014, combined hybrid CS with clustering, that is, inside the cluster data gathering is done without CS and between CHs data gathering is done with CS, and analytically found the optimal size of the cluster that leads to the minimum number of transmissions possible. A data gathering tree spanning all CHs is constructed to transmit data to the sink by using the CS method. Minh, N, et al., 2013 also calculated the optimal number of clusters and proved that consumed power is a decreasing function of the number of clusters, i.e. more clusters result in more power saving. Lan. K.C et. al., 2017 proposed a Compressibility Based Cluster Algorithm (CBCA) that enables fewer data transmissions than the random clustering method used in the real integration of CS and clustering. In CBCA, the network topology is converted first into a logical chain similar to the concept used in PEGASIS and then the spatial correlation of the cluster nodes readings is employed for CS. In the work, it is shown that CBCA enables less data transmission than the Random Clustering method. While this method is applicable to many sensor networks such as chain or mesh networks, it might fail for some particular types of networks. Various clustering algorithms are proposed by Xiaoronga, C., et. al. 2012, but the volume of work on a spatially correlated cluster in a dense WSN is less. It is observed that nearby sensor nodes monitoring a phenomenon typically register similar values. Hence grouping such sensor nodes together and appointing a node to represent the entire group value to the sink, can reduce energy consumption and hence prolong the network lifetime. By exploiting inherent spatial and data correlation in WSN some researchers have systematically discussed spatial correlation, which can be categorized based on the spatial distance between nodes or spatial correlation of sampled data. In literature, the judgment of the spatial correlation is based on the geographic distance of sensors (Yuan, J. et. al., 2009), tolerance error of different sensor reading (Liu, C. et. al, 2007) or area of overlap between sensors (Shakya, R. K., et. al, 2013), and the combination of error tolerance range and spatial correlation range (Liu, Z., et. al, 2013) to form the correlated cluster. Instead of using the conventional circular shape we assume the shape of the cluster as hexagon. Hexagon has the highest coverage area and satisfies all required criteria. Wang, D. et. al, 2011 assumed a hexagonal clustered wireless sensor network and applied subdivision of the hexagon for overall power saving. The deeper the subdivision, the more the power saving was achieved and prolonged the lifetime of sensor nodes. Thus analytical results have shown that subdivision will yield considerable savings in overall power consumption of the cluster. The saving is heavily dependent on the node's transmission range and their deployment density. Fig 2 below shows the subdivision used by the author.

Figure 2. Number of subdivision in hexagon

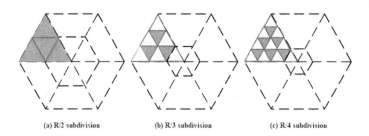

(a) R/2 subdivision   (b) R/3 subdivision   (c) R/4 subdivision

Existing CS-based data gathering integrated with clustering neglect the inherent spatial correlation in a dense WSN while forming a cluster and form clusters randomly. Also, a spatially correlated cluster in literature has no specific or strict requirement on the similarity measure between nodes or distance. By considering these facts, we propose to estimate the spatial correlation between sensor nodes on the geographical proximity within them and propose a Spatially Correlated Cluster using Compressive Sensing (SCC_CS).

Problem Formulation

In, clusters integrated with CS, all nodes in the cluster send data to the CH and CH applies CS further transmitting compressed data to BS. Thus, we have intra-cluster communication between CH and Member Nodes (MN) and inter-cluster communication between CHs of different clusters.

Let WSN consist of N sensor node with C clusters. Let the ith cluster be formed by S_i number of sensor nodes and

$$\sum_{i=1}^{C} S_i = N \tag{2.1}$$

In each cluster, one node is CH and (S_i-1) nodes are member nodes. These member nodes transmit their data to CH in the transmission phase and then CH performs CS. Therefore, the number of transmissions in the network are

$$\sum_{i=1}^{C} S_i - 1 + \sum_{i=1}^{C} M_i \tag{2.2}$$

i.e. the number of transmissions of an ith cluster is the number of transmissions of (S_i-1) *SNs* to its *CH* and CS performed by i^{th} *CH*. If each cluster contains the same number of nodes S and each cluster head performs CS to M number of measurements, then the communication load of the whole network is:

$$C*(S - 1) + C*M = N + C(M - 1) \tag{2.3}$$

To reduce the communication load of the network, we propose correlated clusters with all sensor nodes that have similar information. Hence, only CH transmits the data to the BS and other nodes are in sleep mode, so no transmission takes place within the cluster i.e. no intra-cluster communication cost. Only CH transmits data to the BS. So, the number of transmissions is reduced to $C*M$. Hence, with correlated cluster and CS, the communication load reduces significantly

SPATIALLY CORRELATED CLUSTER WITH COMPRESSIVE SENSING

In this section, we present the design of the proposed Spatially Correlated Cluster with Compressive Sensing (SCC_CS). Before that, we present the model used for building the cluster architecture and proof of the same in the subsections.

Spatial Correlation Model

Consider the sensing range of a sensor node as circular using the unit disk sensing model. The correlation between sensory data of nodes related to the spatial correlation between them is estimated based on sensory coverage of nodes. Also, their reading association character can describe the correlation of different sensor nodes, which is covariance. The covariance between two measured values from the node n_i and n_j at location R_i and R_j respectively can be expressed as

$$Cov\{R_i, R_j\} = \tilde{A}_s^2 K_g(d) \tag{3.1}$$

Where

σ_s^2 = variance of sample observation from sensor nodes
$Cov(.)$ = mathematical covariance
$K_g(.)$ = denotes correlation function

A model depicting the correlation between sensor nodes and spatial proximity is discussed in the next section

Assume two sensor nodes with a sensing range r_s located d distance apart as shown in Fig. 3.

Using geometry to set up the correlation model with the meaning of symbols explained as follows:

R_i and R_j denote the location of the node n_i and n_j of the disk with radius r_s.
A_i: Area R_i denoting the area of R_i
A_j: Area R_j denoting area of R_j
R_i^j : Region delimitate by the perpendicular bisector of R_i and R_j and belongs to R_i
A_i^j : Area R_i^j denoting the area of R_i^j
A_j^i : Area R_j^i denoting the area of R_j^i
R: denotes the sensing region

Figure 3. Correlation model

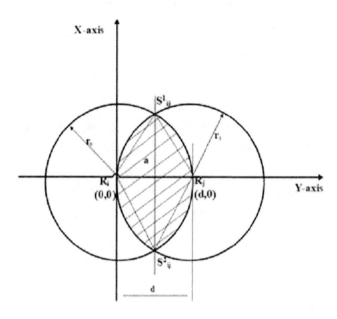

A: denotes the area of R

d:distance between R_i and R_j

S_{ij}^1 and S_{ij}^2 : are intersection points of the two nodes

a: length of the common chord joining S_{ij}^1 and S_{ij}^2

$$a = \sqrt{4r_s^2 - d^2} \tag{3.2}$$

The intersection of two nodes with radius r_s and distance d is

$$O(d, r_s) = \begin{cases} 0 \\ \pi r_s^2 \\ 2r_s^2 \cos^{-1}\left(\dfrac{d}{2r_s}\right) - \dfrac{d}{2}\sqrt{4r_s^2 - d^2} \end{cases} \tag{3.3}$$

1st case corresponds to non-intersecting discs

2nd case corresponds to one disc entirely contained in the other

3rd case corresponds to the non-trivial intersection.

If $d < 2r_s$ then R_i overlaps with R_j and correlation is defined as

$$K_9(d) = \frac{A_i^j + A_j^i}{A} \tag{3.4}$$

Due to symmetry, $A_i^j = A_j^i = A^{int}$ (say). The area of the asymmetry lens which intersects the sensing range of two sensor node and is calculated using the formula of the circle segment of radius R' and triangle height d'

$$A^{int}(R', d') = R'^2 \arccos(\frac{d'}{R'}) - d'\sqrt{R'^2 - d'^2} \tag{3.5}$$

Then, eqn.3.3 is

$$K_9(d) = \frac{2A^{int}}{A} \tag{3.6}$$

Depending on the location of the node n_j at R_j concerning node n_i at R_i, whether it is outside, inside or on the sensing radius, the area of the intersection will vary. Eqn. 3.6 gives a direct relationship between correlation and area of intersection. If location R_j is outside the sensing range that is the radius r_s, the intersecting region is less; hence correlation is weak. On the other hand, if the location R_j is inside the sensing range that is the radius r_s, then the intersecting area is more; hence correlation is robust. At the boundary, that is on the circumference of the circle with the area of $\pi r_s 2$ the area of overlap is 50%, i.e. $\pi/2$. If the circle has a unit radius then,

$$K_9(d) = \frac{2 \times \pi/2}{\pi} = 1 \tag{3.7}$$

Which means the correlation is strong. After this point, the correlation starts becoming weak. Hence, the impact area where correlation is strong corresponds to the area of the circle.

Spatial Data Correlation Proof

In order to prove that spatial correlation exists we replicate the performance of CS at CH, the dataset used is the real data set obtained by a WSN deployed at

Intel Laboratory Berkeley (Intel labs Berkeley data), and simulation for this was developed in MATLAB R2015. This data set contains temperature, humidity, light and voltage values periodically collected with 54 distributed Mica2Dot sensor nodes from 25th February - 5th April 2004. Fig. 4 shows the distribution of sensor nodes in Berkeley Lab.

Figure 4. Intel Berkeley lab

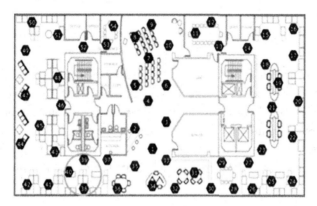

The communication radius of the sensor node in the Intel Berkeley Lab is set to 6m and the sensing radius is 3m since the sensing radius is half the communication radius. The correlated cluster is formed with Node numbers 38, 39 and 40 and depicted in the Berkeley lab by a green circle as shown in Fig. 4 To prove that the spatial readings of the nodes are correlated, we tabulate their spatial coordinates, i.e. x and y coordinates (in meters relative to the upper right corner of the lab) and plot a graph of the temperature reading at the same time instance in Fig. 5. Table 1 gives the spatial coordinates of sensor Node 38, Node 39 and Node 40. As seen in the graph, readings are highly correlated. Hence, instead of all the nodes sending the same readings to the BS, we can suppress correlated information in the geographical proximity [11] and save the number of transmissions in the network.

Table 1. Node coordinates

Node	(x location, y location)
38	(30.5,31)
39	(30.5,26)
40	(33.5,28)

Figure 5. Spatial correlation between readings of Node38, 39 and40

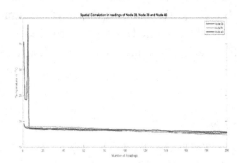

But with the primary objective of data gathering in WSN with an accurate approximation of the signal, in the proposed cluster with CS, we evaluate the performance of our method with respect to the reconstruction accuracy. Hence, we calculated the RMSE value for the three nodes in the WSN and tabulate in table 2. From the table, we observe that the RMSE values are very much similar. Hence, one node can transmit the data from the correlated region to the sink.

Table 2. RMSE values

Node	RMSE Value
38	0.244194
39	0.228896
40	0.244613

System Design

Following assumptions are made in this work

i. The sensed information is highly correlated.
ii. The sensor data is K sparse.
iii. The network contains N sensor nodes and the information of $\varnothing$ and Ψ are stored in the sink.
iv. Sensor nodes know the geographic location via the attached GPS or other localization techniques.
v. The sensing range r_s is same and half the transmission range r_t i.e. $r_t = 2r_s$ for each sensor.
vi. The sensor data can be reconstructed with high probability when $M = 3K \sim 4K$

The process of clustering is mainly composed of three phases: CH selection, cluster formation, and data transmission. As sensor nodes are stationary, cluster formation is a one-time process and CH selection is performed during each interval. The initial clusters are formed immediately after the first interval of CH selection by BS. The selected CH is analyzed at each interval for energy remaining and it varies based on the energy threshold. Hence, every node associated with the cluster will get the chance of becoming CH once. As the current CH change to new CH, it leads to a change in cluster position.

Randomly deployed sensor network that has to be partitioned into clusters, can be based on the order of whether the cluster is formed first and then CH (leader node) is selected or CH is selected first and then cluster is formed. Two approached are devised for cluster formation

- Cluster First approach
- Leader First approach

In our base station assisted cluster formation method, we work on both these approaches and show that the leader first approach coupled with the cluster first approach for a hexagonal shape.

Cluster First Approach

In this base station assisted clustering approach the following modules are involved viz: Hexagonal base area division, cluster head selection and sub-cluster head selection

Hexagonal Deployment

Hexagon is a geometric shape with the largest coverage area and resembling the radiation pattern of an omnidirectional antenna is chosen for sensor deployment. Hexagons can be overlaid without overlapping each other, this has been proven in cellular geometry and is well applicable to wireless sensor networks also as shown in Fig. 6a. Hence the area of interest i.e. sensing field is divided by the base station into tessellating hexagons as shown in Fig. 6b.

Once all the nodes are deployed in the area of interest, they inform their location and energy to the Base Station (BS). Considering only one part of the network that is a hexagon, BS divides the area into six regions as triangles as shown in Fig. 7a with each region contains sensor nodes. The formula for dividing the areas into six triangles is given is:

Figure 6. (a) Hexagon around the base station (b)Tessellating hexagons

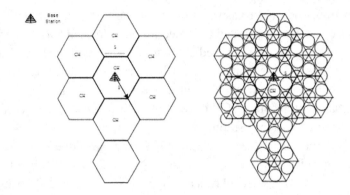

$$Area = \frac{\left(x_1\left(y_2 - y_3\right) + x_2\left(y_3 - y_1\right) + x_3\left(y_1 - y_2\right)\right)}{2} \tag{3.8}$$

Selection of Cluster Head

Considering BS assisted cluster formation, the BS selects the CH with the highest residual energy and appropriate location. The CH selection process is mathematically described by a parameter W_{CH}, which depends upon the average residual energy of the CH and their position and is expressed as:

$$W_{CH} = w_1 R_{CH}^L + w_2 R_{CH}^E \tag{3.9}$$

Where R_{CH}^L is location and R_{CH}^E is an energy factor of CH while w_1 and w_2 indicate the contribution of both the parameters in the expression of W_{CH}. After dividing the field into six regions the BS calculates the geographic center point of each region, this is the value R_{CH}^L. The center point of the cluster is calculated using the formula:

$$O_x = \frac{A_x + B_x + C_x}{3} \quad O_y = \frac{A_y + B_y + C_y}{3} \tag{3.10}$$

The sensor nodes do not know who is the closest to the central point of a cluster area, in this scenario, cluster of all nodes within the range from the center be the CH candidates of the cluster. The candidate that has the smallest distance to the center of the cluster among the other candidates becomes the CH of the cluster. BS geocast the tuple (R_{CH}^L, R_{CH}^E) into the arca. All nodes receive this tuple, but the one

Figure 7. (a) Hexagon divided into six regions (b) Location of cluster head in the region of interest (c) Formation of sub-cluster heads

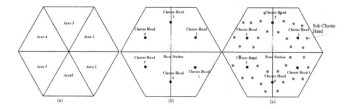

with the closest value announced in the message is elected as CH. As shown in Fig. 7b.

Selection of Sub-Cluster Head

After the identification of CH, the sub-cluster heads of each cluster-area has to be elected. Following the same procedure, each region is further subdivided into six sub-regions as given by Eqn. 3.9 and then SCH is calculated using the formula in Eqn. 3.10. Six SCH are elected in each region. These sub-clusters from the correlated regions; hence only the SCH is required to gather and transmit data for this region to the CH of that region. Fig. 7c shows the location of these sub-cluster heads.

Leader First Approach

As against the cluster being formed first in this leader first approach, the CH is first chosen by the BS and then a cluster is formed around it, in this approach.

Cluster Head Selection

As discussed in the above section, BS selects the CH with the highest residual energy and most appropriate location. The parameters R_{CH}^{L} and R_{CH}^{E} are to be computed as follows: R_{CH}^{L} is evaluated at the intersection of distance $2r_s$ and 60^0 angle line as illustrated in Fig. 8a

Let this location be denoted by coordinate (x_i, y_i) at R_i. R_{CH}^{E} is measured by taking the ratio of average residual energy of CH and the average residual energy of non-CHs. Thus, BS calculates W_{CH} and geocasts the tuple $\left(R_{CH}^{L}, R_{CH}^{E} \right)$. All nodes receive this tuple but the one with the closest value announced in the message is elected as CH and forms a cluster around. These selected CHs can expand the network by

Figure 8. (a) Location of cluster head in the region of interest (b) Cluster formation

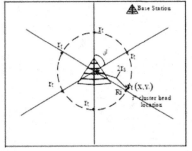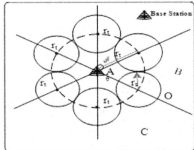

selecting CHs at twice their sensing range thus achieving scalability. The next section explains the cluster formation.

Cluster Formation

Considering, spatial correlation of sensory information between sensor nodes geographically proximal to each other, a cluster with a radius equal to the sensing range of the CH. Hence, after the CHs election, they announce their availability as a CH within the area of its sensing radius r_s by transmitting announcement packets to form a cluster. Nodes associated to CHs via sending a packet, which includes its ID, location information and residual energy. These nodes remain permanent members of the cluster and will become CH once in their lifetime. Fig. 8b illustrates cluster formation according to the disk-sensing model. The average number of nodes in a correlated cluster will depend on the sensing range r_s and node density in the area.

The Pseudocode for cluster head selection and cluster formation is given:

I.   *Cluster Head Selection Phase*
     Input: Sensing range r_s, Sensor nodes location and Initial Energy
     Weights: w_1 and w_2
     Output: CHs selected $\text{ÇH}', \text{ÇH}'', \ldots\ldots\ldots$
          Step 1: BS calculate CHs among the nodes randomly deployed using

$$W_{CH} = R_{CH}^L + R_{CH}^E \text{ where}$$

$$R_{CH}^L = \frac{E_0}{mean(N.E_0)}, R_{CH}^E = \left(2.r_s, angle\left(60^0\right)\right)$$

Step 2: Geocast tuples $S\left(R_{CH}^{L}, R_{CH}^{E}\right)$

Step 3: For each node j; $j \in N$ {/* Parallel process for each j */}

Node j= CH if $dist\left(j\right) = pdist(tuple\left(R_{CH}^{L}, R_{CH}^{E}\right)$

Step 4: $ÇH', ÇH'', \ldots = \varnothing$

II. *Cluster Formation*

Input: list of CHs: $ÇH', ÇH'', \ldots\ldots\ldots$

Output: Clusters formed

Step1: Each CH'_j broadcasts announcement packets with its identity embedded into it to all the nodes i within $d(j) - d(i) \leq r_s$

Step 2: Node i chooses a cluster to join:

a. Node i receives only one message from a CH'_j, then it joins a cluster of j

b. If i receives $n\left(2 \leq n \leq \left|CH'_j\right|\right)$ message from cluster head then j chooses a CH'_j to join if it satisfies $RSSI = \max\left\{CH'_j\right\}$

Step 3: Node i associate to CH'_j by sending a join message.

Step 4: Cluster is formed.

Step 5: Check if all nodes are associated with a cluster, if no then go back to step 1

Data Transmission

After clustering, CS-based data gathering is done by CH and CS measurement is transmitted to the BS.

Let CH_i denote the CH of the cluster i, and f_i represent the sensor reading at CH_i. CH_i has N_i readings which can be denoted as,

$$f_i = \left[f_1, f_2, \ldots\ldots\ldots\ldots f_{N_i}\right] \tag{3.11}$$

The CH multiplies this by a random matrix $\varnothing_i$ and then sends the product Y_i to the BS. The BS collects each measurement from one cluster at a time, and a Block Diagonal Matrix (BDM) as a sensing matrix is built.

$$\begin{bmatrix} Y_1 \\ Y_2 \\ \vdots \\ Y_M \end{bmatrix}_{M \times 1} = \begin{bmatrix} \varnothing_1 & 0 & 0 & 0 \\ 0 & \varnothing_2 & 0 & 0 \\ 0 & 0 & \ddots & 0 \\ 0 & 0 & 0 & \varnothing_M \end{bmatrix}_{M \times N} \begin{bmatrix} f_1 \\ f_2 \\ \vdots \\ f_{N_i} \end{bmatrix}_{N \times 1} \tag{3.12}$$

This BDM with only one nonzero entry in each row and column is the sparsest measurement matrix. Finally, the BS receives, $Y = \bigcup\limits_{i=1}^{C} Y_i$, the compressed information of all clusters at the BS. The original data can be reconstructed from Y by using l_1 minimization the compressed information of all clusters at the BS, the original data can be reconstructed from Y by using l_1 minimization.

Cluster Head for Next Round

Since the cluster head is solely sensing, transmitting and communicating, it will drain out of its energy. Hence when its energy reaches a threshold, E_{th}, another member has to take its position as cluster head. The member is in sleep mode and the sleep schedule is distributed in the beacon (Zhen, C., et. al. 2014) by the cluster head. This schedule is not the same for all the members, it is based on the area of overlap between the cluster head and the member nodes. Degree of correlation increases with internodes proximity in a densely deployed network. Also, correlated variability is directly related to the amount of information present. Hence the area of overlap will decide the next cluster head for the cluster.

Let N_i and N_j be two sensor nodes with a sensing range r_s. Let, d_{ij}, be the distance between the two nodes. If distance $d_{ij} \leq 2r_s$ then the sensing area will overlap. The area of intersection of the two circles, in terms of distance d_{ij} and radius r_s is

$$A = r_s^2 \cos^{-1}\left(\frac{d_{ij}}{2r_s}\right) - \frac{d_{ij}}{4}\sqrt{4r_s^2 - d_{ij}^2} \tag{3.13}$$

The graph shown in Fig. 9 gives the relation between d_{ij} and A.

If the two nodes are at a distance of twice the sensing range they are uncorrelated. Flowchart calculating the CH in the next round is given in Fig. 10

Figure 9. Graph of distance vs Area

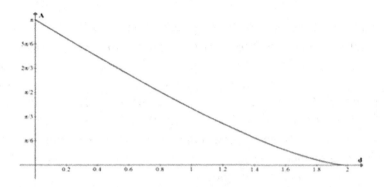

Figure 10. Flowchart of selection of next cluster head

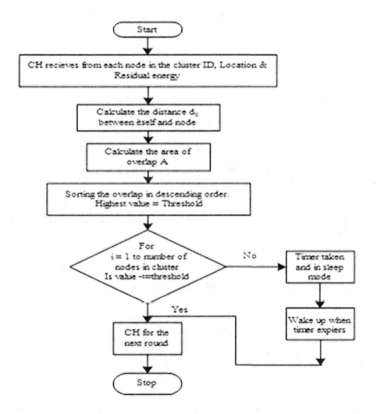

EXPERIMENTAL ANALYSIS

We evaluate the performance of the proposed cluster with CS applied at CH. The experimental design and analysis of SCC_CS are performed using the MATLAB tool. We designed by varying number of sensor nodes and evaluated the performance of proposed protocol compared to existing state-of-art methods such as random clustering without CS, CCS (Nguyen, M. T., et.al. 2016), Xie's method (Xie, R., et. al.,2014) and CBCA (Lan, K. C., et. al.,2017). Performance metrics such as the number of transmissions, energy consumption and network lifetime are compared with the existing method. The scenario is kept the same for both methods that are the number of nodes and energy model.

Energy Model

The energy model used in this work is as shown in Fig. 11(Heinzelman, W. B. 2000)

Figure 11. Radio energy dissipation model

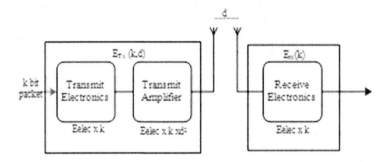

To transmit k bit to a distance d, the radio expends is given by

$$E_{Tx}(k,d) = \begin{cases} \{E_{elec}k + \mathcal{E}_{fs}kd^2 & d \le d_0 \\ \{E_{elec}k + \mathcal{E}_{mp}kd^4 & d > d_0 \end{cases} \tag{4.1}$$

E_{elec} = Electronics energy depends on factors such as digital coding, modulation filtering and spreading of signal.

$\mathcal{E}_{fs}d^2$ or $\mathcal{E}_{mp}d^4$ amplifier energy depends on the distance to the receiver and acceptable bit-error-rate

$$E_{Rx}(k) = E_{elec}k \qquad\qquad (4.2)$$

Simulation Results

The simulation parameters are tabulated in Table 3. MATLAB R2015 is used for simulation, with a network area of dimensions 20 X 10 Square units with the sink node at the corner as depicted in Fig. 12. The nodes are randomly deployed within the given area. The simulation scenario is kept the same for all the methods Viz. clustering without CS, CCS, Xie's, CBCA and SCC_CS method.

Figure 12. Deployment frame with 500 nodes in the sensor field

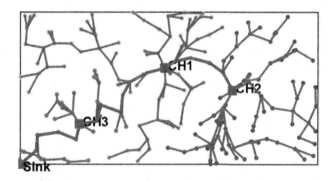

Table 3. Simulation Parameters

Name	Value
Number of Sensor Nodes	500
Compression Ratio	10%
Number of Iterations	10
Initial Energy	0.2 Joule
Transmitter Energy	0.05 Joule
Receiver Energy	0.05Joule

Clustering in WSN is a periodic process in which at each interval the re-clustering is performed in which the clusters position and CH may change according to the clustering method. In this work, we performed 10 rounds and for each round, we measure the parameters of the result. At each round, the deployment of sensor nodes is changed and hence the results are varying. The rounds performed in simulation to verify the reliability of the proposed method at different network conditions. Constructing a base station assisted SCC_CS, we observe that since only the CH is sensing and transmitting data on behalf of the entire cluster the number of transmissions is very less as compared to that of the other methods. Compressive sensing is applied in all the methods with a compression ratio in 0.1.

Figure 13. (a) Average number of transmission with the number of rounds (b) Residual energy graph

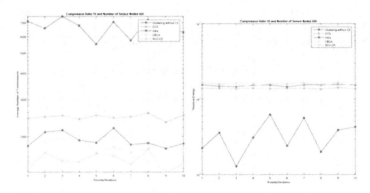

The plot of an average number of transmission with the number of iterations or rounds is given in Fig. 13 (a). The figures demonstrate the performance of a number of transmissions and network lifetime at each clustering round for 500 sensors network respectively. As observed in figure 13, the number transmissions required for clustering which is random and hence we name it Random Clustering (RC) method without CS is more as compared to other methods with CS. As the number of transmission in a wireless network is directly associated with the energy consumption, as the number of transmissions reduces the energy of the network is saved and so the network lifetime performance is poor for other methods in comparison to our proposed method of clustering namely, Spatially Correlated Clustering with CS(SCC_CS). More the transmission rounds, the less the network lifetime as more energy consumption is consumed in wireless transmission. In our work, lifetime is defined as the time interval when the first node in the network dies. Since wireless transmission is a major contributor to power consumption in every sensor node reducing the number of transmissions ultimately leads to less power consumption in a battery-operated sensor node. The residual energy of the node is enhanced and depicted in the graph plotted in Fig. 13 (b). The enhancement of the residual energy in a node will lead to an increase in the network lifetime of the wireless sensor network. A comparison of the network lifetime for all the methods is given in Fig. 13 (b). From the above results, at each round of clustering the SCC_CS method shows superior performance as compared to the benchmark clustering with CS methods especially the recent methods Xie's and CBCA. The number of transmissions is significantly reduced at each round for the SCC_CS method and hence improves the overall network energy efficiency as compared to state-of-art methods.

Density Evaluation

As per our study, most of the existing clustering methods deliver the worst performances as the number of sensor nodes increases in terms of a number of transmissions and energy efficiency as a large number of sensor nodes becomes the part of data collection and aggregation processes in WSN. As the number of nodes increases, the number of transmissions increases in a network (Fig. 14 (a)). To mitigate such challenges, we designed the novel clustering method to work effectively for the dense networks as discussed in the above section. In this section, we compare the SCC_CS method with all the investigated methods. Like any other compression technique measuring the accuracy of the reconstruction is an important parameter. One of the most popular ways to do it by calculating the root mean square error (RMSE) value. The expression of which is given by:

$$RMSE = \frac{\|s - \hat{s}\|_2}{s_2} \tag{4.3}$$

Where s is the original signal, $\hat{s}$ is the approximated signal and $s_2 = \left(\sum_{i=1}^{n} |s_i|^2 \right)^{\frac{1}{2}}$ is the 2-norm or Euclidean length of s.

As observed in Fig. 14(a), it is obvious that the number of transmissions of SCC_CS method is significantly smaller than that of the RC method without using CS. The reason is that data are compressed using the CS method at the CHs in the SCC_CS method. SCC_CS method shows the less number of transmissions at each network as compared to other CS-based methods. The reason is that novel spatial correlation-based methods do not allow all the sensor nodes to become the part of data capturing and transmission to the CH, rather put them into the sleep mode. This can lead to a significant reduction in energy consumption and hence increases the energy efficiency (Fig. 14 (b)) and network lifetime (Fig. 14 (c)) performances.

Figure 14. (a) Average number of transmissions performance evaluation (b) Energy efficiency performance evaluation (c) Average network lifetime

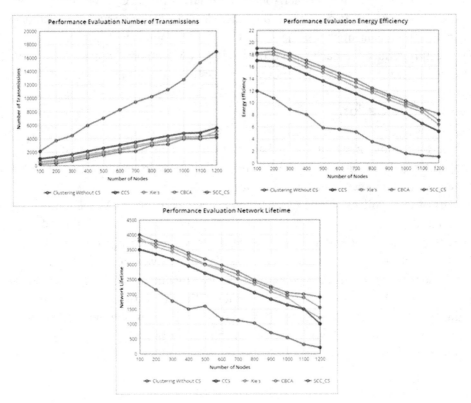

In the proposed clustering method, SCC_CS, we used the matrix-based CS technique with l_1-norm reconstruction approach, which is different from other investigated data gathering methods. Thus we evaluate the performance of CS methods to recover the correct data at the sink node using the RMSE metrics. As observed, the RMSE performance of RC with basic CS approach shows the worst performance, whereas the proposed method and CBCA method shows the mixed performances but better than Hybrid CS. As some networks, the performance of CBCA is better as compared to proposed, but the overall performance of RMSE is good for the proposed method.

CONCLUSION AND FUTURE SCOPE

A massive amount of data generated by a dense WSN can be economically compressed with CS. Usually, CS is integrated with clustering to gain its benefits neglecting the inherent spatial correlation in WSN. In this work, we exploit spatial correlation in the cluster at the same time reducing the number of transmissions making the network transmission efficient. Different from other clusters a spatially correlated BS assisted cluster is proposed SCC_CS. Two approaches, cluster first and leader first, are used to form the correlated clusters were correlation is estimated based on the sensing range of the CH. From results obtained, we believe that due to the proposed clustering method we can able to achieve the reduction in the total number of transmissions which leads to energy efficiency for all WSNs. In benchmark methods, Hybrid CS and CBCA, the random clustering and compressibility-based clustering (PEGASIS chain formation) are used for the data aggregation and transmissions. In both methods, the cluster members also become part of this process and hence increase the number of transmissions. The CBCA shows the better performance as compared to Hybrid CS as it forms the logical chain based on the spatial correlation technique. But in comparison to CBCA, the performance of our method is superior in terms of all the parameters evaluated which is the number of transmissions, energy consumption and network lifetime. In the proposed method, we not only reduce the intra-cluster communications, since only cluster head is the sole node sensing the data in the entire cluster and all other nodes are put in the cluster in sleep position so that energy consumption minimizes. The outcomes prove the effectiveness of the proposed method. For future work, it will be interesting to investigate the new CH selection using the energy parameter of other SNs in-network (if the current CH drops its energy level below the threshold value).

REFERENCES

Chen, J., & Chang, Z. (2016). Hierarchical Data Gathering Scheme for Energy Efficient Wireless Sensor Network. *International Journal of Future Generation Communication and Networking, 9*(3), 189–200. doi:10.14257/ijfgcn.2016.9.3.18

Donoho David, L. (2006). Compressed sensing. *IEEE Transactions on Information Theory, 52*(4), 1289–1306. doi:10.1109/TIT.2006.871582

Duarte, M. F., Shen, G., Ortega, A., & Baraniuk, R. G. (2012). Signal compression in wireless sensor networks. *Philosophical Transactions of the Royal Society A: Mathematical, Physical and Engineering Sciences, 370*(1958), 118-135.

Heinzelman, W. B. (2000). *Application-specific protocol architectures for wireless networks* (Doctoral dissertation). Massachusetts Institute of Technology.

Intel labs Berkeley data. (n.d.). http://db.csul.mit.edu/www.select.cs.cmu.edu/data/labapp3/

Lan, K. C., & Wei, M. Z. (2017). A compressibility-based clustering algorithm for hierarchical compressive data gathering. *IEEE Sensors Journal, 17*(8), 2550–2562. doi:10.1109/JSEN.2017.2669081

Leinonen, M., Codreanu, M., & Juntti, M. (2015). Sequential compressed sensing with progressive signal reconstruction in wireless sensor networks. *IEEE Transactions on Wireless Communications, 14*(3), 1622–1635. doi:10.1109/TWC.2014.2371017

Liu, C., Wu, K., & Pei, J. (2007). An energy-efficient data collection framework for wireless sensor networks by exploiting spatiotemporal correlation. *IEEE Transactions on Parallel and Distributed Systems, 18*(7), 1010–1023. doi:10.1109/TPDS.2007.1046

Liu, Z., Xing, W., Zeng, B., Wang, Y., & Lu, D. (2013, March). Distributed spatial correlation-based clustering for approximate data collection in WSNs. In *Advanced Information Networking and Applications (AINA), 2013 IEEE 27th International Conference on* (pp. 56-63). IEEE.

Luo, C., Wu, F., Sun, J., & Chen, C. W. (2009, September). Compressive data gathering for large-scale wireless sensor networks. In *Proceedings of the 15th annual international conference on Mobile computing and networking* (pp. 145-156). ACM. 10.1145/1614320.1614337

Luo, J., Xiang, L., & Rosenberg, C. (2010, May). Does compressed sensing improve the throughput of wireless sensor networks? In Communications (ICC), 2010 IEEE international conference on (pp. 1-6). IEEE. doi:10.1109/ICC.2010.5502565

Miller, H. J. (2004). Tobler's first law and spatial analysis. *Annals of the Association of American Geographers, 94*(2), 284–289. doi:10.1111/j.1467-8306.2004.09402005.x

Nguyen, M. T., & Rahnavard, N. (2013, November). Cluster-based energy-efficient data collection in wireless sensor networks utilizing compressive sensing. In *Military Communications Conference, MILCOM 2013-2013 IEEE* (pp. 1708-1713). IEEE. 10.1109/MILCOM.2013.289

Nguyen, M. T., Teague, K. A., & Rahnavard, N. (2016). CCS: Energy-efficient data collection in clustered wireless sensor networks utilizing block-wise compressive sensing. *Computer Networks, 106*, 171–185. doi:10.1016/j.comnet.2016.06.029

Shakya, R. K., Singh, Y. N., & Verma, N. K. (2013). Generic correlation model for wireless sensor network applications. *IET Wireless Sensor Systems, 3*(4), 266–276. doi:10.1049/iet-wss.2012.0094

Wang, D., Lin, L., & Xu, L. (2011). A study of subdividing hexagon-clustered WSN for power saving: Analysis and simulation. *Ad Hoc Networks, 9*(7), 1302–1311. doi:10.1016/j.adhoc.2011.03.001

Wu, X., Xiong, Y., Yang, P., Wan, S., & Huang, W. (2014). Sparsest random scheduling for compressive data gathering in wireless sensor networks. *IEEE Transactions on Wireless Communications, 13*(10), 5867–5877. doi:10.1109/TWC.2014.2332344

Xiang, L., Luo, J., & Vasilakos, A. (2011, June). Compressed data aggregation for energy efficient wireless sensor networks. In *Sensor, mesh and ad hoc communications and networks (SECON), 2011 8th annual IEEE communications society conference on* (pp. 46-54). IEEE. 10.1109/SAHCN.2011.5984932

Xiang, L., Luo, J., & Vasilakos, A. (2011, June). Compressed data aggregation for energy efficient wireless sensor networks. In *Sensor, mesh and ad hoc communications and networks (SECON), 2011 8th annual IEEE communications society conference on* (pp. 46-54). IEEE. 10.1109/SAHCN.2011.5984932

Xiaoronga, C., Mingxuan, L., & Suc, L. (2012). Study on clustering of wireless sensor network in distribution network monitoring system. *Physics Procedia, 25,* 1689–1695. doi:10.1016/j.phpro.2012.03.296

Xie, R., & Jia, X. (2014). Transmission-efficient clustering method for wireless sensor networks using compressive sensing. *IEEE Transactions on Parallel and Distributed Systems, 25*(3), 806–815. doi:10.1109/TPDS.2013.90

Yuan, J., & Chen, H. (2009, September). The optimized clustering technique based on spatial-correlation in wireless sensor networks. In *Information, Computing and Telecommunication, 2009. YC-ICT'09. IEEE Youth Conference on* (pp. 411-414). IEEE.

Zhang, C., Zhang, X., Li, O., Yang, Y., & Liu, G. (2017). Dynamic clustering and compressive data gathering algorithm for Energy-efficient wireless sensor networks. *International Journal of Distributed Sensor Networks, 13*(10), 1550147717738905. doi:10.1177/1550147717738905

Zhen, C., Liu, W., Liu, Y., & Yan, A. (2014). Energy-efficient sleep/wake scheduling for acoustic localization wireless sensor network node. *International Journal of Distributed Sensor Networks, 10*(2), 970524. doi:10.1155/2014/970524

Chapter 3
Study of Self-Organizing Coordination for Multi-UAV Systems

Meena T.
VIT-AP University, India

Ravi Sankar Sangam
VIT-AP University, India

ABSTRACT

In recent days, the usage of drones was increased and extended to various domains such as surveillance, photography, military, rescue, etc. Drones are small flying computers with on-board sensors and camera with a limited battery and coverage area. Due to the limited coverage area, usage of standalone drones in the above-mentioned domains such as rescue and military is restrictive. Multi-drones with self-organizing network can help to solve the above discussed issues. Hence, this chapter presents an extensive review on drone networks in which the core areas such as coverage, connectivity, link establishment, etc. are discussed. Finally, this chapter concludes by leveraging the challenges in state-of-the-art technologies in drone networking.

1.INTRODUCTION

Unmanned Aerial Vehicles (UAVs), also called drones, have gotten expanding enthusiasm for ecological and cataclysmic event observing, fringe observation, crisis help, inquiry and safeguard missions, and transfer correspondences. Small

DOI: 10.4018/978-1-5225-9493-2.ch003

multicopters are quite compelling practically speaking because of their simplicity of sending and low acquisition and support costs. Innovative work in small multicopters began with tending to control issues, for example, flight security, mobility, and heartiness, trailed by planning self-ruling vehicles equipped for waypoint flights with negligible client mediation. With progresses in innovation and financially accessible vehicles, the intrigue is moving toward collective UAV frameworks. Consideration of small vehicles for the previously mentioned applications normally prompts organization of multiple aerial vehicles that are arranged. Particularly, for missions that are time basic or that traverse a substantial land zone, a solitary little UAV is deficient because of its constrained vitality and payload. A multi-UAV framework, be that as it may, is more than the aggregate of numerous single UAVs. Notwithstanding permitting scope of bigger territories, numerous vehicles give decent variety by watching and detecting a zone of enthusiasm from various perspectives, which expands the unwavering quality of the detected information. In addition, the natural excess expands adaptation to internal failure. A few undertakings investigated the outline difficulties of UAV frameworks in various applications. The general plan standards of a multi-UAV framework in common applications still needs examination and remains an open issue. In this article, we outline a few difficulties for the plan of an arrangement of various little UAVs. These UAVs have a restricted flight time, are furnished with on-board sensors and implanted handling, speak with each other over remote connections, and have constrained detecting scope.

We recognize the fundamental building squares of a multi-UAV framework as sensing, communication, and coordination modules. Our primary objective is to give an outline of the coveted usefulness inside these plan squares and to pick up understanding toward a general framework engineering. We imagine that such an engineering can be misused in the outline of multi-UAV frameworks with various vehicles, utilizations of intrigue, and goals. To delineate the talked about standards, we present an agent system of collective UAVs and give a few true contextual analyses examining brought together and circulated approaches and the related difficulties. In particular, we utilize our multi-UAV elevated observing framework to help firefighters amid a calamity, to give expansive region scope no mission time imperatives, and for inquiry and protect with continuous video bolster. We show that diverse applications have distinctive coordination, sensing, and communication limitations. For time-basic missions with evolving goals, distributed coordination and reliable sensing and networking are required. For large area coverage, for example, ecological checking with no time imperatives, the way design can be produced before the mission in a unified station, and the detected information can be handled disconnected, unwinding the requirements on correspondence. In the spite of fact that not researched in this article, conveyance of merchandise by UAVs require concentrated or decentralized coordination, though correspondence and detecting should be dependable to adjust to dynamic requests and to maintain a strategic distance from deterrents and crash in urban conditions for safe conveyance. The assorted variety of utilization requests underpins the investigation of multi-UAV frameworks from coordination, sensing, and communication perspectives and we imagine that the exercises learned in our examinations will control the exploration network toward accomplishing a compelling multi-UAV framework for a large number of common applications. The rest of the article is organized as follows. In Section 2 and 3, we present multi-UAVs overview and architecture, respectively. Section 4 describes collaborative drone network and finally Section 5 draws conclusion. Key contributions in this paper,

- We discuss detailed description of UAV
- We explain Architecture of UAV
- We discuss major issues and challenges in drone networks

2.SYSTEM OVERVIEW

Critical properties of a multi-UAV framework are robustness, adaptivity, asset productivity, versatility, helpfulness, heterogeneity, and self-configurability (Szafir et al., 2017). To accomplish these properties, the physical control of individual UAVs and in addition their route and correspondence abilities should be incorporated (Rajappa et al., 2017). Outline and usage of these functionalities, independent from anyone else, constitute surely understood research subjects. Calculations and plan standards proposed by explore networks in remote specially appointed and sensor systems, mechanical technology, and swarm knowledge give important bits of knowledge into at least one of these functionalities and additionally blends of them (Rathinam et al., 2004).

The previous two decades saw a few nonmilitary ventures on UAVs (Cole et al., 2010). A grouping of these tasks can be made as takes after: First, we can recognize the sort of vehicles utilized, for example, helicopters, airships, or settled wing UAVs. These vehicles have diverse sizes, payloads, or flight times, and these distinctions influence the system lifetime, removes that can be voyage, and in addition the correspondence ranges. Second, an order can be made on the focal point of research, for example, outline of the vehicles or plan of calculations. To wrap things up, the applications for which these systems are sent additionally contrast. Prerequisites from the applications include diverse limitations the framework plan and they have as of late been investigated (Hayat et al., 2016).

While these activities begin from various suppositions, center around various functionalities, and mean to address distinctive limitations and objectives, on a fundamental level, they fulfill some normal plan standards (Andre et al., 2014). In like manner, one can think of a natural theoretical graph that catches the embodiment of multi-UAV frameworks in the writing. Figure 1 represents the abnormal state building squares of a multi-UAV framework. The UAV stage in this outline alludes to the utilized vehicles, the product and equipment related with the low level and high-level controls of these vehicles, and locally available processors. The Sensing block is in charge of watching nature and examining the gathered information from the earth as well as different vehicles, while Communication and Networking square empowers scattering of data between gadgets in the system. The basic leadership is taken care of by Coordination block, which forms input and imperatives from the rest of the building blocks. The connections between the blocks and the required usefulness from each block are subject to the objective of the framework. Existing multi-UAV frameworks center around the plan of at least one of these squares for various applications. This dynamic portrayal improves the plan contemplations for multi-UAV frameworks and should be refined further to concoct particular outline standards. On a fundamental level, one can treat these blocks freely when building a multi-UAV framework and address the difficulties forced by each block decoupled from the others. This natural and basic decoupling approach permits bringing in calculations from the comparing research network. An all the more intriguing yet difficult approach is to send a coordinated plan that considers associations and impacts between squares. The technique for incorporating these blocks, outlining the fundamental collaboration and input systems, and designing a perfect group of various UAVs are critical issues to be tended to.

3.SYSTEM ARCHITECTURE

A multi-UAV framework can work in a centralized or decentralized way. In a centralized framework, a substance on the ground gathers data, settles on choices for vehicles, and updates the mission or undertakings. In a decentralized framework, the UAVs need to expressly participate on various levels to accomplish the framework objectives and trade data to share assignments and settle on aggregate choices. Free of whether task is centralized or decentralized,what makes a gathering of single UAVs into a multi-UAV framework is the certain or express collaboration among the vehicles.

The UAVs need to

- Observe the environment
- Evaluate their own observations and information received from other UAVs, and reason from them, and
- Act in an effective way.

Thinking should be possible at the centralized control element or locally available the UAVs with full or fractional data. The conceivable activities are dictated by the capacities of the UAVs and the objective of the multi-UAV framework. In the accompanying, we relate the sensing, communication and networking, and coordination obstructs in a multi-UAV framework to the Observe-Reason- Act (ORA) cycle and condense wanted usefulness and related assignments in these building squares. We don't portray the UAV stage and accept that framework can contain a heterogeneous arrangement of little scale UAVs.

Figure 1. Multi UAV System Design Block

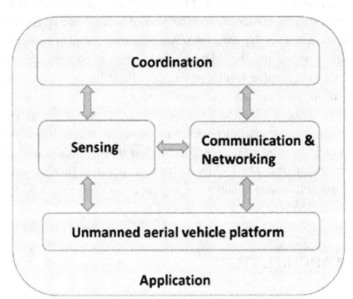

The diagram of Multi UAV System Design Block contain three major blocks. Sensing block is responsible for observing the environment. Communication and Networking block empowers scattering of data between gadgets in the system. The basic leadership is taken care of by Coordination block. Below given detailed description about this diagram.

3.1 Sensing

The sensing block goes about as eyes noticeable all around. Solid and precise detected information is basic for meeting the objectives of the mission. Contingent upon the application, an assortment of sensors might be utilized locally available the UAVs. While cameras as aloof sensors are usually utilized with the end goal of aeronautical checking, dynamic sensors can likewise be utilized for perception. These sensors should be lightweight with an effortlessly available interface for correspondence and in the meantime have the capacity to give adequate nature of tactile information to fulfill mission prerequisites. Generally, during UAV task in a wide region sensor arrange, there are few limitations such as time, add up to vitality utilization and disallowed flight regions so determining the ideal UAV flight way and the strategy used to get the detecting data from a devoted arrangement of sensors are NP-Hard issues (Yang & Yoo, 2018). Some particular issues that should be tended to in this block are:

Robust sensing: The capacities and attributes of sensors may influence the arranging, coordination, and correspondence design. For example, the UAV waypoints are arranged considering the field of perspective of the sensors locally available. Be that as it may, a UAV may tilt because of the flying progression or wind. Some UAV producers mount the sensors on dynamic suspensions to make up for the tilting impact, gave that as far as possible isn't come to.

Sensor fusion: A UAV is regularly outfitted with an assorted arrangement of sensors, for example, GPS, whirligig, accelerometer, and gauge. Subsequently, a hearty strategy for information combination from numerous heterogeneous sensors is required. This tangible information should be additionally synchronized and investigated to accomplish data combination and more elevated amount objectives, for example, coordination and impediment crash shirking.

3.2 Communication

The communication and networking block is in charge of the data stream in the ORA cycle. This block should be vigorous against vulnerabilities in nature and rapidly adjust to changes in the system topology. Correspondence isn't basic for spreading perceptions, assignments, and control data, yet it is expected to organize the vehicles all the more successfully toward a worldwide objective, for example, checking a given territory or identifying occasions in the briefest time, which are particularly critical in a fiasco circumstances. In particular, the 3-D geometry channel demonstration is planned as a mix of the UAV development state data and the channel gain data, where the previous antenna can be gotten by the sensor combination of the flight control framework, while the last antenna can be assessed through the pilot transmission (Zhao et al., 2018). Some particular issues that should be tended to in this block are as per the following:

Connectivity: On the off chance that correspondence framework is inadequate with regards to, the utilization of UAVs as transfers between detached ground stations will end up basic. UAVs have restricted correspondence ranges, are exceedingly versatile, and have rare vitality assets. This square needs to look after availability, and the utilized systems administration and booking conventions need to adjust to the dynamic condition.

Routing and scheduling: Past keeping up availability and meeting nature of administration necessities, conventions that can deal with or, all the more attractively, that fuse three dimensional controlled portability should be outlined.

Communication link models: Multicopters have particular designs and imperatives not quite the same as settled wing UAVs. Models that catch the attributes of UAV-UAV and UAV-ground joins are required.

Data transmission: Transmission of the payload information, e.g., control data, sensor readings, pictures, and videos, must be performed with the end goal that the QoS necessities of the application are met under fluctuating system conditions. This may incorporate adjustment of the payload information.

3.3 Coordination

The coordination block is the reasoning and basic leadership element, which is in charge of utilizing perceptions, mission necessities, and framework requirements to compose the UAVs. Basically, it needs to figure the directions of the UAVs and settle on choices on the best way to dispense undertakings to accomplish group conduct. Coordination can mean accomplishing and managing unbending arrangements or can be errand dissemination among vehicles in a self-sorting out way. So also, it should be possible at a nearby or worldwide level, contingent upon the mission and capacities of the vehicles. Adaptability and heterogeneity are additionally wanted in a multi-UAV framework, since countless with various capacities are normal. In this way, the coordination block needs to deal with developing quantities of heterogeneous UAVs, errands, and perhaps mission zones. Some particular issues that should be tended to inside this block are:

Task allocation: Thinking and basic leadership is expected to ideally circulate undertakings to individual UAVs or gatherings of UAVs that can deal with dubious or deficient data and dynamic missions. Instruments to characterize and adjust assignments to the mission necessities or vehicle capacities should be composed.

Path planning: There are a few way arranging procedures for ground robots and direction outlines for arrangements of robots. More assignment enhanced, correspondence mindful, three-dimensional way arranging techniques are wanted for multi-UAV frameworks that can deal with rare vitality assets and heterogeneous vehicles (Perazzo et al., 2017). The crash likelihood of UAV in development flying is broke down by methods for the likelihood computation and impact evasion technique for satellites or other rockets. Following the impact likelihood investigation in satellites, while breaking down the crash issues of UAVs flying in development, we expect that the UAVs move directly. Therefore, the two UAVs move in uniform direct motion, and their speed can be estimated, so the area mistake ellipsoids keep up unaltered when they meet. We expect that the two gatherings of UAVs will impact or have just impacted when the separation between the two gatherings of UAVs is littler than their real radius. The crash likelihood refers to the probability that the distance between two groups of UAVs is less than the sum of their radius. The crash likelihood isn't just a substantial standard for diminishing and maintaining a strategic distance from UAV impacts and demonstrating the hazard coefficient when UAVs are meeting yet in addition a critical reason for keeping away from impediments (Wu et al., 2018).

This general outline and the portrayal in Figure 1 can be viewed as an underlying deliberation of the parts of a multi-UAV framework and can give a few rules in the outline of multi-UAV frameworks with various abilities and with various requirements forced by various applications. In the accompanying, we refine this theoretical portrayal and actualize a community oriented multi-UAV framework that can be sent for different applications.

4.COLLABORATIVE DRONE NETWORK

The framework studied in this paper is to screen a specific zone in a given day and age and with a given refresh recurrence to help protect work force in a debacle circumstance. It is intended to (i) catch ethereal pictures and give a review picture of the checked zone, and (ii) recognize and record the status of an object in real time. The fundamental task begins with a client characterized errand depiction, which is utilized to register courses for the individual UAVs. For the zone observing application, the UAVs y over the zone of intrigue and procure pictures. The pictures are sent to the ground station and move toward becoming mosaicked to a huge diagram picture. For the Search and Rescue (SAR) application, the UAVs look through the protest of enthusiasm for a given territory. Once the protest is identified, they reposition themselves, shaping a correspondence transfer chain, to convey ongoing video of the objective to the ground station. The abnormal state modules in this design are:(i) the user interface; (ii) the ground station containing mission control, mission arranging, and sensor information investigation; i.e., coordination; (iii) a correspondence foundation; and (iv) the UAVs with their on-board preparing and detecting abilities.

In this diagram the associations between the design blocks are demonstrated by directional connector in Figure 2. For example, the sensing and coordination blocks are connected through sensing capacities, wanted sensor inclusion, and asset confinements of the UAVs. The sensor information to be conveyed impacts the communication

& networking hinder amid booking of transmissions. We additionally think about elective dimensions of collaborations among coordination and communication & networking blocks, where we have the alternative of centralized coordination with no cooperation or decentralized coordination with correspondence subordinate UAV movement. We bolster distinctive kinds of UAVs gave they have some base usefulness, for example, independent flight and intends to determine the route waypoints. The processed courses are given in a stage autonomous arrangement and the UAVs' ready control makes an interpretation of these conventional directions into the UAV-explicit low-level directions. With more UAVs, the multifaceted nature for coordination and arranging additionally increments. Along these lines, we require a strong disseminated engineering for programming advancement which gives a helpful structure to low-level gadget control and message going between the hubs. The Robot Operating System (ROS) has been abused in our framework for this reason. Utilizing the UAVs with a preparing ability on-board and furnished with ROS, every element can get the present status from or send directions to different elements advantageously. We consider both centralized and decentralized coordination and correspondence modules. In the decentralized case, arranging usefulness is moved starting from the ground station to the UAVs.

Figure 2. System Architecture

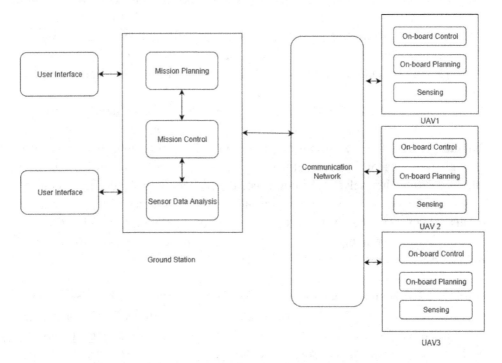

User interface: The user interface has two principles; one is to begin with the client to characterize the abnormal state errands to be proficient by portraying the zone to be checked on a computerized delineate. Other, the client can characterize certain properties, for example, the required picture determination or on the other hand refresh interims. Second, it furnishes the client with the produced mosaicked picture with the present places of the UAVs. A mid mission execution, the client can change the assignments as required (Merino et al., 2006).

Sensing: For the outline of the sensing square, the necessities of the applications within reach and mean to address the difficulties forced by the restricted assets accessible on-board the UAVs. Previously or amid the mission, the flight courses are sent to the UAV's ready Control. The on-board control isn't in charge of the low level control to balance out the UAV's height, yet in addition to explore proficiently to the registered waypoints. The Sensing module is in charge of catching pictures what's more, preparing the picture information on-board before transmission to the ground station. preparing incorporates highlight extraction, comment with meta-information, quality checks, furthermore, multi-determination encoding.

Contingent upon the application, distinctive sorts of cameras are utilized for Contingent upon the application, distinctive sorts of cameras are utilized for sensing. The study utilizes visual camera's for catching pictures amid sunlight, while a warm camera might be utilized for night vision, seeing through smoke or haze, vegetation checking, fire and warm discovery utilizing infrared examples, and so on. On the off chance that the caught pictures will be utilized for additionally, preparing, they will be exchanged to the base station (Yahyanejad & Rinner, 2015). Something else, assuming all the pictures are not important to be exchanged, they can be prepared on-board to separate a particular example or highlight and after that the UAV responds in like manner or potentially begins spilling the caught pictures from an objective region (Scherer et al., 2015).

Little scale UAVs have restricted assets, and it is basic to relegate these assets painstakingly. The study utilizes picture pressure for lessening the extent information before exchanging them. Utilizing a picture pyramid is another approach to spare information transmission stack and handling power. All the time, pictures with low determination and quality are prepared or exchanged first; we continue to higher quality later when assets wind up accessible.

Communication: The examination with respect to the communication and networking square core interests on accomplishing and looking after network, investigation of air-air what's more, air ground channels by means of genuine tests, deciding the impediments of existing remote correspondence advances, proficient information transmission, and examination of correspondence requests from an application perspective.

To set up a dependable multi-UAV framework, it is important to consider the requests postured by systems administration of the UAVs and base stations. An elevated system with three dimensional versatility benefits from radio wires with almost isotropic radiation power designs. Moreover, to empower conveyed online basic leadership, it is important to have ongoing correspondence. In versatile application situations, as in one of our utilization cases, where the mission assignments shift after some time, such correspondence might be required to spread data and errands. Moreover, SAR is a time-basic application where consistent availability to ground staff is obligatory. Accordingly, industrious system network is alluring to engender data productively.

The decision to restrict the remote channel in an aeronautical system, we have moved our concentration to correspondence requests of a multi-UAV framework. The requests on organize network rely upon the application (Hayat, 2016). Consequently, to decide QoS requests from an aeronautical system, taken an application driven approach (Andre et al., 2014) and have recognized the building squares of a multi-UAV framework as far as correspondence needs of the usefulness requested by a given application. Few general guidelines for various application classes, displayed an examination of systems administration execution from genuine world tests in the writing, and gave understanding into the eventual fate of ethereal systems (Chen et al., 2002).

Coordination: The coordination square designs the flight courses of the UAVs, adjusting to the necessities of the use of intrigue. The exploration centers around the outline of way arranging and errand portion systems, considering detecting and correspondence limitations.

The coordination contains three principle parts. Mission Control is the center module of this framework. It takes the client's information what's more, dispatches it to alternate segments. The Mission Planning part separates the abnormal state errands to flight courses for individual UAVs. A flight course contains an arrangement of focuses to visit in world directions and certain activities for each waypoint. At long last, the Sensor Data Analysis part mosaics the pictures from the UAVs into a solitary extensive review picture, which is then exhibited to the client. Since mosaicking is a computationally escalated process, abuse an incremental approach that speedily demonstrates a diagram picture to the client while the UAVs are as yet executing their central goal (Yahyanejad & Rinner, 2015).

Concentrated and conveyed coordination techniques to deal with static and dynamic situations. Like correspondence and systems administration requests, coordination of a multi-UAV framework, the basic leadership process, and the level of data trade among gadgets rely upon the errands identified with every application. For the region scope application, it considers pre-characterized UAV ways created at the ground station (Quaritsch et al., 2010). To this end, the territory of intrigue is separated into cells, relating to picture focuses, with the end goal that the quantity of pictures to cover the territory is limited and the quality necessities for picture sewing are fulfilled. At that point, utilizing a different voyaging salesperson issue approach, briefest ways over the photo focuses are produced considering the quantity of vehicles and flight time confinements. In the event that the ways should be refreshed amid the mission, the new ways are produced at the ground station and conveyed to the UAVs in a concentrated way (Khan et al.,2015). An examination of this proposed pre-characterized and disseminated coordination approaches for the region scope application regarding mission time and arranging multifaceted nature can be found in (Yanmaz et al.,2011).

5.CONCLUSION

This paper presented an extensive study on drone network. From the study it is inferred that the drone network suffers in various aspects such as connectivity, coverage, node stabilization etc. Further this paper presented the experimental evaluation of few techniques which gives insight for the above claim. The presented study infers that there are many open issues yet to be resolved in this area. The need for dynamic application using drones are still on-demand and focus should be given in better way so that the real world constraints can be sorted at ease. In order to achieve the discussed points, a novel model is required to be formulated for drone network.

REFERENCES

Andre, T., Hummel, K. A., Schoellig, A. P., Yanmaz, E., Asadpour, M., Bettstetter, C., Grippa, P., Hellwagner, H., Sand, S., & Zhang, S. (2014). Application-driven design of aerial communication networks. *IEEE Communications Magazine*, *52*(5), 129–137. doi:10.1109/MCOM.2014.6815903

Chen, B., Jamieson, K., Balakrishnan, H., & Morris, R. (2002). Span: An energy-efficient coordination algorithm for topology maintenance in ad hoc wireless networks. *Wireless Networks*, *8*(5), 481–494. doi:10.1023/A:1016542229220

Cole, D. T., Thompson, P., Göktoğan, A. H., & Sukkarieh, S. (2010). System development and demonstration of a cooperative UAV team for mapping and tracking. *The International Journal of Robotics Research*, *29*(11), 1371–1399. doi:10.1177/0278364910364685

Hayat, S., Yanmaz, E., & Muzaffar, R. (2016). Survey on Unmanned Aerial Vehicle Networks for Civil Applications: A Communications Viewpoint. *IEEE Communications Surveys and Tutorials, 18*(4), 2624–2661. doi:10.1109/COMST.2016.2560343

Khan, A., Yanmaz, E., & Rinner, B. (2015). Information exchange and decision making in micro aerial vehicle networks for cooperative search. *IEEE Transactions on Control of Network Systems, 2*(4), 335–347. doi:10.1109/TCNS.2015.2426771

Merino, L., Caballero, F., Martínez-de Dios, J. R., Ferruz, J., & Ollero, A. (2006). A cooperative perception system for multiple UAVs: Application to automatic detection of forest fires. *Journal of Field Robotics, 23*(3-4), 165–184. doi:10.1002/rob.20108

Perazzo, P., Sorbelli, F. B., Conti, M., Dini, G., & Pinotti, C. M. (2017). Drone Path Planning for Secure Positioning and Secure Position Verification. *IEEE Transactions on Mobile Computing, 16*(9), 2478–2493. doi:10.1109/TMC.2016.2627552

Quaritsch, M., Kruggl, K., Wischounig-Strucl, D., Bhattacharya, S., Shah, M., & Rinner, B. (2010). Networked UAVs as aerial sensor network for disaster management applications. *Elektrotechnik und Informationstechnik, 127*(3), 56-63.

Rajappa, S., Bülthoff, H., & Stegagno, P. (2017). Design and implementation of a novel architecture for physical human-UAV interaction. *The International Journal of Robotics Research, 36*(5–7), 800–819. doi:10.1177/0278364917708038

Rathinam, S., Zennaro, M., Mak, T., & Sengupta, R. (2004). An architecture for UAV team control. *IFAC Proceedings Volumes, 37*(8), 573-578.

Scherer, J., Yahyanejad, S., Hayat, S., Yanmaz, E., Andre, T., Khan, A., & Rinner, B. (2015, May). An autonomous multi-UAV system for search and rescue. In *Proceedings of the First Workshop on Micro Aerial Vehicle Networks, Systems, and Applications for Civilian Use* (pp. 33-38). ACM. 10.1145/2750675.2750683

Szafir, D., Mutlu, B., & Fong, T. (2017). Designing planning and control interfaces to support user collaboration with flying robots. *The International Journal of Robotics Research, 36*(5–7), 514–542. doi:10.1177/0278364916688256

Wu, Z., Li, J., Zuo, J., & Li, S. (2018). Path Planning of UAVs Based on Collision Probability and Kalman Filter. *IEEE Access: Practical Innovations, Open Solutions, 6*, 34237–34245. doi:10.1109/ACCESS.2018.2817648

Yahyanejad, S., & Rinner, B. (2015). A fast and mobile system for registration of low-altitude visual and thermal aerial images using multiple small-scale UAVs. *ISPRS Journal of Photogrammetry and Remote Sensing, 104*, 189–202. doi:10.1016/j.isprsjprs.2014.07.015

Yang, Q., & Yoo, S. (2018). Optimal UAV Path Planning: Sensing Data Acquisition Over IoT Sensor Networks Using Multi-Objective Bio-Inspired Algorithms. *IEEE Access: Practical Innovations, Open Solutions, 6*, 13671–13684. doi:10.1109/ACCESS.2018.2812896

Yanmaz, E., Kuschnig, R., Quaritsch, M., Bettstetter, C., & Rinner, B. (2011, April). On path planning strategies for networked unmanned aerial vehicles. In *Computer Communications Workshops (INFOCOM WKSHPS), IEEE Conference on* (pp. 212-216). IEEE. 10.1109/INFCOMW.2011.5928811

Zhao, J., Gao, F., Kuang, L., Wu, Q., & Jia, W. (2018). Channel Tracking with Flight Control System for UAV mmWave MIMO Communications. *IEEE Communications Letters, 22*(6), 1224–1227. doi:10.1109/LCOMM.2018.2824800

KEY TERMS AND DEFINITIONS

Drone: A drone, in a technological context, is an unmanned aircraft. Essentially, a drone is a flying robot. The aircrafts may be remotely controlled or can fly autonomously through software-controlled flight plans in their embedded systems working in conjunction with onboard sensors and GPS.

Self-Organizing Coordination: A self-organizing is an automation technology designed to make the planning, configuration, management, optimization. Self-organizing networks are commonly divided into three major types. distributed, centralized, hybrid.

Unmanned Aerial Vehicle Networks: UAVs are a component of an which include a UAV, a ground-based controller, and a system of communications between the two. The flight of UAVs may operate with various degrees of autonomy, either under remote control by a human operator or autonomously by onboard computers

Wireless Sensor Network (WSN): WSN is a wireless network that contains spatially distributed self-governing devices using sensors to monitor environmental and physical conditions. The WSN system incorporates a gateway that supply wireless connectivity back to the wired world and distributed nodes.

Chapter 4
Vehicular Networks in the Eyes of Future Internet Architectures

Hakima Khelifi
Beijing Institute of Technology, China

Senlin Luo
Beijing Institute of Technology, China

Boubakr Nour
Beijing Institute of Technology, China

Hassine Moungla
Paris Descartes University, France

Syed Hassan Ahmed
Georgia Southern University, USA

ABSTRACT

The challenging characteristics of the vehicular environment such as high mobility, diversity of applications, dynamic topologies, unreliable broadcast channels, and short-lived connectivity call into the need to extend the IP-based network to fulfill the user and VANETs requirements. Researchers are developing new network communication models to transfer the future internet. The information-centric networking (ICN) paradigm is a promising solution that may overcome the issues mentioned above. ICN involves a named content, name-based routing, in-network caching, and content-based security, which make it a suitable architecture for VANET applications. In this chapter, the authors present recent advances in VANET solutions that rely on named-data networking (NDN), which is the most active ICN implementation. The issues of the current host-centric model, mapping between NDN and VANET, is also discussed along with future research directions.

DOI: 10.4018/978-1-5225-9493-2.ch004

1. INTRODUCTION

During the past two decades, Vehicular Ad hoc Networks (VANETs) (Al-Sultan, Al-Doori, Al-Bayatti, & Zedan, 2014) have attracted a vast number of research efforts and projects funding with a dedicated aim to provide a comfortable life, efficient transportation & safety services, and secure data sharing. However, ensuring such requirements in a vehicular environment which is characterized by a high and dynamic mobility of vehicles, that affects the network topologies and the reliability of communication, especially in case of very short-lived wireless connections i.e. vehicles moving so fast and rapidly changing their positions on the roads that it is hard to predict the long-term possible connectivity among nodes. Moreover, today's Internet users and applications requirements are shifting from the connectivity-oriented communications towards the data-oriented paradigm (Koponen et al., 2007). End-users are more related about the requested data regardless of the address of the network host.

Having said that, the Internet Protocol (IP) was initially developed over more than forty years ago to allow communication between two end-users using IP addresses. Altering this model to fulfill users' requirements ends by developing numerous protocol patches and add-ons on top of IP stack, tending to support and enhance security, mobility, management and other networking aspects (Campista, Rubinstein, Moraes, Costa, & Duarte, 2014). Hence, the research community is exploring new architectures for the future Internet (Pan, Paul, & Jain, 2011) that may satisfy the users' demands with a clean and straightforward state design.

In the recent years, Information-Centric Networking (ICN (Ahlgren, Dannewitz, Imbrenda, Kutscher, & Ohlman, 2012) appears as a promising paradigm to replace the current host-centric model and overcome the aforementioned issues. The communication in ICN is based on the content name without using any IP/host addresses. Thus, the content is decoupled from its original location, that by consequence allows the network to cache the content and serves it for future requests. Also, all security related-information are traversed with the content itself, which make the network more distributed and responsible for users' needs. However, due to the characteristics of vehicular networks from both networking and application perspectives, bringing ICN in VANET is a challenging task, that needs tweakings and customization in the networking level.

In this chapter, we aim to discuss the Internet model and its limitations to reply to today's needs, overview the existing architectures for future Internet, and present one of the very actively investigated projects known as Named Data Networking (NDN) in the context of vehicular networks. Also, we tend to discuss the mapping of VANET with NDN, from the requirement perspectives, existing solutions and efforts, and highlighting the existing issues and providing future directions. Table I lists all the acronyms and their explanations used in this chapter.

2. ICN: THE FUTURE INTERNET ARCHITECTURE

The focus of recent research has been on proposing preliminary architectures for the future Internet. In the following, we discuss the limitation of the current Internet model, follow by an overview on some known future Internet architecture. Later, we focus on the Information-Centric Communication Model.

2.1. Limitations of Current Host-Centric Model

The current IP protocol has been designed in the late of 1970s, aiming to provide connectivity between two hosts using IP addresses. This model was kept in development to provide different features including multicast, security support, mobility management, and fast content sharing & distributed. However, adding such add-ons on the TCP/IP model make it more complicated protocol. In the following, we present some limitation of TCP/IP protocol.

Table 1. List of Acronyms Used

Acronyms	Definitions	Acronyms	Definitions
CS	Content Store	NetInf	Network of Information
C2CCC	Car to Car Communication Consortium	PIT	Pending Interest Table
DONA	Data-Oriented Network	PURSUIT	Publish-Subscribe Internet
DSRC	Dedicated Short Range Communications	QoS	Quality of Service
		RSUs	Roadside Units
FCC	Federal Communication Commission	SAIL	Scalable and Adaptive Internet Solutions
FIB	Forwarding Information Base	SeVeCOM	Secure Vehicle Communication
GPS	Global Positioning System	SDN	Software-Defined Networks
GUID	Global Unique Identification	URIs	Uniform Resource Identifiers
ICN	Information-Centric Networking	VANETs	Vehicular Ad hoc Networks
ISO	International Organization for Standardization	VII	Vehicle Infrastructure Integration
ITS	Intelligent Transportation Systems	V2V	Vehicle-to-Vehicle
IVC	Inter-Vehicle Communication	V2I	Vehicle-to-Infrastructure
MANETs	Mobile Ad hoc Networks	V2U	Vehicle-to-Uniform
NDN	Named Data Networking	WAVE	Wireless Access for Vehicular Environment

Routing & Addressing: The Internet continues growing, where new devices and services are connected. Due to the exponential growth and functional evolution, a radical change in Internet addressing is required, not only to address and identify devices, but also services and content in the whole network. As a result of this, designing efficient and scalable routing algorithms that can deal with such is indispensable.

Multi-Protocol Architecture: Due to the existing of different protocols and architectures, the Internet infrastructure is facing a critical problem in which it should deal with all of the existing protocols and systems. Consequently, allowing applications in different domains to communication cross heterogeneous network platforms. Then, various middleware architectures have been proposed tending to adopt multiple protocols for communication. However, adaptation and scalability were the most critical issue in such a step.

Security & Privacy: The security is one of the crucial pillars of the Internet. Due to the heterogeneity in Internet users, services, and protocols and the exchanged information; building trust models, preserving user and data privacy, and ensuring authentication mechanisms became more complex and complicated. The distributed security system can help to achieve better scalability in security and privacy. However, it is not easy to build them on top of complex TCP/IP suite.

Traffic Control: During the data delivery path, all the TCP/IP packets are treated in the same manner, using the same forwarding plane, in which the forwarding decision is independent of the previous packets within the same session, or the same requested application or user, or even the application/traffic class. Although IP header offers Type-of-Service and Precedence bits to customize the quality of service, they are not used in all traffics.

Applications & User Requirements: Today's Internet is witnessing a colossal development of potential applications, including social media, content sharing, online shopping, and video gaming. Although these applications cannot be classified into the same categories, they share the same concept by addressing the user needs and the required content or information regardless of the host addresses.

3. FUTURE INTERNET ARCHITECTURES

To achieve the highest goal of ICN, many research projects have been proposed in the literature under the umbrella of ICN (Xylomenos et al., 2014) such as Data-Oriented Network (DONA) (Koponen et al., 2007), Publish-Subscribe Internet (PURSUIT) (Fotiou, Nikander, Trossen, Polyzos, & others, 2010), Network of Information (NetInf) (Dannewitz et al., 2013), Content-Centric Networking (CCN) (Oehlmann, 2013), and Named Data Networking (NDN) (Zhang et al., 2010) Although these projects have different architectures, they share the same concept which is addressing the content by its name rather than its network address. In the following, we provide a quick overview of each one.

3.1. Data-Oriented Network Architecture (DONA)

DONA is considered as one of the first complete ICN architectures. It uses persistent flat names to identify information objects; in particular, names are in the form *P:L*, where *P* is the ciphered hash of the public key of the content owner, and *L* uniquely identifies one of the contents with respect to the same owner. Publishers use a cryptographic hash as an object identifier, while subscribers verify the content integrity easily by hashing it and comparing the results.

3.2. Publish-Subscribe Internet Technology (PURSUIT)

PURSUIT is based on its predecessor Publish-Subscribe Internet Routing Paradigm (PSIRP) (Fotiou et al., 2010), both are funded by the EU FP7 project. PURSUIT adopts a complete clean-state approach in designing its ICN architecture, by using publish/subscribe stack instead of IP protocol stack. It adopted self-certifying flat names consisting of scope and rendezvous parts.

3.3. Scalable and Adaptive Internet Solutions (SAIL)

SAIL, and its predecessor 4WARD (Architecture and Design for the Future Internet) ("4WARD," 2008) ("FP7 SAIL Project," 2010), inherit aspects both from PURSUIT and from NDN. SAIL uses self-certifying flat names with possible explicit aggregation in the form *ni://A/L*, where *A* is the authority part, and *L* is the local part with respect to the authority, each part can by any type of string, from an URL to a hash value.

3.4. Convergence

Convergence is an EU FP7 project, that inherits a consistent number of features from the NDN architecture. By using self-certifying flat names in the form *namespaceID:name*, resembling the *P:L* pair of DONA, or they can be hierarchical as in NDN.

3.5. MobilityFirst

MobilityFirst mainly focuses on the mobility issue, that adopts a self-certifying flat name, which is a global unique identification (GUID). Each GUID is detached with its original location, i.e. IP address for Uniform Resource Identifiers (URIs). Although that GUID and network address are separated from each other, the MobilityFirst architecture still maintains the mapping between the two; therefore, it implements two routing scheme, GUID and network address based.

4. OVERVIEW OF INFORMATION-CENTRIC MODEL

Regarding the aforementioned limitations of IP protocols, and due to the qualitative and quantitative transformation of the current Internet. Researchers are focusing on developing new architectures and proposing new communication paradigms (Pan et al., 2011) to succeed the existing shortcoming.

Information-Centric Networking (ICN) (Xylomenos et al., 2014) has been proposed as a new future Internet paradigm. In compared to IP-based networks, ICN consists of using the content name as the primary, essential element in the networking layer. This paradigm aims to assign to each content a unique name, and to decouple it from its original location, hereby the network infrastructure might cache the content and serve it for future requests (Figure 1). What makes ICN a strong candidate for the future Internet, is that it does not care about how much the communication channel is secure, it secures the content itself, and all security related-information are traversed with the content packets.

Figure 1. Content exchange scenario in ICN

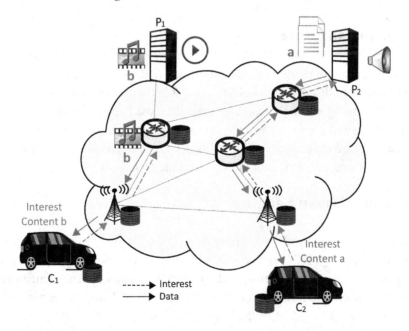

4.1. Content Naming

The content name is the pillar element in ICN. ICN names must be persistence, uniquely, and globally identify each piece of content. Various naming schemes have been proposed including hierarchical, flat, attribute-value based, and hybrid scheme names (Nour, Sharif, Li, Moungla, & Liu, 2017).

Hierarchical Naming Schemes: Similarly, to URIs, hierarchical names are composed of multiple components separated by a delimiter such as "/". Usually, the combination of these components describes the offered services or content. Hierarchical names might enhance the network scalability by offering name prefix aggregation and reducing the routing table size.

Flat Naming Schemes: Flat names can be obtained by performing hash algorithms on existing content or a part of it. Because hash algorithms are involved, neither a naming structure or semantic can be found behind a flat name, nor they are human-friendly. Hereby, they cannot scale as no aggregation rules can be used.

Attribute-value based Naming Schemes: A set of attributes are collected altogether to identify different properties and represent a single content. Attribute-value names are suitable to support searching process using keywords. However, they may provide different content for one search query. Thereby, they cannot ensure naming uniqueness.

Hybrid Naming Schemes: A hybrid name combines at two or more of the previously discussed schemes. Thereby the name consists of taking the best features provided by the base scheme to improve the network scalability, performance, enhance the security and privacy.

4.2. Routing and Forwarding

Due to the use of content name instead of host addresses. ICN uses name-based routing concept to discover and deliver content between requesters, producers and cache store, only by using the content name. In most ICN architectures, a content consumer triggered a request packet by specifying the required content name, nodes in the path forward the request until reaching the original producer or a replica cache store, where a data packet is sent back toward the consumer.

4.3. Content-Based Security

In contrast to the host-to-host communication model where the security of the channel is mandatory, ICN is promising to guarantee intrinsic security and provide privacy by securing the content itself rather than the channel. All the security-related information are coupled with the content and traverse with him in the network, this concept is known as Content-based Security (Afanasyev et al., 2016).

4.4. In-Network Caching

By leveraging unique content names, ICN tends to decouple the content from its original location. Also, integrating all security mechanisms in the content rather than the communication channel, give the network layer the opportunity to cache the content and serve for other requests. This feature is known as in-network caching. in-network caching aims to improve the content availability and distribution in the network by caching the popular content closer to consumers and reduce communication delay (Nour et al., 2018).

5. NAMED DATA NETWORKING (NDN)

Named Data Networking (NDN) (Shemsi & Kadam, 2017) is one of ICN architecture that implements content-centric paradigm, it is the most active project and well-managed ICN implementation. NDN implements hierarchical, human-readable names to identify any particular content instead of identifying the host.

5.1. NDN System Architecture

NDN architecture implements a pull-based communication model based on Interest-Data exchange. Interest packets are triggered by content consumer to discover the content in the network by specifying its name. The request may be satisfied either by the original data producer or any replica node. Hence a data packet is delivered back carrying the content and its name.

The content name is embedded in both interest and data packets. Interest packets are forwarded hop-by-hop using name-based routing until they reach the content producer or a replica-node. Data packets carried the same requested name and delivered to the requester using the reverse path made by interest packets. NDN is a receiver-driven architecture, as shown in Figure 2, where each node maintains three data structures:

Content Store (CS): The use of content names instead of host addresses allows a decoupling of the content and its original location, also applying security mechanisms into the content rather than the communication channel five the network layer the opportunity to cache the content. The intermediate nodes may cache the processing data and make them available for future requests instead of forwarding them to the original producer. The cache store aims to store the cached content along with its name, maintaining caching strategy by deciding which content should be cached, and deciding which content should be evicted in case of memory limitations. The in-network caching increases the data sharing & availability, saves more bandwidth, and reduces content retrieval time.

Figure 2. NDN Architecture

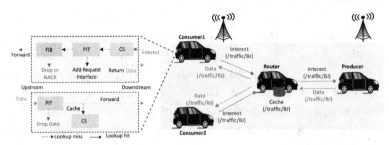

Pending Interest Table (PIT): Pending Interest Table is the pillar of the receiver-driven design, where it tracks information about the incoming interest packets until they will be satisfied by forwarding the associated data packets. Information such as the requested content name, list of interfaces at which the said interest requests have been received (interest aggregation), and timeout to keep the PIT table fresh, are stored in each PIT entry. NDN uses the PIT entry to deliver the data back to consumers in a multicast manner.

Forwarding Information Base (FIB): Forwarding Information Base table stores information about each next-hop relay nodes for each reachable destination name prefixes. It acts as a database for the destination name prefixes for the current intermediate node that are populated by the routing protocol. NDN engine performs longest-name prefix to find the suitable next-hop for the processing interest packet.

6. VEHICULAR AD HOC NETWORKS (VANETS) OVERVIEW

Vehicular Ad hoc Network (Cunha et al., 2016) is a specific type of Mobile Ad hoc Networks (MANETs) in which mobile nodes are mainly vehicles. Vehicles in VANET can exchange data with each other (V2V: Vehicle-to-Vehicle) or with infrastructure (V2I: Vehicle-to-Infrastructure) as well as with charging stations, personal communication devices, and smart grids (V2U: Vehicle-to-Uniform).

VANET has specific characteristics than MANET, that makes vehicular network featured, at the same time more challenging (Laroui et al., 2018). Vehicles in VANET move and run in high speed, which makes the VANET topology quickly change as well as frequent connect and disconnect in a very short period. Also, and due to the high movement of vehicles, it is hard of predicting the mobility of nodes which affect the routing and security tasks, and the delay of data delivering that is a critical factor in such applications, especially in safety applications. In additional, VANET is considered as an unbounded network that has grown quicker and faster which make standardized rules, policies and security enforcement a challenging task. Another characteristic that helps VANET is that each vehicle uses Global Positioning System (GPS) that may help to deliver data efficiency. Furthermore, VANET has unlimited power and storage to support complex algorithms that can be fitted with these factors to make VANET network more reliable and efficient.

6.1. State of VANET Projects

Over the past several decades, VANET attracted many researchers trying to improve human safety and driving conditions as well as provide other different services such as online video streaming and internet access. In the beginning, of 1980, JSK (Association of Electronic Technology for Automobile Traffic and Driving) in Japan have started Inter-Vehicle Communication (IVC) project (Luo & Hubaux, 2004). Later, California Partners for Advanced Transportation TecHnology (PATH) (Hedrick, Tomizuka, & Varaiya, 1994) and Chauffeur of Europe (Gehring & Fritz, 1997) have proved the coupling the vehicles together electronically to determine a train. Then in Europe, the International Organization for Standardization (ISO) evolved a framework for Intelligent Transportation Systems (ITS) including VANET standards (Booysen, Zeadally, & Van Rooyen, 2011). To improve safety and comfortable driving, European project CarTALK 2000 (Reichardt, Miglietta, Moretti, Morsink, & Schulz, 2002) was developed to overcome the related problems to different VANET applications. For automobile manufacturers side, and based on inter-vehicle communications to create efficient and safety traffic, Audi, BMW, Daimler Chrysler, Fiat, Renault, and Volkswagen have dealt to create a non-profit organization called Car to Car Communication Consortium (C2CCC) ("Car to Car Communication Consortium, CAR 2 CAR Communication Consortium Manifesto," n.d.). Moreover, IEEE created the IEEE 802.11p standard in order to provide Wireless Access for Vehicular Environment (WAVE) ("IEEE Standards Association, Wireless access in vehicular environments," n.d.).

In December 2003, Dedicated Short Range Communications (DSRC) (Toor, Muhlethaler, & Laouiti, 2008) was required to be the first wide-scale vehicular environment by the US Federal Communication Commission (FCC) to support the communication from vehicles to and vehicles/infrastructures for safety applications. Also, other projects were created for VANET such PReVENT (Amditis et al., 2010) project (Europe), Internet ITS Consortium ("Initernet ITS Consortium," n.d.), Advanced Safety Vehicle project (Takahashi & Asanuma, 2000) (Japan), Vehicle Infrastructure Integration program (VII) ("Vehicle Infrastructure Integration (VII) program of US Federal and State departments of transportation and automobile manufacturers," n.d.), Secure Vehicle Communication (SeVeCOM) (Kargl et al., 2008), and Network on Wheels project (Germany) (Abusch-Magder et al., 2007).

6.2. VANET Applications

In VANET-enabled networks, vehicles are equipped with different on-board devices, network interface, sensors and GPS receivers. These embedded devices allow vehicles to collect, process and broadcast information concerning itself and its environment to other vehicles and Roadside Units (RSUs), which improve safety and traffic condition as well as the provision of passenger comfort.

VANET application can be divided into three main class: safety, traffic and comfort applications.

Safety Applications: This category provides safety of drivers and passengers and clean environment by exchanging emergency messages via different V2V and V2I communications. Examples of this emergency messages are an accident, emergency warning system, lane-changing assistant and road-condition warning. Usually, these applications require V2V communication due to the critical and sensitive time requirements.

Traffic Applications: This kind of application improves the traffic conditions on roads by informing vehicles about it in order to avoid congestion and providing comfortable driving for passengers. Regularly this application uses unicast mode to forward messages to other nodes.

Comfort Applications: This type of application is also known as non-safety applications that provides suitable convenience to the drivers and passengers like optimized the route to a destination, weather and traffic information, location of gas stations, restaurants and hotels. As well as provide online video streaming, online games, and internet access while vehicles connect to the infrastructure.

6.3. IP-Based VANET Issues

Although VANET is a sub-class of MANET, its requirements and characteristics are fundamentally different. These characteristics caused many issues and challenges for the vehicular network that based on IP solution network. We summarized this issues in the following:

- **Standardization:** VANET is growing faster and bigger to an unbounded network which made the IEEE 802.11 standard cannot fit the requirement of scalable and robust network connectivity which needs more effort about the definition the roles of standardization.
- **Mobility:** Due to the high speed and mobility of vehicles that caused continuously changing VANET topology that affects routing protocols. The intermediate nodes may not found between source and destination, and the end-to-end communication may not be established.

- **Routing Protocols:** Another challenge that faced in VANET is designed suitable routing protocols that can deal with high mobility of vehicles and can run with different standards
- **Connectivity:** Connectivity between vehicles or vehicles and infrastructures is the most critical challenges in VANETs. It needs to create and design a network that provides an optimal delay under different constraints such as vehicles speed, high dynamic topology, and channel bandwidths.
- **Security & Privacy:** Because of VANET is a scalable and an unbounded network, the security enforcement needs careful design attention. Where the most problems exist are the trust in data with protecting the privacy of the users. The network should be designed with establishing trust between vehicles and protect the user's privacy.
- **Broadcasting:** Vehicles exchange a huge of amount of data with each other which caused collisions then lead to frequent retransmissions by vehicles that affect the delivery task and increase the transmission time of the messages. These issues need the effort to be created intelligent flooding schemes that can deal with broadcast storms.

7. NDN-BASED VEHICULAR NETWORKS

Various efforts have been shown to enhance the core NDN architecture or proposing new guidelines to adapt NDN for VANET. Figure 3 depicts a generic VANET architecture on top of NDN, where different VANET applications and scenarios can be modeled using NDN as a communication enabler.

7.1. Mapping NDN with VANETs

In the last decade, considerable attention has been paid to vehicular networks from both academic and industry sectors as a result of which different solutions have been proposed and commercialized. Due to the fact that most of the communications in VANETs follow a content-centric fashion, efforts have been made to check the applicability of ICN as a communication model for VANET.

The use of content name instead of host addresses and adopting a receiver-driven forwarding scheme regardless of knowing the address of requester of provider make NDN as a robust network architecture for VANET application, that can scale without decreasing the overall performance. In addition, the in-network caching and the simple content discovery mechanism based on Interest-Data exchange helps in the content access and dissemination. Similarly, introducing the content-based security

Figure 3. Generic NDN Architecture in VANET

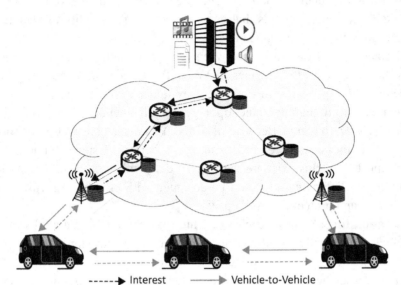

rather than communication-channel security adds more advantages on using NDN as a communication architecture in a vehicular environment.

This is demonstrated by the results that are shown in Figures 4 and 5. Where we have compared the host addresses with NDN in term of throughput and end-to-end delay. We can notice that NDN achieves a better throughput compared to the TCP/IP, as well as NDN provide an optimal and small end-to-end delay rather than TCP/IP. These results are due to NDN has a simple content discovery and delivery, and by using interest aggregation may help to make the communication delay smaller. Furthermore, by using in-network caching, the content will be cached in a near cache store, then the delay will be smaller rather than get it from the provider like TCP/IP, where requests are delivered to the original producer.

7.2. Existing Efforts on NDN-VANET

Toward the merger of NDN with VANET networks, various research efforts have been proposed by enhancing the core NDN architecture or targeting different features and aspects (Saxena, Raychoudhury, Suri, Becker, & Cao, 2016).

Figure 3 illustrates different vehicular communication scenarios through NDN model. In such an environment, vehicles might act as a content consumer, producer, or forwarder. Similarly, RSU and other core network elements such as switches and routers can serve as intermediate forwarders with the capabilities to cache the content.

Figure 4. Average End-to-End Delay

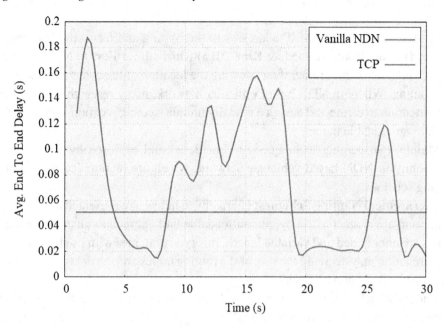

Figure 5. Network throughput Measurement

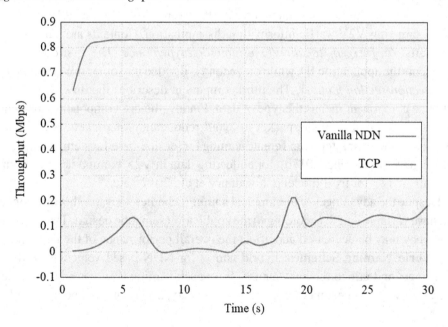

7.2.1. Naming Schemes

NDN uses names instead of IP addresses in the routing and forwarding processes (Safdar Hussain Bouk, Ahmed, & Kim, 2015) which allow users to get content by only giving their names rather than knowing the location of the content from where it is coming. Where in NDN based vehicular network, many research efforts show their attention to create and design a globally unique, secure, location-independent, and human-readable names.

Mainly, two naming technique are considered and explored by the research community in NDN based vehicular network, which are hierarchical and hybrid naming schemes:

Hierarchical Naming Schemes: Hierarchical names are presented as URL-like names and are characteristic by human-readable and aggregation names. Because names are unbounded and variable length, this type of names with its characteristic of aggregation may decrease the size of the routing tables and the delay of searching names as well as support a scalable network.

Under this scheme, many works have been implemented taking the benefits of hierarchical naming to identify both application types and traffic information (Lucas Wang, Wakikawa, Kuntz, Vuyyuru, & Zhang, 2012) (X. Liu, Nicolau, Costa, Macedo, & Santos, 2016) (Modesto & Boukerche, 2017b) (Chowdhury, Gawande, & Wang, 2017). The first naming for vehicular networks was proposed in (Lucas Wang et al., 2012), targeting V2V traffic information dissemination. Contents are named with five parts: */traffic/geo- location/timestamp/datatype/nonce*. The first component represents the application ID, while the second is divided into other sub-component: *ID/direction/section number*. The third component describes the time period, and datatype component means the type of data. Finally, the last component represents the sequence number of the packet to avoid redundancy of packets. Then many research shows their interest and create naming for service-based system architecture (Modesto & Boukerche, 2017b), for collecting data in V2X communication (Drira & Filali, 2014), for trust model (Chowdhury et al., 2017), etc.

Despite the advantages of hierarchical naming schemes, they may become longer and have variable length according to the application semantic design. Thus, lookup efficiency may be decreased and affect the overall performance of the network.

Hybrid Naming Schemes: hybrid names for NDN based vehicular network, refer to get and utilize the best features from each naming schemes and combined to create hybrid names in order to improve security, scalability and better mobility and heterogeneous.

A few hybrid naming schemes were proposed for NDN-based VANET. In (Quan, Xu, Guan, Zhang, & Grieco, 2014), a scalable naming scheme is proposed for multimedia contents in order to improve the quality of experience. The content name is divided into three parts: *Hierarchical Routable Prefix*, which is used for routing and name aggregation; *Flat Content Identifier* that used in caching aspect; and *Primary Attribute Labels*, which is used for multiple aggregations requests. Another work (Safdar H Bouk, Ahmed, & Kim, 2014) combines the hierarchical and flat names to design hybrid scheme. However, this scheme suffers from the same issues of the latter naming category.

7.2.2. Forwarding Solutions

The routing process in NDN uses only names with two packets which are Interest and Data packets to discover and deliver the content in the network. Many forwarding and routing schemes have been proposed that take the benefits of NDN, and the applicability in VANET to create a reliable and efficient network and limited the issue routing that faced from IP-based solution. These proposed schemes can broadly be divided into four classes and are as follow:

Geolocation-based Forwarding Schemes: The interest packets are guided and forwarded by the sender based on using the location information. This category of schemes may overcome the high-speed and mobility issues. Example of this class is introduced in (Grassi, Pesavento, Pau, Zhang, & Fdida, 2015), and namely Navigo. Navigo combines names with the geolocation producers' information to guide and forward interest packets in order to reduce the transmitted packets. Moreover, work in (X. Liu et al., 2016) also uses geographic location of interest and vehicle trajectories with designing new format packets in order to select the best next replay nodes. Similarly, work in (Bian, Zhao, Li, & Yan, 2015b) (Bian, Zhao, Li, & Yan, 2015a) use a new Neighbor Table that contains the position of vehicles in order to guide interest packets geographically; this table is updated periodically using beacon messages.

The benefits of using geolocation information are that they may limit flooding storm and packet retransmission. However, such scheme uses beacon message to refresh their geolocation information may lead overhead and collision as well as define other structures and change the nature of NDN is not preferable.

Neighbor-based Forwarding Schemes: In this category, the forwarding of interest packets is based on using the information of neighbors' nodes such as mobility factor or caching probability in order to select the best forwarder such as work in (Ahmed, Bouk, Yaqub, Kim, & Song, 2017) that used neighbor's speed, direction, and position to select potential forwarders. Work in (Ahmed, Bouk, & Kim, 2015) allows the consumer to select only one relay node to forward interest packets. The information of neighbors is updated based on exchange beacon message periodically. Another work proposed in (Kuai, Hong, & Yu, 2016) that uses the distance between vehicles and its neighbors as the metric to select the relay node aiming to overcome the mobility issues and provide a reliable end-to-end connection.

This scheme may reduce retransmission packet by selecting the best forwarder. However, it leads to the collision and large overhead because of using the beacon messages periodically.

Hybrid Forwarding Schemes: A hybrid forwarding scheme combines several techniques and methods to perform a forwarding decision. Different solutions have been proposed under this category, such as (Deng, Xie, Shi, & Li, 2015) that introduces a hybrid forwarding strategy for location dependent and location independent information by using opportunistic and probabilistic forwarding strategy to select only one wireless interface is used in FIB and PIT tables. In (Yu, Gerla, & Sanadidi, 2015), another hybrid scheme had been discussed, that use Bloom-Filter to define name prefix for each partition in the network area.

Although, hybrid forwarding schemes may provide better forwarding by taking each good benefit from each forwarding strategy, a careful study is needed to select best metrics and chooses for this combination.

7.2.3. Caching Strategies

Initially, the Internet was designed to forward all requests to the original producer where it has to be continuously connected. Thus, a vast network load is created because of the popular content demands, in which will be served with long retrieval delay with large bandwidth is consumed. Also, a single point failure affects the content availability.

By applying in-network caching (Abdullahi, Arif, & Hassan, 2015), an NDN intermediate node may cache the content and serve it for future requests. In-network caching reduces the network load and decreases the overhead on the producer side. Also, by providing a ubiquitous content in the network, the data availability is extremely increased.

From VANETs point of view, the in-network caching (Modesto & Boukerche, 2017a) play an important role to disseminate the content towards and among vehicles. The existing caching efforts can be classified into the following classes:

Probabilistic Caching Schemes: A probabilistic caching scheme aims to increase the content diversity in the network to improve the content dissemination. The caching decision is probabilistic using different metrics.

Deng et al. (Deng, Wang, Li, & Li, 2016) designed a distributed probabilistic caching strategy in VANETs, in which different metrics are involved such as demand and preference of vehicles, the importance of vehicles in the ego network, and the relative movement of the receiver and the sender. In same way, work in (H Khelifi et al., 2018) proposed a caching-based mobility prediction scheme in order to cache the content proactively in the predicted next hop, aiming to solve the mobility and replicate data issues.

Similarly, Mauri et al. (Mauri, Gerla, Bruno, Cesana, & Verticale, 2017) designed a scheme to maximize the average content retrieval probability by studying the users in the communication system, the available bandwidth in the access point, the propagation latency, and the caching capacity.

Using a probabilistic method, that can be presented as a random-fashion manner, to decide the caching operation may not be a suitable solution. Based on the used metrics, a non-popular content can be treated as a popular one and cache in the network.

Cooperative Caching Schemes: A cooperative scheme may be applied when the caching process is applied under more than one administrative authority. Quan et al. (Quan et al., 2014) proposed an innovative highway cooperative scheme to improve the quality of service of multimedia streaming. While Liu et al. (L. C. Liu, Xie, Wang, & Zhang, 2015) discussed caching placement issue, and proposed an algorithm to select the caching vehicular and overcome the vehicle mobility problem.

Although these schemes may enhance the content access and distribution in high mobility scenarios, they require major modifications in NDN's working nature and may violate its primitives.

Content Popularity Caching Schemes: In this class, the caching decision is based on the content popularity, the more popular content and frequently requested, is preferable of the caching. Zhao et al. (Zhao, Qin, Gao, Foh, & Chao, 2017) designed a community similarity scheme for V2V communications. This scheme aims to estimate the community similarity of the content and selects the caching vehicle based on the content popularity.

Caching scheme based on content popularity may improve the content diversity in the network and reduce the content retrieval delay. However, they are hard to be applied in dynamic content or use-once data.

7.2.4. Mobility Solutions

In contrast to IP networks where a mobile node is required to get a new IP address whenever connects to a new network. In NDN, and due to the use of the content name, a node is not required to ask for an address. It uses only the content name to send a request. Moreover, the handover and mobility management became easier; a mobile node can re-issue the same request the has not been satisfied during the mobility, without any complicated management routine.

However, VANET environment is characteristic by the high speed of vehicles and their movement in the different directions for a very short time make which make the data retrieval in such mobility environment challenging. To overcome the mobility challenges, various solutions have been presented.

MobiCCN (Liang Wang, Waltari, & Kangasharju, 2013) is a routing protocol that addresses the mobility and mobile content issue, by using standard CCN protocol and a greedy one. Also, by assigning a Virtual Coordinate to each NDN router, that is embedded into the data packet, intermediate nodes can maintain Neighbors' Coordinates Table to select the closest neighbor to the destination.

While Duarte et al. (Duarte, Braun, & Villas, 2018) evaluated the effect of data source mobility, and analyzed the existing solutions. Hence, they designed a new mobility scheme based on Floating Content and Home Repository tending to improve the content delivery in mobility scenarios.

7.2.5. Data Security and User Privacy

The security of today's Internet is based on how much the communication channel is secure, but one of the major drawbacks of this concept is that the traversed data cannot be used when the session gets expired. In NDN, a content-based security concept is used (S H Bouk, Ahmed, Hussain, & Eun, 2018), that consists of securing the content itself regardless of the used communication channel. Also, and because nodes in NDN are not directly addressable, the DoS attack is effectively reduced.

As NDN is still in its earlier design, other security and privacy challenges can appear when merging it with VANET (Bernardini, Asghar, & Crispo, 2017).

As far as data is concerned in NDN while the mobility still a critical issue that may affect its dissemination, Wang et al. (J. Wang, Wakikawa, & Zhang, 2010) designed a secure mechanism for data collection from mobile vehicles. The proposed mechanism allows manufacturers to verify the integrity and authenticity of incoming content and protect the privacy of mobile users from malicious attacks and nodes.

AutoNDN, a trust model for autonomous vehicular applications, has been designed in (Chowdhury et al., 2017). AutoNDN aims to preserve data and user privacy, by preventing the dissemination of false data and vehicle tracking. Thus, a hierarchical trust model is used that has the following levels: *Autonomous-Vehicle*, *Manufacturers*, *Vehicles*, and *Data*, combined with a pseudonym and proxy-based scheme to make vehicles tracking more difficult for attackers. Similarly, work in (Hakima Khelifi, Luo, Nour, Moungla, & Ahmed, 2018) proposed a reputation-based blockchain scheme in order to secure the cache store by caching only the trusted data.

8. FUTURE VIEW OF ICN-BASED VANETS

The Information-Centric model is still on its earlier design, and still taking shape. Many efforts need to be investigated to apply it as future Internet architecture (Zalak, Aemi, Raval, Ukani, & Valiveti, 2017). In the following, we present a future view of ICN-based vehicular networks, issues, challenges, and identify future research.

8.1. Content/Data Naming

The content/data naming (Afanasyev et al., 2017) is the fundamental ICN/NDN element, that can identify content or service using either flat, hierarchical, or attribute-value based scheme. However, hierarchical and hybrid naming schemes have attracted more attention from researchers. The former scheme because of its simplicity and name aggregation feature that provides better scalability. While the latter, hybrid names, provides an efficient and scalable lookup, aggregation rules, optimized lookup with less memory usage, and privacy and security support; all of this because of the combination of many features.

Notwithstanding, naming convention and agreements on which scheme should be used is still an open challenge. Also, the use of long variable unbounded names is a serious issue that affects the lookup performance, and consequently the quality of service, network scalability \& performance.

8.2. Routing and Data Forwarding

The NDN stack has not a dedicated transport layer (Chen et al., 2016), all the transport features have been included in the forwarding plane. Most of the existing VANET-based NDN forwarding schemes are based on geographical or location techniques, by exploring neighbor status and performing QoS and distributed content discovery.

We believe that an efficient forwarding scheme should integrate various features when selecting the most appropriate next-hop such as load balancing, congestion avoidance, link failure, detect and security attacks.

8.3. In-network Caching

In compared to traditional host-centric networks, the in-network caching is promising to enhance content dissemination and achieve better performance (Din et al., 2018). Toward this, most of the existing caching strategy in VANET focuses on the content popularity and probabilistic caching, aiming to cache the popular content closer to consumers. However, and due to the diversity of VANET applications and traffic, selecting the cached content based the traffic-class or application needs more investigation. Furthermore, and due to the high mobility of vehicles, nearby consumers may not stay the same asking the same content, where a large content replacement in the cache store will be applied. Hence, designing a suitable caching replacement scheme according to the mobility and diversity of applications is required.

8.4. Quality of Service

NDN is receiver-driven architecture, where the content delivery follows the same content discovery path in reverse. Hence, providing the quality of service support and provisioning (Qu et al., 2018) should be taken into the design during the content discovery. Traffic prioritization in the forwarding plane is not a suitable solution to low latency application and enhance the scalability and reliability especially with the high mobility.

8.5. Mobility Support

To overcome the high mobility issue in vehicular networks, most researchers are trying to take benefits of in-network caching and prediction on vehicle destination, neglecting the mobility of replica cache stores (Feng, Zhou, & Xu, 2016). One of the promising solutions to enhance the data delivery in VANET environment is the integration of software-defined networks (SDN) to make a global vision on the network and using mobile/edge computing to ensure a distribution content caching.

8.6. Data Security and Privacy

NDN uses a content-based security paradigm where all security-related information are embedded into the content that is identified by a name. Hence, various security and privacy threats can be launched from any network level exploring the plain content name, poising the cached content, and polluting the cache stores. Furthermore, maintaining public keys in high mobility scenarios is very challenging (Hakima Khelifi, Luo, Nour, & Shah, 2018). Finally, a trust model for VANET applications is required to handle the distributed access-content, authentication, key management, and enforcing the content/user privacy.

9. CONCLUSION

In this chapter, we first overview the Information-Centric communication model, by showing the limitation of the current host-centric paradigm and providing a brief introduction of some future Internet architectures. Furthermore, we discuss the

Named Data Networking architecture and its working principle. Besides, we also focus on the primary objective of this chapter, that is, to map and provide recent advances in the vehicular networks on top of NDN, covering different features such as naming, forwarding, caching, mobility, and security. In the end, we also highlight few of the existing challenges and research direction in the era of VANET-NDN.

ACKNOWLEDGMENT

The work of Dr. Luo was supported by the National 242 Project under Grant No. 2017A149. Dr. Luo is the corresponding author.

REFERENCES

Abdullahi, I., Arif, S., & Hassan, S. (2015). Survey on caching approaches in Information Centric Networking. *Journal of Network and Computer Applications, 56*(2015), 48–59. doi:10.1016/j.jnca.2015.06.011

Abusch-Magder, D., Bosch, P., Klein, T. E., Polakos, P. A., Samuel, L. G., & Viswanathan, H. (2007). 911-NOW: A network on wheels for emergency response and disaster recovery operations. *Bell Labs Technical Journal, 11*(4), 113–133. doi:10.1002/bltj.20199

Afanasyev, A., Halderman, J. A., Ruoti, S., Seamons, K., Yu, Y., Zappala, D., & Zhang, L. (2016). Content-based security for the web. In *New Security Paradigms Workshop* (pp. 49–60). Academic Press.

Afanasyev, A., Jiang, X., Yu, Y., Tan, J., Xia, Y., Mankin, A., & Zhang, L. (2017). NDNS: A DNS-like name service for NDN. *International Conference on Computer Communications and Networks (ICCCN).* 10.1109/ICCCN.2017.8038461

Ahlgren, B., Dannewitz, C., Imbrenda, C., Kutscher, D., & Ohlman, B. (2012). A Survey of Information-Centric Networking. *IEEE Communications Magazine, 50*(July), 1–26. doi:10.1109/MCOM.2012.6231276

Ahmed, S. H., Bouk, S. H., & Kim, D. (2015). RUFS: RobUst forwarder selection in vehicular content-centric networks. *IEEE Communications Letters, 19*(9), 1616–1619. doi:10.1109/LCOMM.2015.2451647

Ahmed, S. H., Bouk, S. H., Yaqub, M. A., Kim, D., & Song, H. (2017). DIFS: Distributed Interest Forwarder Selection in Vehicular Named Data Networks. *IEEE Transactions on Intelligent Transportation Systems.*

Al-Sultan, S., Al-Doori, M. M., Al-Bayatti, A. H., & Zedan, H. (2014). A comprehensive survey on vehicular ad hoc network. *Journal of Network and Computer Applications, 37*, 380–392. doi:10.1016/j.jnca.2013.02.036

Amditis, A., Bertolazzi, E., Bimpas, M., Biral, F., Bosetti, P., Da Lio, M., Danielsson, L., Gallione, A., Lind, H., Saroldi, A., & Sjogren, A. (2010). A holistic approach to the integration of safety applications: The INSAFES subproject within the European framework programme 6 integrating project PReVENT. *IEEE Transactions on Intelligent Transportation Systems, 11*(3), 554–566. doi:10.1109/TITS.2009.2036736

Bernardini, C., Asghar, M. R., & Crispo, B. (2017). *Security and privacy in vehicular communications: Challenges and opportunities.* Vehicular Communications.

Bian, C., Zhao, T., Li, X., & Yan, W. (2015a). Boosting named data networking for data dissemination in urban VANET scenarios. *Vehicular Communications, 2*(4), 195–207. doi:10.1016/j.vehcom.2015.08.001

Bian, C., Zhao, T., Li, X., & Yan, W. (2015b). Boosting named data networking for efficient packet forwarding in urban VANET scenarios. In *IEEE International Workshop on Local and Metropolitan Area Networks (LANMAN)* (pp. 1–6). 10.1109/LANMAN.2015.7114718

Booysen, M. J., Zeadally, S., & Van Rooyen, G.-J. (2011). Survey of media access control protocols for vehicular ad hoc networks. *IET Communications, 5*(11), 1619–1631. doi:10.1049/iet-com.2011.0085

Bouk, S. H., Ahmed, S. H., Hussain, R., & Eun, Y. (2018). Named Data Networking's Intrinsic Cyber-Resilience for Vehicular CPS. *IEEE Access.* Retrieved from https://ieeexplore.ieee.org/abstract/document/8492520/

Bouk, S. H., Ahmed, S. H., & Kim, D. (2014). Hierarchical and hash-based naming scheme for vehicular information centric networks. In *International Conference on Connected Vehicles and Expo (ICCVE)* (pp. 765–766). 10.1109/ICCVE.2014.7297653

Bouk, S. H., Ahmed, S. H., & Kim, D. (2015). Hierarchical and hash based naming with Compact Trie name management scheme for Vehicular Content Centric Networks. *Computer Communications, 71*, 73–83. doi:10.1016/j.comcom.2015.09.014

Campista, M. E. M., Rubinstein, M. G., Moraes, I. M., Costa, L. H. M. K., & Duarte, O. C. M. B. (2014). Challenges and research directions for the future internetworking. *IEEE Communications Surveys and Tutorials, 16*(2), 1050–1079. doi:10.1109/SURV.2013.100213.00143

Car to Car Communication Consortium. (n.d.). *CAR 2 CAR Communication Consortium Manifesto.* Author.

Chen, Q., Xie, R., Yu, F. R., Liu, J., Huang, T., & Liu, Y. (2016). Transport control strategies in named data networking: A survey. *IEEE Communications Surveys and Tutorials, 18*(3), 2052–2083. doi:10.1109/COMST.2016.2528164

Chowdhury, M., Gawande, A., & Wang, L. (2017). Secure Information Sharing among Autonomous Vehicles in NDN. In *International Conference on Internet-of-Things Design and Implementation (IoTDI)* (pp. 15–26). 10.1145/3054977.3054994

Cunha, F., Villas, L., Boukerche, A., Maia, G., Viana, A., Mini, R. A. F., & Loureiro, A. A. F. (2016). Data communication in VANETs: Protocols, applications and challenges. *Ad Hoc Networks, 44*, 90–103. doi:10.1016/j.adhoc.2016.02.017

Dannewitz, C., Kutscher, D., Ohlman, B., Farrell, S., Ahlgren, B., & Karl, H. (2013). Network of Information (NetInf) - An information-centric networking architecture. *Computer Communications, 36*(7), 721–735. doi:10.1016/j.comcom.2013.01.009

Deng, G., Wang, L., Li, F., & Li, R. (2016). Distributed Probabilistic Caching strategy in VANETs through Named Data Networking. *IEEE INFOCOM,* 314–319. doi:10.1109/INFCOMW.2016.7562093

Deng, G., Xie, X., Shi, L., & Li, R. (2015). Hybrid information forwarding in VANETs through named data networking. In *International Symposium on Personal, Indoor, and Mobile Radio Communications (PIMRC)* (pp. 1940–1944). 10.1109/PIMRC.2015.7343616

Din, I. U., Hassan, S., Khan, M. K., Guizani, M., Ghazali, O., & Habbal, A. (2018). Caching in Information-Centric Networking: Strategies, Challenges, and Future Research Directions. *IEEE Communications Surveys and Tutorials, 20*(c), 1–1. doi:10.1109/COMST.2017.2787609

Drira, W., & Filali, F. (2014). NDN-Q: An NDN query mechanism for efficient V2X data collection. In *IEEE International Conference on Sensing, Communication, and Networking Workshops (SECON Workshops)* (pp. 13–18). 10.1109/SECONW.2014.6979698

Duarte, J. M., Braun, T., & Villas, L. A. (2018). Source Mobility in Vehicular Named-Data Networking: An Overview. In Ad Hoc Networks (pp. 83–93). Springer.

Feng, B., Zhou, H., & Xu, Q. (2016). Mobility support in Named Data Networking: A survey. *EURASIP Journal on Wireless Communications and Networking, 2016*(1), 220. Advance online publication. doi:10.118613638-016-0715-0

Fotiou, N., Nikander, P., Trossen, D., Polyzos, G. C., & Associates. (2010). Developing Information Networking Further: From PSIRP to PURSUIT. In Broadnets (pp. 1–13).

Gehring, O., & Fritz, H. (1997). Practical results of a longitudinal control concept for truck platooning with vehicle to vehicle communication. In *Intelligent Transportation System, 1997. ITSC'97., IEEE Conference on* (pp. 117–122). 10.1109/ITSC.1997.660461

Grassi, G., Pesavento, D., Pau, G., Zhang, L., & Fdida, S. (2015). Navigo: Interest forwarding by geolocations in vehicular Named Data Networking. In *International Symposium on World of Wireless, Mobile and Multimedia Networks (WoWMoM)* (pp. 1–10). 10.1109/WoWMoM.2015.7158165

Hedrick, J. K., Tomizuka, M., & Varaiya, P. (1994). Control issues in automated highway systems. *IEEE Control Systems, 14*(6), 21–32. doi:10.1109/37.334412

IEEE Standards Association. (n.d.). *Wireless access in vehicular environments*. IEEE.

Kargl, F., Papadimitratos, P., Buttyan, L., Müter, M., Schoch, E., Wiedersheim, B., Thong, T.-V., Calandriello, G., Held, A., Kung, A., & Hubaux, J.-P. (2008). Secure vehicular communication systems: Implementation, performance, and research challenges. *IEEE Communications Magazine, 46*(11), 110–118. doi:10.1109/MCOM.2008.4689253

Khelifi, H., Luo, S., Nour, B., Moungla, H., & Ahmed, S. H. (2018). Reputation-based Blockchain for Secure NDN Caching in Vehicular Networks. In *IEEE Conference on Standards for Communications and Networking (CSCN)* (pp. 1–6). Paris, France: IEEE. 10.1109/CSCN.2018.8581849

Khelifi, H., Luo, S., Nour, B., Sellami, A., Moungla, H., & Naït-Abdesselam, F. (2018). An Optimized Proactive Caching Scheme based on Mobility Prediction for Vehicular Networks. In *IEEE Global Communications Conference (IEEE GLOBECOM)* (pp. 1–6). 10.1109/GLOCOM.2018.8647898

Khelifi, H., Luo, S., Nour, B., & Shah, C. S. (2018). *Security & Privacy Issues in Vehicular Named Data Networks: An Overview*. Mobile Information Systems. doi:10.1155/2018/5672154

Koponen, T., Chawla, M., Chun, B.-G., Ermolinskiy, A., Kim, K. H., Shenker, S., & Stoica, I. (2007). A data-oriented (and beyond) network architecture. *Computer Communication Review, 37*(4), 181–192. doi:10.1145/1282427.1282402

Kuai, M., Hong, X., & Yu, Q. (2016). Density-Aware Delay-Tolerant Interest Forwarding in Vehicular Named Data Networking. In *IEEE Vehicular Technology Conference (VTC-Fall)* (pp. 1–5). 10.1109/VTCFall.2016.7880953

Laroui, M., Sellami, A., Nour, B., Moungla, H., Afifi, H., & Boukli-Hacéne, S. (2018). Driving Path Stability in VANETs. *IEEE Global Communications Conference (IEEE GLOBECOM)*.

Liu, L. C., Xie, D., Wang, S., & Zhang, Z. (2015). CCN-based cooperative caching in VANET. In *International Conference on Connected Vehicles and Expo (ICCVE)* (pp. 198–203). 10.1109/ICCVE.2015.24

Liu, X., Nicolau, M. J., Costa, A., Macedo, J., & Santos, A. (2016). A geographic opportunistic forwarding strategy for vehicular named data networking. *Intelligent Distributed Computing*, *9*, 509–521.

Luo, J., & Hubaux, J.-P. (2004). *A survey of inter-vehicle communication*. Academic Press.

Mauri, G., Gerla, M., Bruno, F., Cesana, M., & Verticale, G. (2017). Optimal Content Prefetching in NDN Vehicle-to-Infrastructure Scenario. *IEEE Transactions on Vehicular Technology*, *66*(3), 2513–2525. doi:10.1109/TVT.2016.2580586

Modesto, F. M., & Boukerche, A. (2017a). An analysis of caching in information-centric vehicular networks. In *IEEE International Conference on Communications (ICC)* (pp. 1–6). 10.1109/ICC.2017.7997019

Modesto, F. M., & Boukerche, A. (2017b). SEVeN: A novel service-based architecture for information-centric vehicular network. *Computer Communications*.

Nour, B., Sharif, K., Li, F., Moungla, H., Kamal, A. E., & Afifi, H. (2018). NCP: A Near ICN Cache Placement Scheme for IoT-based Traffic Class. In *IEEE Global Communications Conference (GLOBECOM)* (pp. 1–6). 10.1109/GLOCOM.2018.8647629

Nour, B., Sharif, K., Li, F., Moungla, H., & Liu, Y. (2017). M2HAV: A Standardized ICN Naming Scheme for Wireless Devices in Internet of Things. In *International Conference on Wireless Algorithms, Systems, and Applications (WASA)* (pp. 289–301). Springer International Publishing. 10.1007/978-3-319-60033-8_26

Oehlmann, F. (2013). Content-Centric Networking. *Network (Bristol, England)*, *43*, 11–18.

Pan, J., Paul, S., & Jain, R. (2011). A survey of the research on future internet architectures. *IEEE Communications Magazine, 49*(7), 26–36. doi:10.1109/MCOM.2011.5936152

Qu, D., Wang, X., Huang, M., Li, K., Das, S. K., & Wu, S. (2018). A cache-aware social-based QoS routing scheme in Information Centric Networks. *Journal of Network and Computer Applications, 121*, 20–32. doi:10.1016/j.jnca.2018.07.002

Quan, W., Xu, C., Guan, J., Zhang, H., & Grieco, L. A. (2014). Social cooperation for information-centric multimedia streaming in highway VANETs. In *International Symposium on a World of Wireless, Mobile and Multimedia Networks (WoWMoM)* (pp. 1–6). 10.1109/WoWMoM.2014.6918992

Reichardt, D., Miglietta, M., Moretti, L., Morsink, P., & Schulz, W. (2002). CarTALK 2000: Safe and comfortable driving based upon inter-vehicle-communication. In *Intelligent Vehicle Symposium, 2002. IEEE* (Vol. 2, pp. 545–550). IEEE.

Saxena, D., Raychoudhury, V., Suri, N., Becker, C., & Cao, J. (2016). Named Data Networking: A survey. *Computer Science Review, 19*, 15–55. doi:10.1016/j.cosrev.2016.01.001

Shemsi, I., & Kadam, P. (2017). Named Data Networking in VANET: A Survey. *International Journal of Scientific Engineering and Science*, 1–5.

Takahashi, A., & Asanuma, N. (2000). Introduction of Honda ASV-2 (advanced safety vehicle-phase 2). In *Intelligent Vehicles Symposium, 2000. IV 2000. Proceedings of the IEEE* (pp. 694–701). 10.1109/IVS.2000.898430

Toor, Y., Muhlethaler, P., Laouiti, A., & La Fortelle, A. (2008). Vehicle ad hoc networks: Applications and related technical issues. *IEEE Communications Surveys and Tutorials, 10*(3), 74–88. doi:10.1109/COMST.2008.4625806

Vehicle Infrastructure Integration (VII) program of US Federal and State departments of transportation and automobile manufacturers. (n.d.).

Wang, J., Wakikawa, R., & Zhang, L. (2010). DMND: Collecting data from mobiles using named data. In IEEE Vehicular networking conference (VNC) (pp. 49–56). doi:10.1109/VNC.2010.5698270

Wang, L., Wakikawa, R., Kuntz, R., Vuyyuru, R., & Zhang, L. (2012). Data naming in vehicle-to-vehicle communications. In *IEEE Conference on Computer Communications Workshops (INFOCOM WKSHPS)* (pp. 328–333). 10.1109/INFCOMW.2012.6193515

Wang, L., Waltari, O., & Kangasharju, J. (2013). Mobiccn: Mobility support with greedy routing in content-centric networks. In *IEEE Global Communications Conference (GLOBECOM)* (pp. 2069–2075). 10.1109/GLOCOM.2013.6831380

Xylomenos, G., Ververidis, C. N., Siris, V. A., Fotiou, N., Tsilopoulos, C., Vasilakos, X., Katsaros, K. V., & Polyzos, G. C. (2014). A Survey of Information-Centric Networking Research. *IEEE Communications Surveys and Tutorials*, *16*(2), 1024–1049. doi:10.1109/SURV.2013.070813.00063

Yu, Y.-T., Gerla, M., & Sanadidi, M. Y. (2015). Scalable VANET content routing using hierarchical bloom filters. *Wireless Communications and Mobile Computing*, *15*(6), 1001–1014. doi:10.1002/wcm.2495

Zalak, P., Aemi, K., Raval, G., Ukani, V., & Valiveti, S. (2017). Open Issues in Named Data Networking—A Survey. In *International Conference on Information and Communication Technology for Intelligent Systems* (pp. 285–292). Academic Press.

Zhang, L., Estrin, D., Burke, J., Jacobson, V., Thornton, J. D., Smetters, D. K., … others. (2010). *Named Data Networking (NDN) Project.* Relatório Técnico NDN-0001, Xerox Palo Alto Research Center-PARC.

Zhao, W., Qin, Y., Gao, D., Foh, C. H., & Chao, H.-C. (2017). An Efficient Cache Strategy in Information Centric Networking Vehicle-to-Vehicle Scenario. *IEEE Access: Practical Innovations, Open Solutions*, *5*, 12657–12667. doi:10.1109/ACCESS.2017.2714191

Chapter 5

Constrained Average Design Method for QoS–Based Traffic Engineering at the Edge/Gateway Boundary in VANETs and Cyber–Physical Environments

Daniel Minoli
DVI Communications, USA

Benedict Occhiogrosso
DVI Communications, USA

ABSTRACT

Cyber physical systems (CPSs) are software-intensive smart distributed systems that support physical components endowed with integrated computational capabilities. Tiered, often wireless, networks are typically used to collect or push the data generated or required by a distributed set of CPS-based devices. The edge-to-core traffic flows on the tiered networks can become overwhelming. Thus, appropriate traffic engineering (TE) algorithms are required to manage the flows, while at the same time meeting the delivery requirements in terms of latency, jitter, and packet loss. This chapter provides a basic overview of CPSs followed by a discussion of a newly developed TE method called 'constrained average', where traffic is by design allowed to be delayed up to a specified, but small value epsilon, but with zero packet loss.

DOI: 10.4018/978-1-5225-9493-2.ch005

I. INTRODUCTION AND BACKGROUND

Cyber-Physical Systems

The term cyber-physical systems (CPSs) refers to an evolving generation of systems with integrated computational and physical capabilities that can interact with humans in a number of ways (Baheti & Gill, 2011; Song, Rawat et al.,2017; Romanovsky & Ishikawa, 2017). The term CPS was originally introduced in 2006 at a National Science Foundation (NSF) workshop in Austin, TX, where it was defined as "*a system composed of collaborative entities, equipped with calculation capabilities and actors of an intensive connection with the surrounding physical world and phenomena, using and providing all together services of treatment and communication of data available on the network*" (Quintanilla, Cardin et al., 2016). Some researchers now see a CPS as an orchestration of computers and physical systems where embedded computers monitor and control physical processes, usually making use of feedback loops, and where the physical processes affect computations, and vice versa (Lee, 2015; Sivakumar, Sadagopan et al., 2016; Hua, Lua et al., 2016). Thus, CPSs are systems of collaborating computational entities that are intensively connected with the surrounding physical world and its on-going processes, simultaneously providing and using data-accessing and data-processing services available on a cloud and/or on the Internet (Monostori, 2014; Zanero, 2017; Yu & Xue, 2016). CPSs therefore consist of computer networks and devices with built-in controllers that control (possibly, with human participation) physical processes by means of feedback; that is, physical processes exert influence on computations and computations exert influence on the choice and course of physical processes (Letichevsky, Letychevsky et al., 2017; Stankovic, 2017; Müller, Litoiu et al., 2016). Documented applications include but are not limited to: Vehicular Ad hoc Networks (VANETs), Intelligent Transportation Systems (ITSs), automotive systems, biomedical and healthcare systems, smart grid and renewable energy systems, manufacturing process control, military systems, air traffic control and safety systems, aircraft instrumentation, water management systems, physical security systems(access control and monitoring), asset management, and distributed robotics and drones (Baheti & Gill, 2011; Lee, 2015; Zanero, 2017; Letichevsky, Letychevsky et al., 2017; Massey, 2017). While some researchers consider CPSs as distinct from Internet of Things (IoT) systems, others broadly equate the two concepts, (e.g., Wang, Zhu et al., 2018, Thramboulidis & Christoulakis, 2016, Burg, Chattopadhyay et al., 2018, Antonino, Morgenstern et al., 2018, He, Maple et al., 2016, Blasch, Kadar et al., 2017), as a short list of references -- in fact, the National Institute for Standards and Technology (NIST) observes that "*CPS and related systems (including the Internet of Things, Industrial Internet, and more) are widely recognized as having great potential to enable innovative applications and impact multiple economic sectors in the world-wide economy*", thus drawing a relationship between the two concepts (Griffor, 2017).

As seen above, the basic concept of CPSs is to provide intelligent capabilities to dispersed devices in order to support automated (machine-to-machine or people-to-machine) data aggregation, and also, as appropriate, to reliably transmit configuration files or commands to end-systems. CPSs comprise interacting digital, analog, physical, and human components engineered for functionality through integrated physical elements and logic capabilities (Griffor, 2017). The data can be end-system data (e.g., device status, patient monitoring), near-environment sampled data (e.g., site temperature, vehicle or device location and/or proximity, traffic patterns, site parameters), or interactive data (e.g., one- or two-way multimedia streams, video surveillance). The commands can be actuation information, for example, resetting a device or system parameter, or performing some action (e.g., for a road signal, a dam door, a grid transfer switch, a drone function, or a remote robot action). The devices can be stationary or mobile, simple or complex (Wang, Lei et al., 2016; Koutsoukos, Karsai et al., 2018; Minoli, 2013). The applicable CPS technology spans (i) sensors; (ii) access networks -- especially wireless systems for personal area networks, or more generally, Wireless Sensor Networks (WSNs); (iii) edge networks (also called fogs by some), which include VANETs; (iv) core networks -- such as metro Ethernet systems, 5G cellular; (v) cloud and/or Internet services; and (vi) data analytics/big data systems.

Evolving Multimedia Requirements

There is now an evolving requirement to support CPS-oriented video and multimedia (Wang, Wang et al., 2017, Wang, Minoli et al., 2017). These applications include classical video surveillance, drone-based video surveillance, drone-based multimedia sensors, crowdsensing and vehicular crowdsensing, social networking, sensing (including crowdsensing and Smart Environments), streaming, Smart Buildings/Campuses/Cities, and ITSs. Figure 1 depicts graphically the evolution towards multimedia-based CPSs. Just focusing on smartphones, the data traffic is expected to grow ten-fold from about 2,500 PB/month in 2014 to 25,000 PB/month in 2020 (Cisco, 2017).

Wireless channels, which are the norm in most CPS applications, are usually at a premium in terms of bandwidth, especially for traditional 3G/4G systems. Therefore, besides new radio access/transmission technologies (RATs), new and/or improved Traffic Engineering (TE), Quality of Service (QoS), and Mobility Management (MM) techniques are needed (3GPP, 2015) (Figure 2 depicts some available radio technologies.) Given the expected size of the networks and the relatively low-intelligence of some CPSs, a low-complexity TE mechanism is desirable.

Evolving video and multimedia-based applications can utilize the relatively new High Efficiency Video Coding (HEVC) standard, specifically ITU-T Recommendation H.265/ISO IEC 23008-2; this coding standard was ratified by the ITU and ISO in 2013. While HEVC utilizes the same transport structure as H.264/AVC (Advanced Video Coding) and MPEG-2, it provides the same image resolution as H.264/AVC but with a lower coding data rate, or it can provide better image resolution than H.264/AVC but at the same coding data rate. HEVC aims at achieving a compression gain of 50% over H264/AVC (Sullivan, Ohm et al., 2012, ISO/IEC, 2013, Seeling & Reisslein, 2014, Kandris, Tsagkaropoulos et al., 2011, Seema & Reisslein, 2011). A tighter data rate is desirable in CPS environments not only because of transmission limitations (for example in wireless networks, such as Low Power Wide Area [LPWA] networks), but also storage requirements. The reliance of CPSs on wireless technology drives the requirement for efficient use of the channel bandwidth. Regardless of the coding scheme used there is still a need to support real time QoS-based traffic management and TE. The expectation is that within a few years there will be between 20 to 30 billion CPS devices. The traffic generated by these devices, although initially comprised of relatively low rate bursty data flows, is expected to increasingly consist of isochronous streams as multimedia applications are introduced. As a result, the edge-to-core traffic in such networks can be overwhelming. Therefore, Appropriate TE algorithms are required in these CPS systems, particularly low-complexity mechanisms.

Figure 1. CPS/IoT Evolution towards Multimedia-Based Applications

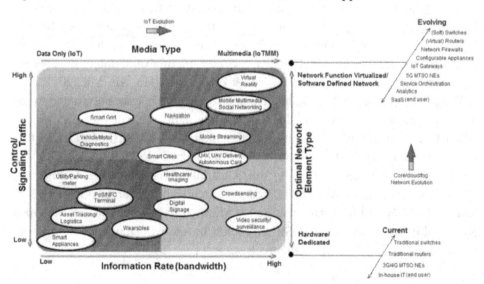

Figure 2. Available Radio Technologies for CPS/IoT

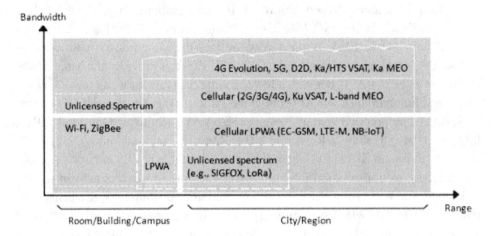

Architectural Requirements for CPSs

In the context of assessing the traffic flows, one should make note that CPSs benefit from architectural formulation and standardization. Architecture deals with how things are assembled, including hierarchy. Standards relate to the ability to deploy the technology in a commodity fashion, assuring simple and reliable end-to-end interoperability – these also include Machine to Machine (M2M) standards and approaches.

A CPS architecture can help define fundamental connectivity and data management building blocks, which, in turn determine information flows (and traffic). Several CPS (and/or IoT) architectural models have been proposed, each addressing or focusing on some specific functionality, formulation, or abstraction of the concept (Quintanilla, Cardin et al., 2016; Weyrich & Ebert, 2016; Hu, Xie et al., 2012; Guo & Zeng, 2019; Jiang, 2018). Open architectures include but are not limited to the following: the architecture formulated by the International Society of Automation (ISA), ISA-95 (International Society of Automation (ISA), 2018); the 5C architecture (Lee & Bagheri, 2015); the IoT Reference Architecture (IoT RA -- ISO/IEC WD 30141); the IoT Architecture (IoT-A); the Industrial Internet Reference Architecture (IIRA); the IEEE P2413 WG's Standard for an Architectural Framework for the IoT; and the ETSI High Level Architecture for M2M. Another architecture, the Open Systems IoT Reference Model (OSiRM) entails the following model layers (Minoli, Occhiogrosso et al., 2017) (see Figure 3):

Layer 1: Things: this layer is comprised of the universe of "things";

Layer 2: Data Acquisition: this layer encompasses the "data acquisition" capabilities;

Layer 3: Fog/Edge Networking: this layer supports "fog networking", localized (site- or neighborhood-specific) networking;

Layer 4: Data Aggregation: this layer supports the "data aggregation", data summarization or protocol conversion;

Layer 5: Data Centralization: this layer supports the "data centralization" function (traditional core networking);

Layer 6: Data Analytics & Storage: this layer encompasses the "data analytics and storage functions";

Layer 7: Applications: this layer is the "applications" layer, a vast array of horizontal and/or vertical applications.

Figure 3. OSiRM Architecture Model

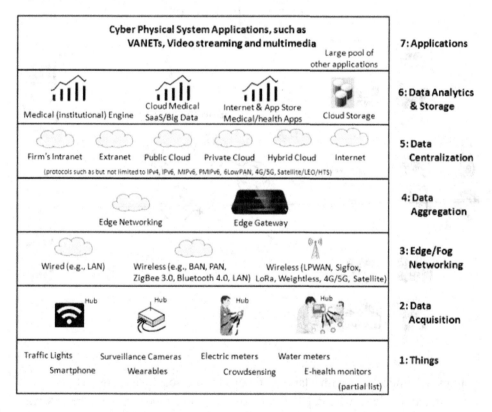

Traffic Engineering Requirements for CPSs

In many CPS environments, such as process control, VANETs, medical systems, and multimedia, real-time QoS-based traffic management and TE are critically important requirements. While some traffic generated by these devices produce low rate bursty data flows, low latency and low packet loss are a key consideration for process control, VANETs, and medical systems. Additionally, many applications give rise to isochronous streams due to multimedia uploading or distribution. The resulting edge-to-core traffic must be appropriately managed. Many networks, but especially networks handling multimedia, almost invariably require mechanisms for traffic control, also known as Traffic Management (TM). Figure 4 depicts the typical CPS environment from a topological perspective. The gateway supports an aggregation function to funnel a large number of edge nodes into the network core.

Figure 4. CPS Topological Environment (also VANETs)

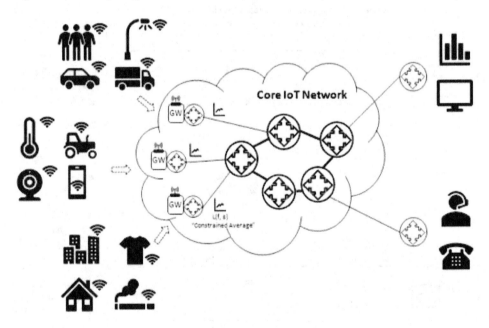

In many applications with a large number of dispersed devices, the edge-to-core concentrated traffic flows can potentially overwhelm supporting networks. Novel, yet simple, TE/TM algorithms are required to manage the flows, while at the same time meeting the delivery requirements in terms of latency, jitter, and packet loss.

This chapter discusses a TE/TM method applicable to CPSs and VANETs called 'constrained average', where traffic is by design allowed to be delayed up to a specified, small value epsilon, but with zero packet loss. This approach is based on some previous traffic engineering work (Minoli, 1983) and is one example of a traffic smoothing algorithm. Its advantages are the relative simplicity (effectively straightforward buffer management under a computational discipline) and its strong mathematical underpinning. The concept also has more general applicability to a number of other traffic management environments: this method can be used in conjunction with, or to enhance, a number of the existing traffic management systems such as: Adaptive Virtual Queue (AVQ), Class-Based Queueing (CBQ), Controlled Delay (CoDel), Credit-Based Fair Queuing (CBFQ), Generic Cell Rate Algorithm (GCRA), Hierarchical Token Bucket (HTB), Fair Queuing (FQ) and Weighted Fair Queuing (WFQ), Random Early Detection (RED) and Advanced Random Early Detection (ARED), Generalized Random Early Detection (GRED), Weighted Random Early Detection (WRED), Round-Robin (RR) and Weighted Round Robin (WRR), and Token Bucket Filter (TBF).

II. BASIC TRAFFIC ENGINEERING

Figure 5 depicts a high-level diagram of the gateway. Traffic arriving at the switch is stored temporarily in the store-and-forward memory. The amount of time that the traffic is (or can be stored) is a function of (a) the arrival/processing rate of the incoming traffic, (b) the QoS requirements of the traffic, and (c) the amount of resources available on the core-facing communication system (e.g., trunks, bandwidth, TDM slots, Ethernet frames, MPLS/IP packets, etc.)

The arrival rate of the traffic is captured here as f(t), where t is time (e.g., time of the day, or an interval of interest [a,b]). As implied, the traffic rate is not a constant f, but a function of time. Classical models have, and often continue to be used, such as the Erlang B model and the Erlang C model, although, in these models, the arrival rate is assumed to be constant. In the Erlang B model, the assumption is that if insufficient resources are unavailable at the instant the traffic arrives, the traffic is dropped. Therefore, the practical design criterion is to look at

$$f = \max_{(a,b)} f(t)$$

and deploy the resources under that value of traffic intensity. This can be wasteful (or at least inefficient), since the number of resources will be based on the absolute busy instance (or busy hour). (In some instances, however, this becomes a rigid design requirement.)

105

Figure 5. High level Gateway Architecture

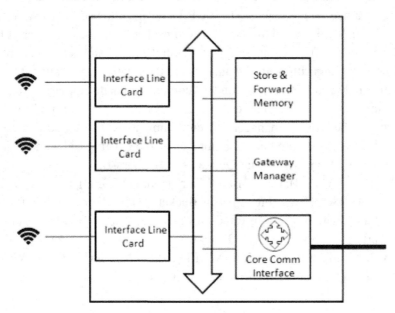

The Erlang C model (M/M/N queue) assumes all traffic (for example, in a classical call center) waits or can be delayed as long as necessary for service or forwarding. In the Erlang C model traffic arrives consistent with a Poisson process at an average rate of λ -- the inter-arrival times are independent and identically distributed exponential random variables with mean 1/λ. Traffic is held in a queue of infinite length and is serviced on a First Come First Served discipline. Service times (for example, transmission) follow an exponential distribution with a mean service time of 1/μ. Traffic (packets) in the queue are serviced by a pool of Nstatistically-identical (homogeneous) resources at an average rate of Nμ. The offered load is referred to as the number of Erlangs, and is by definition the unitless quantity R = λ/μ =f. The traffic intensity (occupancy or utilization) is defined as ρ =λ/(Nμ)= R/N (the traffic intensity must be less than 1, otherwise the system of N resources becomes unstable.)The steady state probability that there will be a queuing delay is

$$P\{Wait > 0\} = 1 - \left(\sum_{m=0}^{N-1} \frac{R^m}{m!}\right) \Big/ \left(\sum_{m=0}^{N-1} \frac{R^m}{m!} + \left(\frac{R^m}{N!}\right)\left(\frac{1}{1 - R/N}\right)\right)$$

The main limitation, again, is that the traffic rate is not a constant, but is typically a function of time (some concentrated high-utilization links may be a D/D/N or M/D/N operation, where the deterministic rate is fairly constant over time; however, edge-level bursty traffic will be time dependent – e.g., consider crowdsensing by people with vehicular traffic sensors: these applications will have high rates during morning rush hour, low rates during the day, and high rates during the evening rush hour).

One novel approach that takes into consideration the time-dependent arrival is to use the concept of the "constrained average" discussed next (see Figure 6).

Figure 6. Traffic level Criteria

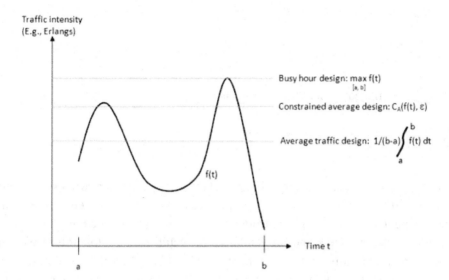

III. CONSTRAINED AVERAGE

Overview

The goal is to compute the relevant number of resources for example trunks, bandwidth, Time Division Multiplexing (TDM) slots, Ethernet frames, Multiprotocol Label Switching/Internet Protocol (MPLS/IP) packets, and so on, to connect a device, such as edge/gateway device in cyber-physical environments, to a core network, when the arrival rate (at a line card or aggregation port) during a time interval [a, b] is f(t) and the traffic can be queued up to a maximum amount of time ε. [a, b] could be a day, a portion of a day, or some other interval. It is assumed that f(t) is an integrable function in the interval in question. To achieve the stated goal, we seek to compute an equivalent traffic level $C_A(f(t), \varepsilon)$, based on the input f(t), representing the design traffic level as a single number which still meets the latency/QoS requirement ε. $C_A(f(t), \varepsilon)$ represents a generalization of the well-known mathematical concept of average of f(t), it being

$$\frac{1}{b-a}\int_a^b f(s)ds$$

If there were no latency constraints in transmitting the traffic ($\varepsilon = \infty$) and the packets could be queued as long as necessary (a gateway with a theoretical infinite [or just very large] memory/buffer), then the average of f(t) over the interval [a,b] is the equivalent traffic level that can be used in the design (using the models described in Section II, or any other more sophisticated model); that is

$$C_A(f(t), \infty) = \frac{1}{b-a}\int_a^b f(s)ds$$

If $\varepsilon = 0$, then the equivalent traffic level is simply

$$C_A(f(t), 0) = \max_{[a,b]} f(t) = \text{busy value}$$

On the other hand, assume that the QoS of the traffic dictates a certain maximum delay ε with $\varepsilon < (b-a)$. Multimedia frames, process control packets, and certain sensing applications with high control/signaling requirements (see Figure 1) require a bounded latency (and some applications, e.g., multimedia, a low jitter.) We aim at developing the equivalent design/traffic level $C_A(f(t), \varepsilon)$ for this case (refer back to Figure 6). This problem is somewhat complex and even after a general process is developed, the development of specific formulas is laborious. In this section, we first provide a general iterative procedure. Then we develop closed-form solutions for two typical categories of input traffic functions f(t).

It can be shown that $C_A(f(t), \varepsilon)$ is the smallest value h satisfying the following constraints:

For any time sequence x_n, n=0, 1, 2, ...M, $x_0=a$, $x_M=b$
    1) $O(x_n) \le he$ for all n;
    2) $O(b)=0$
where

$$O(x_n) = \begin{cases} \max\left\{O(x_{n-1}) + \int_{x_{n-1}}^{x_n} \min(0, f(s) - h)ds, 0\right\} \\ \quad + \int_{x_{n-1}}^{x_n} \max(0, f(s) - h)ds, \text{ for } e \ge 0 \\ \int_{x_{n-1}}^{x_n} \max(0, f(s) - h)ds, \text{ for } e = 0 \end{cases}$$

with $O(a)=0$.

Note that $O(x) \geq 0$ for all x. In place of a formal proof, the following observations help in acquiring an intuitive sense of the process. The term

$$\int_{x_{n-1}}^{x_n} \max(0, f(s) - h)ds$$

represents the new work (e.g., packets, frames) that accumulate between time x_{n-1} and x_n. It represents the amount of service (e.g., transmission time) exceeding the level h; notice that if h is not exceeded, this term (the new work) is zero. The term

$$\int_{x_{n-1}}^{x_n} \min(0, f(s) - h)ds$$

represents the net amount of accumulated work that can be accomplished between time x_{n-1} and x_n (potentially having to postpone some of the new work to a future time, but no later than ε time away.) If f(t) exceeds h, no net new work can be undertaken, but only a postponement of current work, in order to be able to take care of existing work requirement imposed on the gateway. Note that this term is negative, thus it subtracts from the (gateway) work carried over from time x_{n-1}, that is $O(x_{n-1})$, the amount of total work, excluding new work, that can be achieved in the interval (x_{n-1}, x_n). The term

$$O(x_{n-1}) + \int_{x_{n-1}}^{x_n} \min(0, f(s) - h)ds$$

represents the amount of old work that cannot be accomplished in the interval (x_{n-1}, x_n); this work can never be negative (that is, no future work can be done at any point, ever if excess capacity exists at that instance of time). Finally,

$$\max\left[O(x_{n-1}) + \int_{x_{n-1}}^{x_n} \min(0, f(s) - h)ds, 0\right] + \int_{x_{n-1}}^{x_n} \max(0, f(s) - h)ds$$

is the total work that 'spills over' into the interval (x_n, x_{n+1}). In each case that work must be less than $h\varepsilon$.

Figure 7 provides an example. Assume $\varepsilon=1$. Note that $C_A(f(t), 1) = 3$. Work A and B can only be postponed one timeslot over to D and E respectively; C, D, and E must be moved over to F, G, and H. For h=2, $O(1)$ and $O(2)$ exceed $h\varepsilon$, thus making the choice untenable. Also, $C_A(f,1)$ cannot be 4 since we are seeking the smallest h that works. Note that analytical verification of this value can be obtained with the equations derived below.

Figure 7. Example

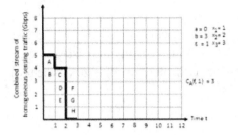

		h = 2			h = 3			h = 4		
		t = 1	t = 2	t = 3	t = 1	t = 2	t = 3	t = 1	t = 2	t = 3
$A = O(x_{n-1})$		0	3	5	0	2	3	0	1	1
$B = \int_{x_{n-1}}^{\delta_1} min$		0	0	-2	0	0	-3	0	0	-4
$C = max(A+B,0)$		0	3	3	0	2	0	0	1	0
$D = \int_{x_{n-1}}^{\delta_1} max$		3	2	1	2	1	0	1	0	0
$E = O(x_n) = C+D$		3	5	4	2	3	0	1	1	0

Single-Peak Function

Assume that f(t) has the shape depicted in Figure 8, which may be typical of some IoT/CPS applications. Applying the mechanisms introduced above we obtain the following two definitions for the overflow work O:

Figure 8. Typical Single-peak Traffic Function

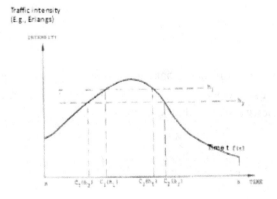

$$O(c_1(h_1)) = 0$$

$$O(c_2(h_1)) = \int_{c_1(h_1)}^{c_2(h_1)} (f(s) - h_1)ds$$

where $c_1(h_1)$ and $c_2(h_1)$ are the points where f(t) intersects h_1. Because of the shape of f(t), $O(c_2(h_1))$ is the largest unfinished work. In fact,

$$O(c_2(h_1) + \varepsilon) = \int_{c_1(h_1)}^{c_2(h_1)} (f(s) - h_1) ds + \int_{c_1(h_1)}^{c_2(h_1)+\varepsilon} f(s) - h_1 ds < \int_{c_1(h_1)}^{c_2(h_1)} f(s) - h_1 ds$$

Since the second term is negative. Therefore, we impose:

$$\int_{c_1(h_1)}^{c_2(h_1)} f(s) - h_1 ds \le h_1 \varepsilon$$

or

$$\frac{1}{c_2(h_1) - c_1(h_1)} \int_{c_1(h_1)}^{c_2(h_1)} f(s) ds \le h_1 \varepsilon$$

We define:

$$h_1 = \frac{1}{c_2(h_1) + \varepsilon - c_1(h_1)} \int_{c_1(h_1)}^{c_2(h_1)} f(s) ds$$

The constants $c_1(h_1)$ and $c_2(h_1)$ are derived by solving the simultaneous (integral) equations described below.

1) $f(c_1(h_1)) = \dfrac{1}{c_2(h_1) + \varepsilon - c_1(h_1)} \int_{c_1(h_1)}^{c_2(h_1)} f(s) ds$

2)   f(c_1(h_1)) = f(c_2(h_1))

Per Figure 8, the second constraint is:

$$O(b) = \max \left[0, \int_{c_1(h_1)}^{c_2(h_1)} (f(s) - h_2) ds + \int_{c_2(h_2)}^{b} (f(s) - h_2) ds \right] = 0$$

or

$$\int_{c_1(h_2)}^{b} (f(s) - h_2) ds \le 0$$

From which one obtains:

$$\frac{1}{b - c_1(h_2)} \int_{c_1(h_2)}^{b} f(s)ds \leq h_2$$

We define:

$$h_2 = \frac{1}{b - c_1(h_2)} \int_{c_1(h_2)}^{b} f(s)ds$$

The constant $c_1(h_2)$ is obtained by solving the integral equation

$$f(c_1(h_2)) = \frac{1}{b - c_1(h_2)} \int_{c_1(h_2)}^{b} f(s)ds$$

Note that because of the shape of $f(t)$, $h_1 \leq h_2$, $c_1(h_2) \geq c_1(h_1)$, $c_2(h_2) \leq c_2(h_1)$; therefore,

$$\int_{c_1(h_2)}^{c_2(h_2)} (f(s) - h_2)ds \leq \int_{c_1(h_2)}^{c_2(h_2)} (f(s) - h_1)ds \leq \int_{c_1(h_1)}^{c_2(h_1)} (f(s) - h_2)ds \leq h\varepsilon$$

namely, the constant

$$\int_{c_1(h_2)}^{c_2(h_2)} (f(s) - h)ds \leq h\varepsilon$$

is met. If $h_1 > h_2$, $c_1(h_2) \leq c_1(h_1)$, $c_2(h_1) \leq c_2(h_2)$, then

$$\int_{c_1(h_1)}^{b} (f(s) - h_1)ds \leq \int_{c_1(h_2)}^{b} (f(s) - h_1)ds \leq \int_{c_1(h_2)}^{b} (f(s) - h_2)ds \leq 0$$

Namely, the constraint

$$\int_{c_1(h_1)}^{b} (f(s) - h)ds = 0$$

is met. Finally,

$$C_A(f(t), \varepsilon) = \max \left(\frac{1}{c_2(h_1) + \varepsilon - c_1(h_1)} \int_{c_1(h_1)}^{c_2(h_1)} f(s)ds, \frac{1}{b - c_1(h_2)} \int_{c_1(h_2)}^{b} f(s)ds \right)$$

= max (h1, h2),

with the constants obtained as indicated above. By the nature of f(t),

$$\frac{1}{b - c_1(h_2)} \int_{c_1(h_1)}^{b} f(s)ds \geq \frac{1}{b - a} \int_{a}^{b} f(s)ds$$

or,

$$C_A(f(t), \varepsilon) \geq \frac{1}{b - a} \int_{a}^{b} f(s)ds$$

which shows that the constrained average traffic (design) level must exceed the average traffic level. An example follows, where the traffic requirements function is:

$$f(s) = \begin{cases} s, & 0 < s \leq 8 \\ 16 - s, & 8 < s \leq 16 \end{cases}$$

and $\varepsilon = 1$. Here a=0 and b=16.

$$c_1(h_1) = \frac{1}{c_2(h_1) + 1 - c_1(h_1)} \times \left(\int_{c_1(h_1)}^{8} f(s)ds + \int_{8}^{c_2(h_1)} f(s)ds \right) = 16 - c_2(h_1)$$

Or,

1) $c_1(c_2 - c_1 + 1) = -64 - c_1^2/2 + 16c_2 - c_2^2/2$
2) $(16 - c_2)(c_2 - c_1 + 1) = -64 - c_1^2/2 + 16c_2 - c_2^2$

   Solving simultaneously for c_1 and c_2 one obtains $c_1(h_1) = 5.64$, $c_2(h_1) = 10.36$. Regarding h_2 we have

$$\frac{1}{16 - c_1(h_2)} \int_{c_1(h_2)}^{16} f(s)ds = c_1(h_2)$$

from which $c_1(h_2) = 4.61$ and $h_2 = 4.67$. Therefore $C_A(f(t), 1) = \max (5.64, 4.67) = 5.64$. Note that $O(5.64) = 0$; $O(8) = 2.72$; $O(10.36) = 5.58$; $O(13.73) = 0$; $O(16) = 0$. Thus, $O(16) = 0$ and the problem is determined by $O(c_2) = h\varepsilon$.

Dual-Peak Function

Assume that the traffic follows the shape of Figure 6, characteristic of many known applications. The constraints to satisfy are:

a) $\quad O(c_2(h_1)) = \int_{c_1(h_1)}^{c_2(h_1)} (f(s) - h_1)ds \le h_1\varepsilon$

b) $\quad O(c_2(h.)) = \begin{cases} \int_{c_3(h_2)}^{c_4(h_2)} (f(s) - h_2)ds \le h_2\varepsilon \\[2mm] \text{if } O(c_3(h_2)) = 0 \text{ or } \int_{c_1(h_2)}^{c_3(h_2)} (f(s) - h_2)ds \le 0 \\[2mm] \int_{c_1(h_3)}^{c_4(h_3)} (f(s) - h_3)ds \le h_3\varepsilon \quad \text{if } O(c_3(h_2)) > 0 \end{cases}$

c) $\quad O(c_2(h.)) = \begin{cases} \int_{c_3(h_4)}^{b} (f(s) - h_4)ds = 0 \\[2mm] \text{if } O(c_3(h_4)) = 0 \text{ or } \int_{c_1(h_4)}^{c_3(h_4)} (f(s) - h_4)ds \le 0 \\[2mm] \int_{c_1(h_5)}^{b} (f(s) - h_5)ds = 0 \quad \text{if } O(c_3(h_4)) > 0 \end{cases}$

Solving for the various values of h one obtains the following (derivation not shown):

$$h_1 = \frac{1}{c_2 + \varepsilon - c_1} \int_{c_1}^{c_2} f(s)ds$$

$$h_2 = \frac{1}{c_4 + \varepsilon - c_3} \int_{c_3}^{c_4} f(s)ds$$

$$h_3 = \frac{1}{c_4 + \varepsilon - c_1} \int_{c_1}^{c_4} f(s)ds$$

$$h_4 = \frac{1}{b - c_3} \int_{c_3}^{b} f(s)ds$$

$$h_5 = \frac{1}{b - c_1} \int_{c_1}^{b} f(s)ds$$

Finally, $C_A(f(t), \varepsilon) = $ max $(h_1, h_2, h_3, h_4, h_5)$ (the various c's have to be solved via simultaneous equations.)

Example

Assume that the traffic arrival at a CPS/IoT gateway follows the simplified profile of Figure 9. For example, this (homogenous) sensing traffic could be generated by crowdsensing elements that monitor vehicular traffic events. The stream shown in the figure is assumed to be fairly isochronous at 5 Gbps in the morning, a 6 Gbps peak, and less in the mid-late afternoon. If the stream could be buffered in the gateway, but must be transmitted within an hour, the nominal outgoing bandwidth would be 4.4 Gbps, due to the constrained average of $C_A(f,1)$. If the traffic could be buffered up to 3 hours, the nominal outgoing bandwidth would be 3.62 Gbps (a saving of about 18%); if the traffic had to be transmitted instantaneously the required bandwidth would be 6 Gbps – compared with $C_A(f,3)$ that is a 65% increase in requisite resources.

Figure 9. Example

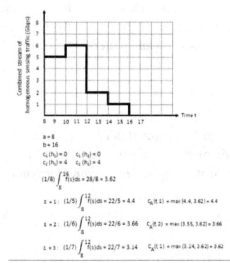

V. CONCLUSION

This chapter discussed a TE method called 'constrained average', where traffic is by design allowed to be delayed up to a specified value epsilon. Many networks, including multimedia-carrying networks and process-control networks, require mechanisms for traffic control. A number of CPS applications can make use of the technique discussed in this chapter (among or in conjunction with other TE/traffic management techniques). For example, Moving Pictures Expert Group (MPEG) frames or Voice over IP (VoIP) packets can only be delayed up to a certain maximum (in many real-time instances of interactive communication) before they become useless. The 'constrained average' approach discussed in this chapter is an example of a traffic smoothing algorithm. Its advantages are the defined per-packet delay, the zero-packet-loss, the relative simplicity and low complexity (effectively straightforward buffer management under a computational discipline), and its strong mathematical underpinning.

Generally, networks allow a 'sustainable' rate of transmission and, if the network could accommodate more traffic (as determined at instantaneous instances), then allowing transmission at a 'peak' rate (for some short bursting interval). The 'constrained average' method shapes the traffic to conform with the sustainable rate of the underlying network infrastructure. Allowing a small delay can reduce the amount of trunking or capacity needed at the edge-to-core convergence point, particularly considering (i) the large number of edge/fog-resident nodes, (ii) the evolution to multimedia-based applications, (iii) the need to maintain QoS, and (iv) the relative scarcity (and/or premium value) of wireless bandwidth.

The 'constrained average' TE concept also has more general applicability to a number of other traffic management environments in conjunction with, or to enhance, existing TE/TM systems. In addition, it is expected that closed-form formulas for many more functional shapes than shown in Section III can be relatively easily developed and applied in edge as well as core switches and routers.

REFERENCES

Antonino, P. O., & Morgenstern, A. (2018, April 30). Straightforward specification of adaptation-architecture-significant requirements of IoT-enabled cyber-physical systems. *2018 IEEE International Conference on Software Architecture Companion (ICSA-C)*. 10.1109/ICSA-C.2018.00012

Baheti, R., & Gill, H. (2011). *Cyber-physical Systems*. In T. Samad & A. M. Annaswamy (Eds.), *The impact of control technology*. IEEE Control Systems Society. Available at www.ieeecss.org

Blasch, E., & Kadar, I. (2017). Panel summary of Cyber-Physical Systems (CPS) and Internet of Things (IoT) opportunities with information fusion. *Proceedings Volume 10200, Signal Processing, Sensor/Information Fusion, and Target Recognition XXVI*. doi:10.1117/12.2264683

Burg, A., Chattopadhyay, A., & Lam, K.-Y. (2018). Wireless communication and security issues for cyber–physical systems and the internet-of-things. *Proceedings of the IEEE, 106*(1), 38–60. doi:10.1109/JPROC.2017.2780172

Cisco. (2017, September 15). *Cisco visual networking index: forecast and methodology, 2016–2021*. Cisco.

3GPP. (2015, September). *RWS-150002: views on next generation wireless access*. 3GPP RAN Workshop on 5G, 9/2015, Lenovo, Motorola Mobility.

Griffor. E. (2017, December 12). *Reference architecture for cyber-physical systems*. NIST (National Institute of Standards and Technology).

Guo, S., & Zeng, D. (Eds.). (2019). *Cyber-Physical Systems: Architecture*. Security and Application, EAI/Springer Innovations in Communication and Computing.

He, H., & Maple, C. (2016, July 24). The security challenges in the IoT enabled cyber-physical systems and opportunities for evolutionary computing & other computational intelligence. *2016 IEEE Congress on Evolutionary Computation (CEC)*. 10.1109/CEC.2016.7743900

Hu, L., & Xie, N. (2012, April). Review of cyber-physical system architecture. *2012 IEEE 15th International Symposium on Object/Component/Service-Oriented Real-Time Distributed Computing Workshops*. doi:10.1109/ISORCW.2012.15

Hua, F., & Lua, Y. (2016, March). Robust cyber–physical systems: Concept, models, and implementation. *Future Generation Computer Systems, 56*, 449–475. doi:10.1016/j.future.2015.06.006

International Society of Automation (ISA). (2018). *ISA-95, Enterprise-control System Integration*. https://www.isa.org/isa95/

ISO/IEC (2013, April). *High Efficiency Video Coding (HEVC), ISO/IEC 23008-2 MPEG-H Part 2/ITU-T H.265*. ISO/IEC Moving Picture Experts Group (MPEG) and ITU-T Video Coding Experts Group (VCEG).

Jiang, J.-R. (2018, June). An Improved cyber-physical systems architecture for Industry 4.0 smart factories. *Advances in Mechanical Engineering*, *10*(6). Advance online publication. doi:10.1177/1687814018784192

Kandris, D., Tsagkaropoulos, M., Politis, I., Tzes, A., & Kotsopoulos, S. (2011). Energy efficient and perceived QoS aware video routing over wireless multimedia sensor networks. *Ad Hoc Networks*, *9*(4), 591–607. doi:10.1016/j.adhoc.2010.09.001

Koutsoukos, X., Karsai, G., Laszka, A., Neema, H., Potteiger, B., Volgyesi, P., Vorobeychik, Y., & Sztipanovits, J. (2018, January). SURE: A modeling and simulation integration platform for evaluation of SecUre and REsilient cyber-physical systems. *Proceedings of the IEEE*, *106*(1), 93–112. doi:10.1109/JPROC.2017.2731741

Lee, E. A. (2015, February). Present and future of cyber-physical systems: A focus on models. *Sensors (Basel)*, *15*(3), 4837–4869. doi:10.3390150304837 PMID:25730486

Lee, J., Bagheri, B., & Kao, H.-A. (2015). A Cyber-physical systems architecture for Industry 4.0-based manufacturing systems. *Manufacturing Letters*, *3*, 18–23. doi:10.1016/j.mfglet.2014.12.001

Letichevsky, A. A., & Letychevsky, O. O. (2017, November). Cyber-physical systems. *Cybernetics and Systems Analysis*, *53*(6), 821–834. Advance online publication. doi:10.100710559-017-9984-9

Massey, D. (2017, November 3). Applying cybersecurity challenges to medical and vehicular cyber physical systems. *Proceeding of SafeConfig '17, 2017 Workshop on Automated Decision Making for Active Cyber Defense*, 39-39. doi:10.1145/3140368.3140379

Minoli, D. (1983). A new design criterion for store-and-forward networks. *Computer Networks*, *7*(1), 9–15. doi:10.1016/0376-5075(83)90003-X

Minoli, D. (2012). *Mobile video with Mobile IPv6*. Wiley., doi:10.1002/9781118647059.

Minoli, D. (2013). *Building the Internet of Things with IPv6 and MIPv6*. Wiley.

Minoli, D., Occhiogrosso, B., & ... . (2017). IoT considerations, requirements, and architectures for insurance applications. In Q. Hassan, A. R. Khan, & S. A. Madani (Eds.), *Internet of Things: challenges, advances and applications*. CRC Press.

Monostori, L. (2014). *Cyber-physical production systems: roots, expectations and R&D challenges. In Procedia CIRP* (Vol. 17). Elsevier. doi:10.1016/j.procir.2014.03.115

Müller, H. A., & Litoiu, M. (2016). Engineering cybersecurity into cyber physical systems. In CASCON 2016. Markham, Canada: ACM.

Quintanilla, F. G., & Cardin, O. (2016). Implementation framework for cloud-based holonic control of cyber-physical production systems. In *IEEE Proceedings of the 14th International Conference on Industrial Informatics*. Poitiers, France: IEEE.

Romanovsky, A., & Ishikawa, F. (Eds.). (2017). *Trustworthy cyber-physical systems engineering*. CRC.

Seeling, P., & Reisslein, M. (2014). Video Traffic characteristics of modern encoding standards: H.264/AVC with SVC and MVC extensions and H.265/HEVC. *TheScientificWorldJournal, 2014*, 189481. Advance online publication. doi:10.1155/2014/189481 PMID:24701145

Seema, A., & Reisslein, M. (2011). Towards efficient wireless video sensor networks: A survey of existing node architectures and proposal for a Flexi-WVSNP Design. *IEEE Communications Surveys and Tutorials, 13*(3), 462–486. doi:10.1109/SURV.2011.102910.00098

Sivakumar, M., & Sadagopan, C. (2016, August). Wireless sensor network to cyber physical systems: addressing mobility challenges for energy efficient data aggregation using dynamic nodes. *Sensor Letters, 14*(8), 852-857. doi:10.11661.2016.3624

Song, H., & Rawat, D. B. (2017). *Cyber-physical systems: foundations, principles and applications*. Academic Press/Elsevier.

Stankovic, J. A. (2017, February). Research directions for cyber physical systems in wireless and mobile healthcare. *ACM Transactions on Cyber-Physical Systems, 1*(1). doi:10.1145/2899006

Sullivan, G. J., Ohm, J.-R., Han, W.-J., & Wiegand, T. (2012). Overview of the High Efficiency Video Coding (HEVC) Standard. *IEEE Transactions on Circuits and Systems for Video Technology, 22*(12), 1649–1668. doi:10.1109/TCSVT.2012.2221191

Thramboulidis, K., & Christoulakis, F. (2016, October). UML4IoT - A UML-based approach to exploit IoT In cyber-physical manufacturing systems. *Computers in Industry, Elsevier, 82*, 259–272. doi:10.1016/j.compind.2016.05.010

Wang, S., Lei, T., Zhang, L., Hsu, C.-H., & Yang, F. (2016, August). Offloading mobile data traffic for QoS-aware service provision in vehicular cyber-physical systems. *Future Generation Computer Systems, Elsevier, 61*, 118–127. doi:10.1016/j.future.2015.10.004

Wang, W., & Minoli, D. (2017, June 6). Multimedia IoT systems and applications. In *Global IoT Summit, GIoTS-2017*. IEEE.

Wang, W., & Wang, Q. (2017, April). Multimedia Sensing As A Service (MSaaS): Cloud-edge IoTs and fogs. *IEEE IoT Journal, 4*(2), 487–495. doi:10.1109/JIOT.2016.2578722

Wang, C., & Zhu, Y. (2018, September). A dependable time series analytic framework for cyber-physical systems of IoT-based smart grid. *ACM Transactions on Cyber-Physical Systems, 3*(1). doi:. doi:10.1145/3145623

Weyrich, M., & Ebert, C. (2016, February). Reference architectures for the Internet of Things. *IEEE Software*.

Yu, X., & Xue, Y. (2016, May). Smart grids: A cyber-physical systems perspective. *Proceedings of the IEEE, 104*(5), 1058–1070. doi:10.1109/JPROC.2015.2503119

Zanero, S. (2017, April). Cyber-physical systems. *Computer, 50*(4), 14–16. doi:10.1109/MC.2017.105

Chapter 6

Resource Allocation Techniques for SC–FDMA Networks:
Advancements, Challenges, and Future Research Directions

Muhammad Irfan
COMSATS University Islamabad, Wah Campus, Pakistan

Ayaz Ahmad
(iD) https://orcid.org/0000-0002-2253-6004
COMSATS University Islamabad, Wah Campus, Pakistan

Raheel Ahmed
COMSATS University Islamabad, Wah Campus, Pakistan

ABSTRACT

Single carrier frequency division multiple access (SC-FDMA) is a promising uplink transmission technique that has the characteristic of low peak to average power ratio. The mobile terminal uplink transmission depends on the batteries with limited power budget. Moreover, the increasing number of mobile users needs to be accommodated in the limited available radio spectrum. Therefore, efficient resource allocation schemes are essential for optimizing the energy consumption and improving the spectrum efficiency. This chapter presents a comprehensive and systematic survey of resource allocation in SC-FDMA networks. The survey is carried out under two major categories that include centralized and distributed approaches. The schemes are also classified under various rubrics including optimization objectives and

DOI: 10.4018/978-1-5225-9493-2.ch006

constraints considered, single-cell and multi-cell scenarios, solution types, and perfect/imperfect channel knowledge-based schemes. The advantages and limitations pertaining to these categories/rubrics have been highlighted, and directions for future research are identified.

INTRODUCTION

Cellular wireless technologies are continuously developing and wireless systems are upgraded to meet the increasing demand of the wireless users and improve the energy as well as spectral efficiency of the systems. Recently, for this purpose, Single Carrier Frequency Division Multiple Accesses (SC-FDMA) has got attention of researchers and industrial analyst for uplink communication. Currently, it is adopted as a multiple access scheme for LTE (Long Term Evolution) uplink transmission (Sofer & Segal, 2005; Wong, Oteri, & McCoy, 2009) and is assumed as a strong candidate for the uplink transmission in the next generation wireless networks.

To fulfill the high demands of capacity in wireless networks, Orthogonal Frequency Division Multiple Access (OFDMA) performs well for downlink whereas SC-FDMA is efficient for uplink wireless networks (Berardinelli, de Temino, Frattasi, Rahman, & Mogensen, 2008). In wireless environment, OFDMA shows robustness in the presence of multipath signal propagation by the parallel transmission of the information on M orthogonal and equally spaced subcarriers. However, on the other hand, OFDMA system exhibits an high envelop variation in its waveform which results an high peak to average power ratio (PAPR). Signals exhibiting high PAPR, require highly linear amplifiers to safeguard the system from intermodulation interference. This linearity is achieved if linear amplifiers at the transmitter operate with large back off from their peak power. This results in low power efficiency and if used for uplink transmission will put a significant burden on the portable wireless station (Fantacci, Marabissi, & Papini, 2004a). Another drawback of OFDMA is that it exhibits a certain offset in the frequency reference between the transmitter and receiver. This frequency offset can destroy the orthogonality of the transmission which introduces inter-carrier interference.

To overcome the above described drawbacks of OFDMA, the Third Generation Partnership Project (3GPP) introduced SC-FDMA for the uplink transmission which is a modified form of OFDMA (Rumney, 2008). As in OFDMA, SC-FDMA system transmit the information signals on different orthogonal frequency subcarriers. However, it transmits the subcarriers in sequence rather than in parallel which reduces the envelop fluctuation of waveform and results in low PAPR than OFDMA system (Ciochina, Castelain, Mottier, & Sari, 2009). In wireless environment with severe multipath, SC-FDMA signals arrive at the base station with inter-symbol interference. However, the base station employs adaptive frequency domain equalization to overcome this inter-symbol interference issue. In this way, the processing burden is shifted from the portable wireless station to the more powerful base station (Huang, Nix, & Armour, 2007; Lafuente-Martinez, Hernandez-Solana, Guio, & Valdovinos, 2011; Raghunath & Chockalingam, 2009). With the evolving wireless technologies, the demand of voice over IP, video conferencing, high data rate demanding wireless gaming, etc is also increasing (Falconer, Ariyavisitakul, Benyamin-Seeyar, & Eidson, 2002). Therefore, in wireless networks, intelligent and efficient resource allocation is very essential for meeting the demands of wireless users and achieving the goals of the service providers. Likewise any other wireless network, with the increasing wireless services, the multiple users in SC-FDMA uplink compete for the limited available radio spectrum. Secondly, the uplink systems are more vulnerable to energy losses as they are generally operated on the batteries with limited energy supply capabilities. Therefore, resource allocation with efficient spectrum utilization and optimized energy consumption is very important for uplink SC-FDMA systems.

In this chapter, we present a survey of resource allocation techniques for the uplink SC-FDMA systems with the objective to review the state-of-the-art techniques for guaranteeing the quality-of- service (QoS) demands of the users and achieving the energy and spectrum efficiencies of the system. The survey on resource allocation techniques is carried out under two major headings that consist of centralized and distributed approaches. The resource allocation schemes are also categorized under various headings/classifications such as spectrum efficient or energy efficient, interference avoidance or QoS guarantee based, faire resource allocation, single cell or multi cell, perfect channel information based or imperfect channel information based, etc. The pros and cons of these classes are discussed and future research directions are outlined. The rest of this survey is organized as follows. In Section 2, we provide the classification of the different resource allocation techniques which are reported in the literature for SC-FDMA. Section 3, provides comprehensive review of the centralized resource allocation techniques. Section 4, describes the distributed resource allocation techniques. In Section 5, we discuss some future research directions and in Section 6, we conclude the paper.

RESOURCE ALLOCATION IN SC-FDMA SYSTEM

In this paper, we survey the resource allocation schemes designed for efficient power and spectrum allocation in uplink SC-FDMA systems. We classify these schemes into two major categories that consist of centralized and distributed schemes. Based on the attainment of different objectives in up-link transmission, these schemes are also classified with respect to the optimization criteria such as energy efficient resource allocation, spectrum efficient resource allocation, QoS guarantee based schemes, fair resource allocation and resource allocation with interference prevention. In addition, the schemes are also classified on the basis of optimization constraints that whether the transmit power or the achieved throughput constraint is considered in the resource optimization problem. The schemes are also divided into single-cell and multi-cell resource allocation approaches. We also categorize these schemes on the basis of channel knowledge. That is, whether perfect or imperfect channel state information is assumed in the design of a resource allocation scheme. The classification tree is shown in Fig. 1 which provides a global view of the different categories and sub-categories of resource allocation in uplink SC-FDMA system along with the relevant references. The references in Fig. 1 provides the information about the amount of work done for each class of resource allocation schemes. In the following, we provide the overview of all the resource allocation classes and highlight the advantages and limitations associated to each of them.

Centralized and Distributed Resource Allocation

Centralized Resource Allocation

For centralized resource allocation in up-link SC-FDMA, the central node/base station (BS) is only responsible for resource allocation. In this case, the users provide their channel state information to the BS and the resource allocation decisions are made at the BS which are then communicated to the end users as illustrated in Fig 2. The decision making BS in centralized resource allocation schemes are central node which has a global view of all the network and due to which centralized schemes have several benefits. Centralized schemes can easily achieve any of the several optimization objectives: power minimization, throughput maximization, minimizing the inter-cell interference, faire resource allocation among multiple users, and allocating resources to users on the basis of their promised priorities. However, there are some drawbacks associated with centralized resource allocation schemes that can severely affect the network performance in practical scenarios. High signaling overhead is the major drawback of these schemes as the BS/central node makes radio resource allocation decision for all the users in the network and communicate these decisions to the users. In scenarios, where users are spread over large area of network, high amount of transmission power is required for the transmission of the control decision to all the users which may lead to inefficient power consumption. In most of the existing works on resource allocation in SC-FDMA systems, centralized schemes have been studied (Afifi, Elsayed, & Khattab,

2013; A. Ahmad, 2015, 2015; Ayaz Ahmad & Assaad, 2011c, 2011b, 2011a; Aijaz, Nakhai, & Aghvami, 2014; Akande, Iqbal, Zerguine, Al-Dhahir, & Zidouri, 2016; Davaslioglu & Ayanoglu, 2014; Fahmi, Astuti, Meylani, Asvial, & Gunawan, 2016; Fan, Lee, Li, & Li, 2015; Fushiki, Ohseki, & Konishi, 2011; Ha, Tuan, & Nguyen, 2013; Hsu, Chao, & Liu, 2013; Hu & Ci, 2015; Kaddour, Pischella, Martins, Vivier, & Mroueh, 2013; Kessab et al., 2012; H. Kim, Chung, Lee, Kim, & Choo, 2015; J. Kim, Hwang, & Kang, 2015; Kiran & Jibukumar, 2016; Lee, Pefkianakis, Meyerson, Xu, & Lu, 2009; Lim, Kim, & Im, 2010; Lu, Yang, Li, Qiu, & Zhang, 2016; Nakada, Obara, Yamamoto, & Adachi, 2012; Pao, Chen, Tsai, Lou, & Chang, 2012; Ruby, Leung, Michelson, 2015; Shah, Gu, Hasan, & Chung, 2015; Sokmen & Girici, 2010; Tsiropoulou & Papavassiliou, 2011a; J. Zhang, Yang, & Hanzo, 2009; M. Zhang & Zhu, 2013).

Figure 1. Types of Resource Allocation in SC-FDMA

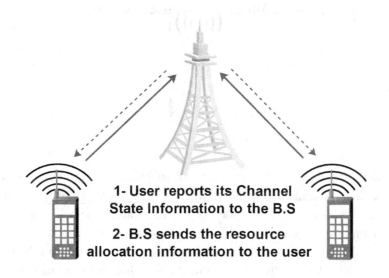

1- User reports its Channel State Information to the B.S

2- B.S sends the resource allocation information to the user

Distributed Resource Allocation

Centralized resource allocation has several advantages as decisions are made at the central entity with global knowledge of the network and users. However, due to the high demand of global and seamless access of the broadband wireless network, distributed resource allocation is very important in uplink SC-FDMA. In distributed resource allocation, the resource allocation decisions are taken at the users end as illustrated in Fig 3.

Figure 2. Centralized Resource Allocation

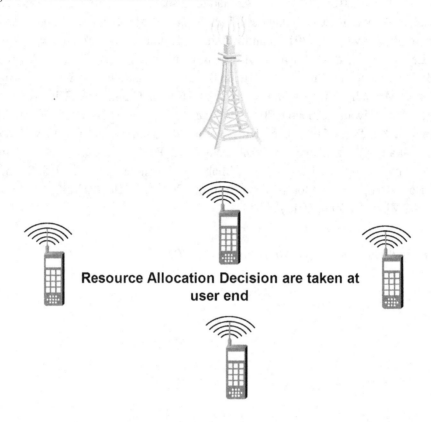

Resource Allocation Decision are taken at user end

An advantage of distributed resource allocation schemes is that they can be adapted to network changes immediately due to which these schemes are effective for time varying wireless environment. Another advantage of these schemes is that they have lower signaling overhead and quick decision making process which is very useful for resource allocation in the uplink. Distributed resource allocation schemes have drawbacks as well. One of their major drawback is that each user makes decision on the basis of local information which may be incomplete. Performing resource allocation using this incomplete information may result in a non-optimal solution. Another issue with distributed schemes is that it cannot provide global fairness among the multiple users. A number of research works dedicated to distributed resource allocation in SC-FDMA systems are available in the literature (Ahmed & Mohamed, 2015; Ahmed, Mohamed, & Shakeel, 2010; Lafuente-Martínez, Hernández-Solana, & Valdovinos, 2011; Noh & Oh, 2009; Tsiropoulou, Ziras, & Papavassiliou, 2015). The advantages and limitations centralized and distributed resource allocation schemes along with the relevant references are summarized in Table 1.

Figure 3. Distributed Resource Allocation

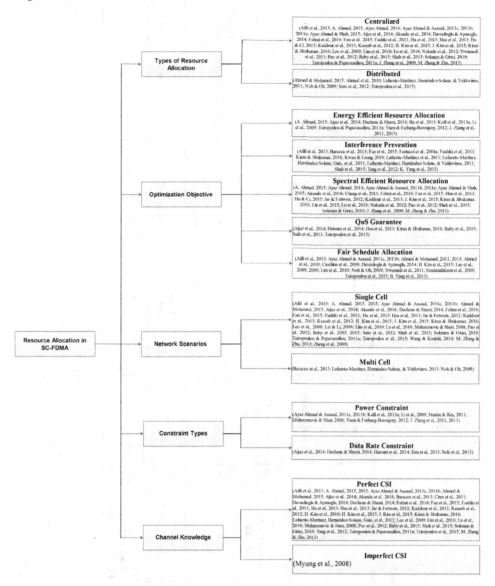

Optimization Objectives

Resource allocation schemes for SC-FDMA with different optimization objectives are overviewed in this subsection.

Table 1. Resource Allocation types for SC-FDMA

Resource Allocation Types	Description	Benefits	Drawbacks	References
Centralized	All the users are connected to BS/central node and get resource allocation decision and other information from this BS/central node	In centralized scheme efficient decisions are made due to the availability of global knowledge about all the users which are connected to the BS/central node	High signaling overhead and increased power consumption for transmitting control signals	(Afifi et al., 2013; A. Ahmad, 2015; Ayaz Ahmad, 2014; Ayaz Ahmad & Assaad, 2011b, 2011c, 2011a; Ayaz Ahmad & Shah, 2015; Akande et al., 2016; Davaslioglu & Ayanoglu, 2014; Fahmi et al., 2016; Fan et al., 2015; Ha et al., 2013; Hsu et al., 2013; Kaddour et al., 2013; Kessab et al., 2012; H. Kim et al., 2015; Kiran & Jibukumar, 2016; Lee et al., 2009; Lu et al., 2016; Nakada et al., 2012; Nwamadi, Zhu, & Nandi, 2011; Pao et al., 2012; Ruby et al., 2015; Shah et al., 2015; Sokmen & Girici, 2010; Tsiropoulou & Papavassiliou, 2011a; J. Zhang et al., 2009; M. Zhang & Zhu, 2013)
Distributed	This type of resource allocation does not involve the BS/central node and the resource allocation decisions are made at each user's end	Low signaling overhead, faster decision making	It use local information for resource allocation which may lead to inefficient solution, Fairness among the users may not be guaranteed	(Ahmed & Mohamed, 2015; Ahmed et al., 2010; Lafuente-Martínez, Hernández-Solana, & Valdovinos, 2011; Noh & Oh, 2009; Sato, Ryusuke, Obara, & Adachi, 2012; Tsiropoulou et al., 2015)

Energy/Power Efficient Resource Allocation

Energy efficient or power efficient resource allocation can prolong the battery life of the mobile terminal by efficient utilization of the energy. Owing to the fact that energy minimization and energy conservation is crucial for battery operated mobile users, this type of schemes are very important and beneficial for uplink SCFDMA systems. The disadvantage of energy efficient resource allocation schemes is that these schemes cannot achieve maximum spectral efficiency and maximum throughput performance. In the literature, a number of energy efficient schemes for resource allocation in SC-FDMA can be found(A. Ahmad, 2015; Aijaz et al., 2014; Dechene & Shami, 2014; Kalil, Shami, & Al-Dweik, 2013a; Li, Yin, & Yue, 2009; Tsiropoulou & Papavassiliou, 2011b; J. Zhang, Yang, & Hanzo, 2011, 2013).

Spectral Efficient Resource Allocation

Spectral efficient resource allocation improves the spectrum usage efficiency and improves the throughput/capacity of a given SC-FDMA uplink network. If the multiple mobile users share and utilize the given spectral bandwidth efficiently than the users' required QoS can be achieved and the network can perform more effectively. Spectral efficient resource allocation schemes has the disadvantage of increasing the power consumption of the users. Another issue associated to these schemes is that they may result in unfair resource allocation among the users. In (Ayaz Ahmad & Assaad, 2011c, 2011b; Ahmed et al., 2010; Fan et al., 2015; Hsu et al., 2013; Jar & Fettweis, 2012; Kaddour et al., 2013; Kessab et al., 2012; Kiran & Jibukumar, 2016; Liu, Chen, & Xiong, 2015; Lu et al., 2016; Pao et al., 2012; Shah et al., 2015; J. Zhang et al., 2009; M. Zhang & Zhu, 2013), the authors have reported spectral efficient resource allocation schemes for SC-FDMA.

QoS Guarantee Based Resource Allocation

For some applications, e.g., the multimedia applications, it may be crucial to guarantee the given QoS requirements while allocating the resource among the users. QoS requirements for SC-FDMA system can include transmission rate and delay requirements, etc. Several articles are published on QoS guarantee based resource allocation for SCFDMA networks (Aijaz et al., 2014; Hatoum, Hatoum, Ghaith, & Pujolle, 2014; Hsu et al., 2013; Lu et al., 2016; Ruby et al., 2015; Safa, El-Hajj, & Tohme, 2013; Tsiropoulou et al., 2015). Though these schemes guarantee the QoS requirement, they cannot achieve the maximum energy efficiency and spectral efficiency.

Fair Resource Allocation

Fair resource allocation among users is important as it accounts for the fair share of resources among users and their priorities and importance. These scheme intelligently decides that how much resources should be allocated to a user such that the promised QoS of the user is assured while the rest of the users are not deprived of the resources. Knowing the importance of fairness in improving the network performance, fair resource allocation in the uplink SC-FDMA networks are presented in (Afifi et al., 2013; Ayaz Ahmad & Assaad, 2011c, 2011b; Ahmed & Mohamed, 2011, 2015; Ahmed et al., 2010; Ciochina et al., 2009; Davaslioglu & Ayanoglu, 2014; Lee et al., 2009, 2009; Noh & Oh, 2009; Nwamadi et al., 2011; Sandanalaksmi, Manivanan, Manikandan, Barathi, & Devanathan, 2009; Tsiropoulou et al., 2015; B. Yang, Niu, He, Xu, & Huang, 2013). Demerit of fairness based schemes is that they may not provide efficient results in terms of network's global performance. Also fair resource allocation may not ensure the QoS requirements of all the users/services.

Interference Prevention

If interference between the users is minimum then the given network has good performance and the required QoS can be achieved. Therefore, some resources allocation schemes in SC-FDMA networks are designed to avoid interference among users and obtain the efficient results. Interference prevention based resource allocation schemes for SC-FDMA networks are reported in a number of works (Afifi et al., 2013; Baracca, Tomasin, & Benvenuto, 2013; Fan et al., 2015; Fantacci, Marabissi, & Papini, 2004b; Fushiki et al., 2011; Kiran & Jibukumar, 2016; Kwan & Leung, 2010; Lafuente-Martinez et al., 2011; Lafuente-Martínez, Hernández-Solana, Guío, & Valdovinos, 2011; Lafuente-Martínez, Hernández-Solana, & Valdovinos, 2011; Shah et al., 2015; Tang, Hong, Xue, & Peng, 2012; K. Yang, Martin, & Yahiya, 2015). One of the major limitations of schemes with interference prevention is that they may not guarantee the QoS requirements of some of the users in the network.

Constraint Types

Different types of constraint are considered in resource allocation for uplink SC-FDMA. Two of the important and widely considered types of constraints are discussed in this paper as follows.

Power Constraint

Power constraints in SC-FDMA contain two types of constraints which are per user transmit power constraint and per sub-channel transmit power constraint. Per user power constraint means that the power transmitted on all sub-channels of a user must be less than or equal to a preset threshold value for that user. On the other hand, the per sub-channel power constraint reflects the fact that the power transmitted on each sub-channel of a user should not exceed a certain maximum value. In (Ayaz Ahmad & Assaad, 2011c, 2011b; Kalil, Shami, & Al-Dweik, 2013b; Li et al., 2009; Muharemovic & Shen, 2008; Sokmen & Girici, 2010; Yuen & Farhang-Boroujeny, 2012; J. Zhang et al., 2011, 2013), the resource allocation schemes for SCFDMA consider the power constraints.

Data Rate Constraint

Data rate constraint in SC-FDMA resource allocation guarantees that the achieved data rate of a user should be greater than or equal to a threshold value. Data rate constraint in SC-FDMA networks are also reported in the literature (Aijaz et al., 2014; Dechene & Shami, 2014; Hatoum et al., 2014; Hsu et al., 2013; Safa et al., 2013).

Single-Cell or Multi-Cell Network Scenarios

In resource allocation schemes for uplink SC-FDMA, either the single-cell network or the multi-cell scenario is considered as discussed in the following.

Single-Cell Scenario

Single-cell scenarios are widely considered in centralized resource allocation of SC-FDMA uplink cellular networks. The resource allocation schemes for single-cell SC-FDMA networks neither consider intra-cell interference nor inter-cell interference. The intra-cell interference is not an issue in SC-FDMA networks thanks to the orthogonal allocation of SCFDMA sub-channel among different users. However, inter-cell interference can severely degrade the system performance and its consideration in the design of resource allocation is necessary. Resource allocation with single-cell assumption are presented in (Afifi et al., 2013; A. Ahmad, 2015, 2015; Ayaz Ahmad & Assaad, 2011c, 2011b; Ahmed & Mohamed, 2015; Aijaz et al., 2014; Akande et al., 2016; Dechene & Shami, 2014; Fahmi et al., 2016; Fan et al., 2015; Fushiki et al., 2011; Ha et al., 2013; Hsu et al., 2013; Jar & Fettweis, 2012; Kaddour et al., 2013; Kessab et al., 2012; H. Kim et al., 2015; J. Kim et al., 2015; Kiran & Jibukumar, 2016; Lee et al., 2009; Lei & Li, 2009; Lim et al., 2010;

Lu et al., 2016; Muharemovic & Shen, 2008; Pao et al., 2012; Ruby et al., 2015, 2015; Sato et al., 2012; Shah et al., 2015; Sokmen & Girici, 2010; Tsiropoulou & Papavassiliou, 2011a; Tsiropoulou et al., 2015; Wang & Konishi, 2010; M. Zhang & Zhu, 2013; Zheng, Wei, Long, Liu, & Wang, 2009).

Multi-Cell Scenario

Resource allocation in multi-cell scenarios are affected by inter-cell interference. To avoid inter-cell interference in multi-cell scenarios different techniques can be used such as fractional frequency reuse (FFR) in which a cell is divided into two region with an inner and an outer area (Baracca et al., 2013). The inner region of all the cells reuses the same frequencies while the outer region reuses a subset of frequencies in a manner that avoids inter-cell interference. A spectrum efficient approach is the dynamic frequency reuse technique in which all the cells are allowed to use all the frequencies and appropriate techniques are employed for interference mitigation (Noh & Oh, 2009). However, the use of interference mitigation techniques augments the computation complexity of the systems. The works in (Lafuente-Martínez, Hernández-Solana, & Valdovinos, 2011), consider multi-cell resource allocation in SC-FDMA networks.

Channel Knowledge

Channel state information (CSI) describes how the wireless channel from the transmitter to the receiver exhibits and how the different parameters of the channel can affect the quality of communication. The channel state information play important role in efficient resource allocation by providing guidelines that how the power and spectrum may be allocated among different users with different channel characteristics. Resource allocation schemes for the uplink SC-FDMA are designed either under the assumption of perfect or imperfect knowledge of the user's channels.

Perfect Channel Knowledge

Perfect channel knowledge based resource allocation in uplink SC-FDMA assumes that both the transmitter and the receiver have complete knowledge of the channel state of the users' channels. Based on this assumption, resource allocation decisions are made and power and spectrum is allocated among users. Most of the resource allocation schemes for SC-FDMA assume the availability of perfect channel state information (Afifi et al., 2013, 2013; A. Ahmad, 2015; Ayaz Ahmad & Assaad, 2011b, 2011c; Ahmed & Mohamed, 2015; Akande et al., 2016; Chen, Song, & Letaief, 2011; Davaslioglu & Ayanoglu, 2014; Dechene & Shami, 2014; Fan et al.,

2015; Fushiki et al., 2011; Ha et al., 2013; Hsu et al., 2013; Kaddour et al., 2013; Kessab et al., 2012; D. Kim, Kim, Kim, Kim, & Han, 2010; H. Kim et al., 2015; J. Kim et al., 2015; Kiran & Jibukumar, 2016; Lafuente-Martínez, Hernández-Solana, Guío, et al., 2011; Lim et al., 2010; Lu et al., 2016; Muharemovic & Shen, 2008; Pao et al., 2012; Ruby et al., 2015; Shah et al., 2015; Tang et al., 2012; Tsiropoulou & Papavassiliou, 2011a; M. Zhang & Zhu, 2013). The assumption of perfect channel knowledge makes the solution of the resource allocation problems quite easy. However, resource allocation performed under this assumption may not be practically applicable as it is not realistic that channel state is fully known to both the transmitter and the receiver.

Imperfect Channel Knowledge

As the transmitter may not have perfect channel state information, and the resource allocation decision taken on the basis of this imperfect channel state information will not be efficient. Therefore, the consideration of imperfect channel knowledge in resource allocation for SC-FDMA systems is very important and a number of works have taken it into account (Myung, Oh, Lim, & Goodman, 2008). The solution of resource allocation problems with the availability of only imperfect channel information is difficult compared to the those with the assumption of perfect channel information.

CENTRALIZED RESOURCE ALLOCATION

In this section, we review centralized resource allocation schemes for uplink SC-FDMA systems. Centralized schemes have been investigated for attaining various types of resource allocation objectives that were described in Section II. Table II summarizes the centralized schemes from the optimization objectives' perspective.

Centralized Energy/Power Efficient Resource Allocation

Energy/power-efficient joint power and sub-channel allocation is investigated in (Tsiropoulou & Papavassiliou, 2011a) whereas only power allocation is considered in (A. Ahmad, 2015) and (Ha et al., 2013). In (Tsiropoulou & Papavassiliou, 2011b), optimal joint power and sub-channel allocation in uplink SC-FDMA network is explored. The authors have proposed an iterative algorithm to solve the problem of joint power and subcarrier allocation. The approach of (Tsiropoulou & Papavassiliou, 2011b) is demonstrated to show effectiveness in terms of power saving and user satisfaction. In (A. Ahmad, 2015), power allocation problem with finite power channel input with arbitrary distribution in uplink SC-FDMA system is investigated. A power allocation algorithm is developed by exploiting the relationship between the mutual information and the minimum mean square error. The simulation results in(A. Ahmad, 2015) have shown that the power allocation to the users in accordance to the employed modulation scheme achieves significant power saving as compared to the power allocation performed on the basis of common assumption of Gaussian channel input. In (Ha et al., 2013), joint source power control and relay beamforming framework for multi-relay and multi-user SCFDMA networks with amplify-and-forward strategy is investigated. This work considers two problems that include the maximization of the users' SINR subject to transmit power constraints and the minimization of sum transmit power under users' target minimum SINR. The problems are converted into difference-of-convex programs and iterative solutions are developed that perform efficiently. In addition, these problems are simplified and solved via branch-and-bound technique in order to find a simple yet acceptable solution.

Table 2. Relevant References of Centralized Resource Allocation Scheme for SC-FDMA

Optimization Objective	Relevant References
Centralized Energy Efficient Resource Allocation	(A. Ahmad, 2015; Ha et al., 2013; Tsiropoulou & Papavassiliou, 2011b)
Centralized Spectral Efficient Resource Allocation	(Ayaz Ahmad, 2014; Ayaz Ahmad & Assaad, 2011b, 2011c, 2011a; Ayaz Ahmad & Shah, 2015; Akande et al., 2016; Fahmi et al., 2016; Fan et al., 2015; Fushiki et al., 2011; Hsu et al., 2013; Hu & Ci, 2015; Kaddour et al., 2013; Kessab et al., 2012; J. Kim et al., 2015; Kiran & Jibukumar, 2016; Lu et al., 2016; Nakada et al., 2012; Pao et al., 2012; Sokmen & Girici, 2010; J. Zhang et al., 2009; M. Zhang & Zhu, 2013)
Centralized Fair Resource Allocation	(Afifi et al., 2013; Davaslioglu & Ayanoglu, 2014; H. Kim et al., 2015; Lee et al., 2009; Lim et al., 2010; Nwamadi et al., 2011; Shah et al., 2015)
Centralized QoS Guarantee	(Aijaz et al., 2014; Ruby et al., 2015)

Centralized Spectral Efficient Resource Allocation

In (Ayaz Ahmad & Assaad, 2011c, 2011b, 2011a, 2011a; Ayaz Ahmad & Shah, 2015; Pao et al., 2012), the authors have studied joint power and sub channel allocation for maximizing the spectral efficiency of the system. In (Pao et al., 2012), the authors have considered power allocation and subcarrier assignment problem in uplink SC-FDMA systems. The authors present a scheme that is comprised of constant power allocation and iterative chunk and subcarrier assignment techniques. The iterative techniques in the first phase allocates a chunk of subcarriers to each users. In the second phase, it reassigns the consecutive subcarriers to the users to achieve high data rate. The authors in (Ayaz Ahmad & Assaad, 2011b)[has studied a weighted sum rate maximization based resource allocation problem in uplink SC-FDMA system. The authors have proposed a polynomial complexity power and sub channel allocation algorithm by using the canonical duality theory. The authors have shown through simulation that their algorithm has near to optimal performance. The same authors have used a similar approach for the solution of resource allocation with the objective of sum-power minimization in uplink SC-FDMA system in (Ayaz Ahmad & Assaad, 2011c). Using the canonical duality theory, joint resource allocation and adaptive modulation with the objective of utility maximization and cost minimization for uplink SC-FDMA networks are studied in (Ayaz Ahmad & Assaad, 2011a) and (Ayaz Ahmad & Shah, 2015), respectively. In (Ayaz Ahmad & Assaad, 2011a)], the authors have considered three different problems of efficient sub-channel and user transmission power allocation for uplink SC-FDMA wireless network. These problems include weighted-sum-rate maximization, sum-power minimization and users rate constraints satisfaction with minimum sub channel usages problems. The authors have developed optimal as well as low-complexity greedy algorithms for the solution of these problems, and have demonstrated via numerical results that the greedy algorithms performs close to their optimal counterparts.

Resource block allocation schemes for achieving high aggregate throughput/cell capacity are presented in (Kaddour et al., 2013; Tsiropoulou & Papavassiliou, 2011b)]. In(Kaddour et al., 2013), the authors have proposed an opportunistic and efficient resource block allocation algorithm which maximize the aggregate throughput of an uplink SC-FDMA network. The authors suggested a solution for the efficient resource block allocation in which a resource block is allocated to the user for which the throughput improvement with this additional resource block is the highest. The authors of (Tsiropoulou & Papavassiliou, 2011b) have proposed a strategy of multiple resource block allocation to each user in uplink SC-FDMA network and have evaluated the gain associated to their scheme. This multiple resource block allocation approach which aims at achieving high cell capacity, is compatible with the basic SC-FDMA specifications, i.e., limited transmission power at user side and the use of the same modulation and coding scheme for all resource blocks of each user. The authors have suggested to first update the signal to interference ratio (SINR) for all users and then allocate the multiple resource blocks to the users. This approach is shown to be efficient for cells with low-to-medium traffic.

Fair sub channel allocation with the objective of spectral efficiency are studied in (Hsu et al., 2013; M. Zhang & Zhu, 2013). In (Hsu et al., 2013), proportional fair scheduling in uplink SC-FDMA system has been studied. In their study, the authors have considered the two important and basic constraints of SC-FDMA systems, i.e.,

1) multiple subcarriers allocated to a user must be contiguous in frequency in each time slot and 2) users should use the same modulation and coding scheme on all its allocated subcarrier. In view of the NP-hard nature of the scheduling problem, the authors have developed an efficient proportional fair heuristic algorithm which results in improved system throughput and thereby improves the overall efficiency of the system. In (M. Zhang & Zhu, 2013), an enhanced greedy two step algorithm is devised. Authors, use this technique to solve the resource allocation problem in uplink SC-FDMA networks with adjacent sub channel allocation constraint. This constraint makes the resource allocation in SC-FDMA network difficult.

In the first step of the enhanced greedy algorithm, the base station selects a number of stronger users and gives them priority to find their first sub-channels while in the second step, the base station allows all the users to find their sub channels among the remaining sub-channels on the basis of the best contribution criterion. The given two step algorithm is shown to result in a high spectral efficiency while improving the proportional fairness among the multiple users.

Spectral efficient subcarrier chunk allocation with average BER consideration is investigated in (Fahmi et al., 2016). In this work, the authors present a solution in which a chunk is allocated to a user for which his/her average BER is low for that chunk. This approach minimizes the average BER of all users and maximizes the throughput of the network. In (Fushiki et al., 2011), the impact on the achieved throughput due to the interference arising from user end to the neighboring cells in SC-FDMA system is studied. This interference degrades the system performance, especially the performance of the users at the cell edge. To overcome this problem, authors have studied the benefits of fractional frequency reuse (FFR) technique for achieving throughput gain in SC-FDMA system. To examine the through-put gain of the FFR in uplink SC-FDMA system, the authors studied its performance relationship with that of the frequency selective scheduling (FSS). This analysis has revealed that the throughput gain of FFR is small due to the degradation of FSS gain. However, the authors have found that under the conditions of (1) small number of users and (2) larger or zero multipath delay spread, the FFR is effective and achieves high throughput gain. In (J. Zhang et al., 2009), the authors investigated that spectral efficiency in SC-FDMA networks can be achieved by using a relay protocol. In the relay protocol presented in (J. Zhang et al., 2009)], a relay can be accessed by multiple users at a time simultaneously. In the relay protocol, the authors have also taken measures to overcome the interference problem in multiuser access of a relay protocol. In (Nakada et al., 2012), the authors have devised a power allocation technique for direct/cooperative SC-FDMA networks with amplify and forward relay scheme. The relaying is used when it can result in higher throughput than the direct communication. Power is allocated to the relay stations and mobile terminals in accordance to the wireless channel condition with the objective of maximizing the system throughput. The technique proposed in (Nakada et al., 2012) is shown to achieve almost same throughput as achieved by the grid search method. Virtual multiple input multiple output (virtual MIMO) systems can exploit the multiplexing capability of the uplink cellular communications to improve the spectral efficiency of the network. As SC-FDMA is a promising multiple access technique in the uplink, scheduling algorithms for virtual MIMO SC-FDMA system has attracted researchers' attention. Consequently, several algorithms for grouping/pairing the multiple users and allocating the same resource to multiple users in virtual MIMO SC-FDMA systems are reported in the literature (Fan et al., 2015; Hu & Ci, 2015; J. Kim et al., 2015; Kiran & Jibukumar, 2016; Lu et al., 2016). In (Hu & Ci, 2015), the authors focus on joint user pairing and resource allocation with the objective of improving the service quality of experience (QoE) in virtual MIMO SC-FDMA system. For this propose, the authors have quantified the QoE of the users by using a No-reference logarithmic model and have maximized the sum of users' mean of score (MOS). The approach proposed in (Hu & Ci, 2015) provides a good balance between the computational complexity and accuracy as well as improves the QoE with increased system throughput. A similar approach of joint resource allocation and user grouping for obtaining a balanced trad-off between transmission performance (average mean squared error (MSE) performance) and system throughput for virtual MIMO SC-FDMA uplink is investigated in (Lu et al., 2016). For hard MSE constraint algorithms, the approach proposed in (Lu et al., 2016) guarantees the required average MSE while achieving maximum system throughput. On the other hand, for elastic MSE constraint algorithms, it achieves alterable trade-off between average MSE and system throughput. In [(Kiran & Jibukumar, 2016), optimal selection of users for a group and dynamic addition of new user to an existing multiuser group in SC-FDMA based virtual MIMO system is studied. In this work, based on the average SINR at the receiver, a limit on the interference from the group users is determined such that the receiver can perform successful detection. According to (Kiran & Jibukumar, 2016), a new user is added to the existing group if the total interference due to the group after the addition of the new user remains in the given limit. In (J. Kim et al., 2015), joint frequency-domain resource allocation and user pairing problem for virtual MIMO SC-FDMA system is considered and a greedy heuristic algorithm is proposed which results in improved performance up to 90% of the upper-bound (optimal) performance. A joint spectrum allocation, power control and user pairing in uplink spatial multiuser SC-FDMA systems with MMSE (frequency domain) equalization is investigated in (Fan et al., 2015). A joint optimization problem with the objective of weighted throughput maximization is formulated and a two-step suboptimal solution is developed. This two-step solution first perform user pairing and spectrum allocation for a fixed transmit power and then adjusts the transmit power for the obtained user pairing and spectrum allocation to maximize the system throughput. Based on interference information sharing among multiple base-stations, a distributed iterative algorithm is also devised. System level simulation are performed to demonstrate the interference mitigation and throughput maximization capability of these algorithms.

In (Akande et al., 2016), the authors have by developed a novel frequency domain multi-modulus blind equalization algorithm for improving the throughput and spectral efficiency of SC-FDMA system. The frequency domain nature of this new algorithm reduces the computational complexity while its blind implementation improves the system throughput as well as its spectral efficiency. In the weight update, the frequency bins are normalized to further improve the convergence of this algorithm.

Centralized Fair Resource Allocation

Centralized fair resource allocation is investigated in (H. Kim et al., 2015; Lim et al., 2010; Nwamadi et al., 2011; Safa et al., 2013; Shah et al., 2015). The authors of (Shah et al., 2015)[, proposed a combined spectrum allocation and power control scheme for device-to-device (D2D) communication with fairness consideration. The authors assume that the uplink resources of the conventional cellular system are reused by the D2D communication and SC-FDMA is adopted as the multiple access technique in the uplink. The scheme proposed in (Shah et al., 2015) uses fractional frequency reuse (FFR) architecture for mitigating the interference between the cellular and D2D users and proportional fair scheduling algorithm for allocating the resources among the users in efficient and fair passion. In addition, the authors have also devised a power control algorithm that achieves a minimum target signal to interference plus noise ratio (SINR) for the cellular as well as D2D users. The schemes are shown to achieve low PAPR and improve the capacity of the overall system. In(Lim et al., 2010), subcarrier allocation in grouped FDMA (GFDMA) system is considered. GFDMA is SCFDMA with iterative multiuser detection which allows a number of users to share a common set of subcarriers. The authors have proposed a proportional fair scheduling algorithm which improves the fairness among users and guarantees the required quality of service (QoS) for the users sharing the same subcarrier sets. Simulation results show that the proposed algorithm in (Lim et al., 2010) not only increases the spectral efficiency of the GFDMA system but also ensures the fairness among users.

The work in (Safa et al., 2013) considered the contiguous sub channel allocation constraint in SC-FDMA systems that limit the scheduling flexibility and studied a rate maximization problem. For this purpose, the authors suggested a mechanism of formulating the conventional time domain proportional fair rate maximization problem in frequency domain. The authors showed the NP-hardness of the frequency domain allocation problem under this contiguous allocation constraint and presented an alternative low-complexity practical algorithms. The authors demonstrated that their algorithm improves the system performance in terms of system throughput as well as fairness among multiple users in the network. In (H. Kim et al., 2015), the authors propose a scheduling algorithm called maximum PF selection with contiguity constraint (MSCC) that satisfies the resource allocation contiguity constraint and makes sure the attainment of high cell capacity and fairness with low complexity. Authors, also examine cell edge user throughput and Peak to average power ratio (PAPR). Simulation result shows that MSCC provide improved fairness and cell throughput than previous schemes. In (Nwamadi et al., 2011), the authors have investigated physical resource block allocation in uplink LTE system with SC-FDMA and have proposed three reduced complexity algorithms named maximum-greedy, mean-enhanced-greedy, and single-mean-enhanced-greedy algorithms. The authors have demonstrated via simulation that the data rate fairness and bit-error-rate performance of all these three algorithm is better than the existing two-dimensional schemes. The effects of channel imperfections caused due to the feedback delay, channel estimation error, and Doppler Spread on the performance of the three algorithms are also investigated.

In (Afifi et al., 2013), the authors have proposed an integrated QoS aware power allocation approach for uplink SC-FDMA systems and have evaluated its performance. The proposed approach exploits the per cell interference limiting concept and a closed loop form power control technique and provides an autonomous local inter-cell interference coordination. The autonomous interference coordination has the benefit of avoiding the interference related information exchange among the neighboring cells. The authors have shown via simulations that their approach guarantees the QoS of the system, limits the interference and ensures fair resource sharing among the users. In (Davaslioglu & Ayanoglu, 2014), the authors have focused on the study of fairness and efficiency tradeoffs in resource allocation and scheduling in SC-FDMA system. The authors have used the set partitioning technique to analyze the performance of different schedulers such as max-min fair, sum rate maximization and proportional fair scheduler. This study established that compared to max-sum maximization schemes, proportional fair schemes offers better fairness and increases the throughput of the cell edge as well as the median users. The study performed in (Davaslioglu & Ayanoglu, 2014) can help the network operators in choosing best trade-off between fairness and efficiency for the growing traffic load on the network.

Centralized Schemes with QoS Guarantee

In (Aijaz et al., 2014), the authors consider efficient power allocation in uplink SC-FDMA system under delay quality of service (QoS) constraint. For this purpose, the authors propose a canonical duality theory based method which minimizes the sum transmitted power of the users in the uplink while satisfying the delay QoS constraint. The methods proposed in (Aijaz et al., 2014) performs closer to the optimal solution. In (Ruby et al., 2015), the authors study resource allocation problem in uplink SC-FDMA networks with both individual user QoS requirement and other standard constraints specific to heterogeneous traffic system. In the analysis, the heterogeneous traffic is considered to be comprised of traffic with delay constraints, traffic with certain throughput requirement, and best-effort traffic. Towards the solution of this problem, the authors used the dual decomposition method and devised optimal solution as well as low-complexity suboptimal algorithm with acceptable performance.

DISTRIBUTED RESOURCE ALLOCATION

This section reviews the distributed resource allocation schemes for uplink SC-FDMA systems in which self-decisions are taken at each user terminal and which provides seamless communication. In Table 3, these schemes are summarized on the basis of the optimization objectives considered in each of them.

Table 3. Relevant References of Distributed Resource Allocation Scheme for SC-FDMA

Optimization Objective	Relevant References
Distributed Fair Resource Allocation	(Ahmed & Mohamed, 2015; Ahmed et al., 2010; Noh & Oh, 2009; Sato et al., 2012)
Distributed QoS Guarantee	(Lafuente-Martínez, Hernández-Solana, & Valdovinos, 2011; Tsiropoulou et al., 2015)

Distributed Fair Resource Allocation

A virtual-clusters based distributed resource allocation scheme with proportional fairness for LTE SC-FDMA uplink has been proposed in (Ahmed et al., 2010). According to this scheme, virtual groups/clusters of users are formed by exploiting the MAC layer's link adaption information. The distributed scheme then allocates resource blocks among the users in proportion to the number of users and the total throughput of each cluster/group. The approach proposed in (Ahmed et al., 2010) is extended by the same authors in (Ahmed & Mohamed, 2015) where a virtual cluster-based proportional fair distributed joint resource block and power allocation scheme is proposed. In (Ahmed & Mohamed, 2015), in each cluster, the scheme guarantees a minimum data rate for all users by allocating the resource blocks among the users proportional to the number of users and the throughput. After the allocation of resource blocks, power is allocated on all the resource blocks of each user by using water-filling algorithm. The scheme performs fair resource allocation with QoS guarantee. In (Sato et al., 2012), a subcarrier allocation scheme is proposed that exploits the fact that assigning a subcarrier to a user with good channel condition can increase the system capacity of SC-FDMA. This scheme also accounts for the trade-off between fairness and system capacity. This trade-off problem can be optimally solved by using cooperative Nash Bargaining Solution (NBS).

However, due to the computational complexity of the optimal NBS, the authors have proposed a suboptimal NBS solution to perform subcarrier allocation while achieving a good balance between the fairness among users and the system capacity. This solution is compared with the proportional pair (PF) solution by applying it to a distributed antenna network which demonstrates that this solution achieves high fairness with a comparable system capacity to PF.

In (Noh & Oh, 2009), the authors presented a distributed algorithm to maximize the utility function of each cell sector for the uplink of 3GPP Long Term Evolution system (SC-FDMA system) while considering fair resource allocation among the users. The maximizing of the utility function defined in (Noh & Oh, 2009) is equivalent to maximizing the sum of average signal to interference plus noise ratio (SINR). The authors have used iterative Hungarian algorithm for the solution of the optimization problem and the proposed distributed algorithm is shown to be able to reduce the inter-cell interference and maximizes the system throughput.

Distributed Schemes with QoS Guarantee

In (Tsiropoulou et al., 2015), the authors have provided a distributed optimal power allocation scheme for uplink multi-service SC-FDMA networks. A utility function is formulated to represent the diverse QoS requirements of the users based on their requested-services. The subcarrier allocation is assumed to be known and user-centric non-cooperative multilateral bargaining approach is used for allocating per subcarrier power to each user. In (Lafuente-Martínez, Hernández-Solana, & Valdovinos, 2011), a per sector based frequency reuse approach is employed for the performance improvement of a tri-sectorized multicellular SC-FDMA network. Towards the achievement of QoS requirements, two types of inter-cell interference coordination techniques are proposed and compared with the cell-based fractional frequency reuse approach. The authors have demonstrated that their techniques can support increased traffic load and guarantees the QoS requirements of the real-time traffic. A summary of the works reviewed in this paper are summarized in Table 4 that provides the global picture of the resource allocation schemes for SC-FDMA networks.

Table 4. Detailed Survey of Different Research Papers For Resource Allocation in SC-FDMA

Relevant References	Centralized	Distributed	Energy Efficient	Spectral Efficient	Fairness	Perfect Channel State	Imperfect Channel State
(A. Ahmad, 2015; Ha et al., 2013; Tsiropoulou & Papavassiliou, 2011b)	✓		✓			✓	
(Ayaz Ahmad & Assaad, 2011b, 2011c, 2011a, 2011a; Ayaz Ahmad & Shah, 2015; Pao et al., 2012)	✓			✓		✓	
(Kaddour et al., 2013; Kessab et al., 2012; Sokmen & Girici, 2010; M. Zhang & Zhu, 2013)	✓			✓		✓	
(Fahmi et al., 2016; Hsu et al., 2013; Hu & Ci, 2015; Nakada et al., 2012; J. Zhang et al., 2009; M. Zhang & Zhu, 2013)	✓			✓		✓	
(Akande et al., 2016; Fan et al., 2015; J. Kim et al., 2015; Kiran & Jibukumar, 2016; Lu et al., 2016)	✓			✓		✓	
(H. Kim et al., 2015; Lee et al., 2009; Lim et al., 2010)	✓				✓	✓	
[(Afifi et al., 2013; Nwamadi et al., 2011)	✓				✓	✓	
(Twyman-Saint Victor et al., 2015)	✓				✓	✓	
(Shah et al., 2015; Fushiki et al., 2011)	✓				✓	✓	
(Ahmed & Mohamed, 2015; Ahmed et al., 2010; Noh & Oh, 2009)		✓				✓	
(Tsiropoulou et al., 2015)		✓				✓	
(Lafuente-Martínez, Hernández-Solana, & Valdovinos, 2011)		✓				✓	
(D. Kim et al., 2010)	✓					✓	
(Sato et al., 2012)		✓				✓	
(Wu, Schober, & Bhargava, 2013)	✓					✓	
(Lafuente-Martínez, Hernández-Solana, Guío, et al., 2011)	✓					✓	
(Baracca et al., 2013; Chen et al., 2011; Davaslioglu & Ayanoglu, 2014; Tang et al., 2012)	✓					✓	
(Aijaz et al., 2014; Dechene & Shami, 2014; Kalil et al., 2013a; Li et al., 2009; Yuen & Farhang-Boroujeny, 2012; J. Zhang et al., 2011, 2013)	✓		✓			✓	
(Myung et al., 2008)	✓						✓
(Chang, Chao, & Liu, 2011)	✓			✓		✓	
(Raghunath & Chockalingam, 2009; Ruby et al., 2015; Sridharan & Lim, 2012)	✓					✓	
(Ahmed & Mohamed, 2011; Lee et al., 2009; Sandanalaksmi et al., 2009; B. Yang et al., 2013)	✓				✓	✓	

FUTURE RESEARCH DIRECTIONS

In this section, we discuss some future research directions in the context of resource allocation in SC-FDMA system.

Imperfect Channel Knowledge Issues

Almost all of the existing schemes for resource allocation in SC-FDMA systems assume the availability of perfect channel state information. However, due to errors in channel estimation and/or channel feedback delay, the information about the channel state at the transmitter may not be perfect. The use of this imperfect channel knowledge in resource allocation may results in inefficient system performance. Therefore, researchers are needed to give appropriate attention to this issue while developing resource allocation algorithms.

Resource Allocation for Delay Sensitive Applications

The existing works on resource allocation for SC-FDMA networks do not consider the delay requirements of the wireless applications. The only works to the best of our knowledge that account for this issue are reported in (Aijaz et al., 2014; Li et al., 2009). However, the modern day wireless networks are proving a number of services such as video conferencing, voice over IP, online gamming, etc., which have stringent delay requirements. Owing to this fact, research is required to explore resource allocation schemes that are applicable to delay sensitive application in SC-FDMA system.

Multi-Tier Networks and Intercell Interference Issues

Due to the high demand of different wireless applications, efficiency and coverage area of the SCFDMA system can be improved by overlaying the macro cell with additional small cells (i.e., forming a multi-tier network). However, this concept of coverage extension creates the problem of inter cell interference which is a challenging hindrance in achieving the benefits of coverage area extension(Han, Chang, Cui, & Yang, 2010; Lafuente-Martínez, Hernández-Solana, & Valdovinos, 2011). Therefore, resource allocation with inter-cell interference management in the multi-tier SC-FDMA network scenarios is an open future research direction.

Resource Allocation in SC-FDMA Cognitive Radio Networks

Cognitive radio networks are considered as the potential solution for resolving the spectrum scarcity problem in the next generation wireless networks. Therefore, the investigation of resource allocation in SC-FDMA systems based cognitive networks is also an important research direction. Except for a couple of works (e.g. (Ayaz Ahmad & Raza Khan, 2017; Wu et al., 2013)), this area of research has not been well explored.

Finite Power Channel Input Consideration

The existing resource allocation schemes in SC-FDMA system considers Gaussian signaling as the input to the channel. However, practical SC-FDMA systems employee finite power symbol as the channel input (Ayaz Ahmad & Anwar, 2016; Al-Imari, Xiao, Imran, & Tafazolli, 2014; Lozano, Tulino, & Verdú, 2008) and the schemes based on Gaussian signaling may not be practically valid. Except for the recent work in (A. Ahmad, 2015) which considers power allocation in uplink SC-FDMA with finite symbol alphabet, no research work has considered this issue in SC-FDMA resource allocation. The consideration of finite power channel input in resource allocation for SC-FDMA systems is an open research area.

CONCLUSION

In this survey, resource allocation schemes for SC-FDMA systems are reviewed in a comprehensive and systematic way. The resource allocation scheme are classified into two major classes and a detailed survey of each of these classes is provided. The two major classes are formed on the basis of resource allocation decisions, i.e., whether the decision are made in a centralized, or a distributed way. The schemes are also categorized on the basis of optimization objectives (i.e., energy efficiency, spectral efficiency, QoS guarantee, fairness consideration, and interference prevention), optimization constraints (i.e., transmit power and data rate constraints), single-cell and multi-cell scenarios, and the perfect/imperfect channel knowledge assumption. The benefits and drawbacks of each of these classes have been discussed and future research directions are identified. The potential future research directions revealed by this survey include the consideration of imperfect channel knowledge in resource allocation, resource allocation schemes for services with stringent delay requirement, resource allocation and inter-cell interference management schemes for multi-tier SCFDMA networks, resource allocation in cognitive SC-FDMA networks, and finite power input based schemes.

REFERENCES

Afifi, A., Elsayed, K. M., & Khattab, A. (2013). Interference-aware radio resource management framework for the 3GPP LTE uplink with QoS constraints. In *2013 IEEE Symposium on Computers and Communications (ISCC)* (pp. 693–698). IEEE. 10.1109/ISCC.2013.6755029

Ahmad, Ayaz, & Assaad, M. (2011b). Polynomial-complexity optimal resource allocation framework for uplink SC-FDMA systems. In *2011 IEEE Global Telecommunications Conference-GLOBECOM 2011* (pp. 1–5). IEEE.

Ahmad, Ayaz, & Assaad, M. (2011c). Power efficient resource allocation in uplink SC-FDMA systems. In *2011 IEEE 22nd International Symposium on Personal, Indoor and Mobile Radio Communications* (pp. 1351–1355). IEEE.

Ahmad, A. (2014). Resource allocation and adaptive modulation in uplink SC-FDMA systems. *Wireless Personal Communications, 75*(4), 2217–2242. doi:10.100711277-013-1464-6

Ahmad, A. (2015). Power allocation for uplink SC-FDMA systems with arbitrary input distribution. *Electronics Letters, 52*(2), 111–113. doi:10.1049/el.2015.1779

Ahmad, A., & Anwar, M. (2016). Resource Allocation for OFDMA Based Cognitive Radio Networks with Arbitrarily Distributed Finite Power Inputs. *Wireless Personal Communications, 88*(4), 839–854. doi:10.100711277-016-3208-x

Ahmad, A., & Khan, R. (2017). Resource allocation for SC-FDMA based cognitive radio systems. *International Journal of Communication Systems, 30*(5), e3046. doi:10.1002/dac.3046

Ahmad, A., & Shah, N. (2015). A joint resource optimization and adaptive modulation framework for uplink single-carrier frequency-division multiple access systems. *International Journal of Communication Systems, 28*(3), 437–456. doi:10.1002/dac.2677

Ahmad, Ayaz, & Assaad, M. (2011a). *Canonical dual method for resource allocation and adaptive modulation in uplink SC-FDMA systems*. ArXiv Preprint ArXiv:1103.4547

Ahmed, I., & Mohamed, A. (2011). Fairness Aware Group Proportional Frequency Domain Resource Allocation in L-SC-FDMA Based Uplink. *International Journal of Communications. Network and System Sciences, 4*(08), 487–494. doi:10.4236/ijcns.2011.48060

Ahmed, I., & Mohamed, A. (2015). Power control and group proportional fairness for frequency domain resource allocation in L-SC-FDMA based LTE uplink. *Wireless Networks*, *21*(6), 1819–1834. doi:10.100711276-014-0845-4

Ahmed, I., Mohamed, A., & Shakeel, I. (2010). *On the group proportional fairness of frequency domain resource allocation in L-SC-FDMA based LTE uplink. In 2010 IEEE Globecom Workshops*. IEEE.

Aijaz, A., Nakhai, M. R., & Aghvami, A. H. (2014). Power efficient uplink resource allocation in LTE networks under delay QoS constraints. In *2014 IEEE Global Communications Conference* (pp. 1239–1244). IEEE. 10.1109/GLOCOM.2014.7036978

Akande, K., Iqbal, N., Zerguine, A., Al-Dhahir, N., & Zidouri, A. (2016). Frequency domain soft-constraint multimodulus blind equalization for uplink SC-FDMA. *EURASIP Journal on Advances in Signal Processing*, *2016*(1), 23. doi:10.118613634-016-0317-3

Al-Imari, M., Xiao, P., Imran, M. A., & Tafazolli, R. (2014). Radio resource allocation for uplink OFDMA systems with finite symbol alphabet inputs. *IEEE Transactions on Vehicular Technology*, *63*(4), 1917–1921. doi:10.1109/TVT.2013.2287809

Baracca, P., Tomasin, S., & Benvenuto, N. (2013). Resource allocation with multicell processing, interference cancelation and backhaul rate constraint in single carrier FDMA systems. *Physical Communication*, *8*, 69–80. doi:10.1016/j.phycom.2012.09.003

Berardinelli, G., de Temino, L. A. M. R., Frattasi, S., Rahman, M. I., & Mogensen, P. (2008). OFDMA vs. SC-FDMA: Performance comparison in local area IMT-A scenarios. *IEEE Wireless Communications*, *15*(5), 64–72. doi:10.1109/MWC.2008.4653134

Chang, C.-H., Chao, H.-L., & Liu, C.-L. (2011). Sum throughput-improved resource allocation for LTE uplink transmission. In 2011 IEEE Vehicular Technology Conference (VTC Fall) (pp. 1–5). IEEE. doi:10.1109/VETECF.2011.6093138

Chen, G., Song, S. H., & Letaief, K. B. (2011). *A low-complexity precoding scheme for PAPR reduction in SC-FDMA systems. In 2011 IEEE Wireless Communications and Networking Conference*. IEEE.

Ciochina, C., Castelain, D., Mottier, D., & Sari, H. (2009). New PAPR-preserving mapping methods for single-carrier FDMA with space-frequency block codes. *IEEE Transactions on Wireless Communications*, *8*(10), 5176–5186. doi:10.1109/TWC.2009.081231

Davaslioglu, K., & Ayanoglu, E. (2014). Efficiency and fairness trade-offs in SC-FDMA schedulers. *IEEE Transactions on Wireless Communications, 13*(6), 2991–3002. doi:10.1109/TWC.2014.042914.131176

Dechene, D. J., & Shami, A. (2014). Energy-aware resource allocation strategies for LTE uplink with synchronous HARQ constraints. *IEEE Transactions on Mobile Computing, 13*(2), 422–433. doi:10.1109/TMC.2012.256

Fahmi, A., Astuti, R. P., Meylani, L., Asvial, M., & Gunawan, D. (2016). A Combined User-order and Chunk-order Algorithm to Minimize the Average BER for Chunk Allocation in SC-FDMA Systems. *KOMNIKA Telecommunication, Computing. Electronics and Control, 14*(2), 574–587.

Falconer, D., Ariyavisitakul, S. L., Benyamin-Seeyar, A., & Eidson, B. (2002). Frequency domain equalization for single-carrier broadband wireless systems. *IEEE Communications Magazine, 40*(4), 58–66. doi:10.1109/35.995852

Fan, J., Lee, D., Li, G. Y., & Li, L. (2015). Multiuser scheduling and pairing with interference mitigation for LTE uplink cellular networks. *IEEE Transactions on Vehicular Technology, 64*(2), 481–492. doi:10.1109/TVT.2014.2321679

Fantacci, R., Marabissi, D., & Papini, S. (2004a). Multiuser interference cancellation receivers for OFDMA uplink communications with carrier frequency offset. In *IEEE Global Telecommunications Conference, 2004. GLOBECOM'04.* (Vol. 5, pp. 2808–2812). IEEE. 10.1109/GLOCOM.2004.1378866

Fushiki, M., Ohseki, T., & Konishi, S. (2011). Throughput gain of fractional frequency reuse with frequency selective scheduling in SC-FDMA uplink cellular system. In *2011 IEEE Vehicular Technology Conference (VTC Fall)* (pp. 1–5). IEEE. doi:10.1109/VETECF.2011.6093053

Ha, H. K., Tuan, H. D., & Nguyen, H. H. (2013). Joint optimization of source power allocation and cooperative beamforming for SC-FDMA multi-user multi-relay networks. *IEEE Transactions on Communications, 61*(6), 2248–2259. doi:10.1109/TCOMM.2013.041113.120480

Han, Y., Chang, Y., Cui, J., & Yang, D. (2010). A novel inter-cell interference coordination scheme based on dynamic resource allocation in LTE-TDD systems. In *2010 IEEE 71st Vehicular Technology Conference* (pp. 1–5). IEEE. 10.1109/VETECS.2010.5494073

Hatoum, R., Hatoum, A., Ghaith, A., & Pujolle, G. (2014). Qos-based joint resource allocation with link adaptation for SC-FDMA uplink in heterogeneous networks. In *Proceedings of the 12th ACM international symposium on Mobility management and wireless access* (pp. 59–66). ACM. 10.1145/2642668.2642673

Hsu, L.-H., Chao, H.-L., & Liu, C.-L. (2013). Window-based frequency-domain packet scheduling with QoS support in LTE uplink. In *2013 IEEE 24th Annual International Symposium on Personal, Indoor, and Mobile Radio Communications (PIMRC)* (pp. 1805–1810). IEEE.

Hu, Y., & Ci, S. (2015). QoE-driven Joint Resource Allocation and User-paring in Virtual MIMO SC-FDMA Systems. *Transactions on Internet and Information Systems (Seoul)*, *9*(10), 3831–3850.

Huang, G., Nix, A., & Armour, S. (2007). Impact of radio resource allocation and pulse shaping on PAPR of SC-FDMA signals. In *2007 IEEE 18th International Symposium on Personal, Indoor and Mobile Radio Communications* (pp. 1–5). IEEE. 10.1109/PIMRC.2007.4394297

Jar, M., & Fettweis, G. (2012). Throughput maximization for LTE uplink via resource allocation. In *2012 International Symposium on Wireless Communication Systems (ISWCS)* (pp. 146–150). IEEE. 10.1109/ISWCS.2012.6328347

Kaddour, F. Z., Pischella, M., Martins, P., Vivier, E., & Mroueh, L. (2013). Opportunistic and efficient resource block allocation algorithms for LTE uplink networks. In 2013 IEEE Wireless Communications and Networking Conference (WCNC) (pp. 487–492). IEEE. doi:10.1109/WCNC.2013.6554612

Kalil, M., Shami, A., & Al-Dweik, A. (2013a). Power-efficient QoS scheduler for LTE uplink. In *2013 IEEE International Conference on Communications (ICC)* (pp. 6200–6204). IEEE. 10.1109/ICC.2013.6655598

Kessab, A., Kaddour, F. Z., Vivier, E., Mroueh, L., Pischella, M., & Martins, P. (2012). Gain of multi-resource block allocation and tuning in the uplink of LTE networks. In *2012 International Symposium on Wireless Communication Systems (ISWCS)* (pp. 321–325). IEEE. 10.1109/ISWCS.2012.6328382

Kim, D., Kim, J., Kim, H., Kim, K., & Han, Y. (2010). An efficient scheduler for uplink single carrier FDMA system. In *21st Annual IEEE International Symposium on Personal, Indoor and Mobile Radio Communications* (pp. 1348–1353). IEEE.

Kim, H., Chung, M. Y., Lee, T.-J., Kim, M., & Choo, H. (2015). Scheduling Based on Maximum PF Selection with Contiguity Constraint for SC-FDMA in LTE Uplink. *Journal of Information Science and Engineering*, *31*(4), 1455–1473.

Kim, J., Hwang, I. S., & Kang, C. G. (2015). Scheduling for virtual MIMO in single carrier FDMA (SC-FDMA) system. *Journal of Communications and Networks (Seoul), 17*(1), 27–33. doi:10.1109/JCN.2015.000006

Kiran, P., & Jibukumar, M. G. (2016). Dynamic Multiuser Scheduling with Interference Mitigation in SC-FDMA-Based Communication Systems. In *Proceedings of the Second International Conference on Computer and Communication Technologies* (pp. 289–297). Springer. 10.1007/978-81-322-2523-2_27

Kwan, R., & Leung, C. (2010). A survey of scheduling and interference mitigation in LTE. *Journal of Electrical and Computer Engineering, 2010*, 1–10. doi:10.1155/2010/273486

Lafuente-Martinez, J., Hernandez-Solana, A., Guio, I., & Valdovinos, A. (2011). Inter-cell interference management in SC-FDMA cellular systems. In *2011 IEEE 73rd Vehicular Technology Conference (VTC Spring)* (pp. 1–5). IEEE. 10.1109/VETECS.2011.5956184

Lafuente-Martínez, J., Hernández-Solana, Á., Guío, I., & Valdovinos, A. (2011). *Radio resource strategies for uplink inter-cell interference fluctuation reduction in SC-FDMA cellular systems. In 2011 IEEE Wireless Communications and Networking Conference.* IEEE.

Lafuente-Martínez, J., Hernández-Solana, Á., & Valdovinos, A. (2011). Sector-based radio resource management for SC-FDMA cellular systems. In *2011 8th International Symposium on Wireless Communication Systems* (pp. 750–754). IEEE. 10.1109/ISWCS.2011.6125369

Lee, S.-B., Pefkianakis, I., Meyerson, A., Xu, S., & Lu, S. (2009). Proportional fair frequency-domain packet scheduling for 3GPP LTE uplink. In *IEEE INFOCOM 2009* (pp. 2611–2615). IEEE. doi:10.1109/INFCOM.2009.5062197

Lei, H., & Li, X. (2009). System level study of LTE uplink employing SC-FDMA and virtual MU MIMO. In *2009 IEEE International Conference on Communications Technology and Applications* (pp. 152–156). IEEE.

Li, Z., Yin, C., & Yue, G. (2009). Delay-bounded power-efficient packet scheduling for uplink systems of lte. In *2009 5th International Conference on Wireless Communications, Networking and Mobile Computing* (pp. 1–4). IEEE. 10.1109/WICOM.2009.5303491

Lim, H.-J., Kim, T.-K., & Im, G.-H. (2010). A proportional fair scheduling algorithm for SC-FDMA with iterative multiuser detection. In *2010 International Conference on Information and Communication Technology Convergence (ICTC)* (pp. 243–244). IEEE. 10.1109/ICTC.2010.5674669

Liu, R., Chen, Y., & Xiong, X. (2015). Efficient Resources Allocation for Femtocells in Heterogeneous Cellular Networks. In *First International Conference on Information Sciences, Machinery, Materials and Energy*. Atlantis Press. 10.2991/icismme-15.2015.255

Lozano, A., Tulino, A. M., & Verdú, S. (2008). Optimum power allocation for multiuser OFDM with arbitrary signal constellations. *IEEE Transactions on Communications*, *56*(5), 828–837. doi:10.1109/TCOMM.2008.060211

Lu, X., Yang, K., Li, W., Qiu, S., & Zhang, H. (2016). Joint user grouping and resource allocation for uplink virtual MIMO systems. *Science China. Information Sciences*, *59*(2), 1–14. doi:10.100711432-015-5514-4

Madan, R., & Ray, S. (2011). Uplink resource allocation for frequency selective channels and fractional power control in LTE. In *2011 IEEE International Conference on Communications (ICC)* (pp. 1–5). IEEE. 10.1109/icc.2011.5963354

Muharemovic, T., & Shen, Z. (2008). *Power Settings for the Sounding Reference signal and the Scheduled Transmission in Multi-Channel Scheduled Systems*. Academic Press.

Myung, H. G., Oh, K., Lim, J., & Goodman, D. J. (2008). *Channel-dependent scheduling of an uplink SC-FDMA system with imperfect channel information. In 2008 IEEE Wireless Communications and Networking Conference*. IEEE.

Nakada, M., Obara, T., Yamamoto, T., & Adachi, F. (2012). Power allocation for direct/cooperative AF relay switched SC-FDMA. In *2012 IEEE 75th Vehicular Technology Conference (VTC Spring)* (pp. 1–5). IEEE.

Noh, J.-H., & Oh, S.-J. (2009). Distributed SC-FDMA resource allocation algorithm based on the Hungarian method. In *2009 IEEE 70th Vehicular Technology Conference Fall* (pp. 1–5). IEEE. 10.1109/VETECF.2009.5378857

Nwamadi, O., Zhu, X., & Nandi, A. K. (2011). Dynamic physical resource block allocation algorithms for uplink long term evolution. *IET Communications*, *5*(7), 1020–1027. doi:10.1049/iet-com.2010.0316

Pao, W.-C., Chen, Y.-F., Tsai, M.-G., Lou, W.-T., & Chang, Y.-J. (2012). A multiuser subcarrier and power allocation scheme in localized SC-FDMA systems. In *2012 IEEE 23rd International Symposium on Personal, Indoor and Mobile Radio Communications-(PIMRC)* (pp. 210–214). IEEE. 10.1109/PIMRC.2012.6362703

Raghunath, K., & Chockalingam, A. (2009). SC-FDMA versus OFDMA: Sensitivity to large carrier frequency and timing offsets on the uplink. In *GLOBECOM 2009-2009 IEEE Global Telecommunications Conference* (pp. 1–6). IEEE.

Ruby, R., Leung, V. C., & Michelson, D. G. (2015). Uplink scheduler for SC-FDMA-based heterogeneous traffic networks with QoS assurance and guaranteed resource utilization. *IEEE Transactions on Vehicular Technology, 64*(10), 4780–4796. doi:10.1109/TVT.2014.2367007

Rumney, M. (2008). *3GPP LTE: Introducing Single-Carrier FDMA*. de Agilent Technologies. Inc.

Safa, H., El-Hajj, W., & Tohme, K. (2013). A QoS-aware uplink scheduling paradigm for LTE networks. In *2013 IEEE 27th International Conference on Advanced Information Networking and Applications (AINA)* (pp. 1097–1104). IEEE. 10.1109/AINA.2013.38

Sandanalaksmi, R., Manivanan, K., Manikandan, S., Barathi, R., & Devanathan, D. (2009). Fair channel aware packet scheduling algorithm for fast UL HARQ in UTRAN LTE. In *2009 International Conference on Control, Automation, Communication and Energy Conservation* (pp. 1–5). IEEE.

Sato, Y., Ryusuke, M., Obara, T., & Adachi, F. (2012). Nash bargaining solution based subcarrier allocation for uplink SC-FDMA distributed antenna network. In *2012 3rd IEEE International Conference on Network Infrastructure and Digital Content* (pp. 76–80). IEEE. 10.1109/ICNIDC.2012.6418715

Shah, S. T., Gu, J., Hasan, S. F., & Chung, M. Y. (2015). SC-FDMA-based resource allocation and power control scheme for D2D communication using LTE-A uplink resource. *EURASIP Journal on Wireless Communications and Networking, 2015*(1), 137. doi:10.118613638-015-0340-3

Sofer, E., & Segal, Y. (2005). *Tutorial on multi-access OFDM (OFDMA) technology*. DOC: IEEE 802.22-05-0005r0.

Sokmen, F. I., & Girici, T. (2010). Uplink resource allocation algorithms for single-carrier FDMA systems. In *2010 European Wireless Conference (EW)* (pp. 339–345). IEEE. 10.1109/EW.2010.5483441

Sridharan, G., & Lim, T. J. (2012). Performance analysis of SC-FDMA in the presence of receiver phase noise. *IEEE Transactions on Communications*, *60*(12), 3876–3885. doi:10.1109/TCOMM.2012.082812.110879

Tang, H., Hong, P., Xue, K., & Peng, J. (2012). Cluster-based resource allocation for interference mitigation in LTE heterogeneous networks. In 2012 IEEE Vehicular Technology Conference (VTC Fall) (pp. 1–5). IEEE. doi:10.1109/VTCFall.2012.6398901

Tsiropoulou, E. E., & Papavassiliou, S. (2011). Utility-based uplink joint power and subcarrier allocation in SC-FDMA wireless networks. *International Journal of Electronics*, *98*(11), 1581–1587. doi:10.1080/00207217.2011.589741

Tsiropoulou, E. E., Ziras, I., & Papavassiliou, S. (2015). Service differentiation and resource allocation in SC-FDMA wireless networks through user-centric Distributed non-cooperative Multilateral Bargaining. In *International Conference on Ad Hoc Networks* (pp. 42–54). Springer. 10.1007/978-3-319-25067-0_4

Twyman-Saint Victor, C., Rech, A. J., Maity, A., Rengan, R., Pauken, K. E., Stelekati, E., ... Odorizzi, P. M. (2015). Radiation and dual checkpoint blockade activate non-redundant immune mechanisms in cancer. *Nature*, *520*(7547), 373–377. doi:10.1038/nature14292 PMID:25754329

Wang, X., & Konishi, S. (2010). Optimization formulation of packet scheduling problem in LTE uplink. In *2010 IEEE 71st Vehicular Technology Conference* (pp. 1–5). IEEE. 10.1109/VETECS.2010.5493797

Wong, I. C., Oteri, O., & McCoy, W. (2009). Optimal resource allocation in uplink SC-FDMA systems. *IEEE Transactions on Wireless Communications*, *8*(5), 2161–2165. doi:10.1109/TWC.2009.061038

Wu, P., Schober, R., & Bhargava, V. K. (2013). Optimal power allocation for wideband cognitive radio networks employing SC-FDMA. *IEEE Communications Letters*, *17*(4), 669–672. doi:10.1109/LCOMM.2013.021913.122708

Yang, B., Niu, K., He, Z., Xu, W., & Huang, Y. (2013). Improved proportional fair scheduling algorithm in LTE uplink with single-user MIMO transmission. In *2013 IEEE 24th Annual International Symposium on Personal, Indoor, and Mobile Radio Communications (PIMRC)* (pp. 1789–1793). IEEE.

Yang, K., Martin, S., & Yahiya, T. A. (2015). LTE uplink interference aware resource allocation. *Computer Communications*, *66*, 45–53. doi:10.1016/j.comcom.2015.04.002

Yuen, C. H. G., & Farhang-Boroujeny, B. (2012). Analysis of the optimum precoder in SC-FDMA. *IEEE Transactions on Wireless Communications*, *11*(11), 4096–4107. doi:10.1109/TWC.2012.090412.120105

Zhang, J., Yang, L.-L., & Hanzo, L. (2009). Multi-user performance of the amplify-and-forward single-relay assisted SC-FDMA uplink. In *2009 IEEE 70th Vehicular Technology Conference Fall* (pp. 1–5). IEEE. 10.1109/VETECF.2009.5378760

Zhang, J., Yang, L.-L., & Hanzo, L. (2011). Energy-efficient channel-dependent cooperative relaying for the multiuser SC-FDMA uplink. *IEEE Transactions on Vehicular Technology*, *60*(3), 992–1004. doi:10.1109/TVT.2011.2104985

Zhang, J., Yang, L.-L., & Hanzo, L. (2013). Energy-efficient dynamic resource allocation for opportunistic-relaying-assisted SC-FDMA using turbo-equalizer-aided soft decode-and-forward. *IEEE Transactions on Vehicular Technology*, *62*(1), 235–246. doi:10.1109/TVT.2012.2220162

Zhang, M., & Zhu, Y. (2013). An enhanced greedy resource allocation algorithm for localized SC-FDMA systems. *IEEE Communications Letters*, *17*(7), 1479–1482. doi:10.1109/LCOMM.2013.052013.130716

Zheng, K., Wei, M., Long, H., Liu, Y., & Wang, W. (2009). Impacts of amplifier nonlinearities on uplink performance in 3G LTE systems. In *2009 Fourth International Conference on Communications and Networking in China* (pp. 1–5). IEEE. 10.1109/CHINACOM.2009.5339757

ADDITIONAL READING

Ahmad, Ayaz, & Assaad, M. (2009). Margin adaptive resource allocation in downlink OFDMA system with outdated channel state information. In *2009 IEEE 20th international symposium on personal, indoor and mobile radio communications* (pp. 1868–1872). IEEE.

Ahmad, A., & Assaad, M. (2010). Optimal resource allocation framework for downlink OFDMA system with channel estimation error. In *2010 IEEE Wireless Communication and Networking Conference* (pp. 1–5). IEEE. 10.1109/WCNC.2010.5506165

Ahmad, A. (2014). Resource allocation and adaptive modulation in uplink SC-FDMA systems. *Wireless Personal Communications*, *75*(4), 2217–2242. doi:10.100711277-013-1464-6

Ahmad, A. (2015). Power allocation for uplink SC-FDMA systems with arbitrary input distribution. *Electronics Letters*, *52*(2), 111–113. doi:10.1049/el.2015.1779

Ahmad, A., & Assaad, M. (2013). Optimal power and subcarriers allocation in downlink OFDMA system with imperfect channel knowledge. *Optimization and Engineering*, *14*(3), 477–499. doi:10.100711081-011-9181-z

Ahmad, A., Khan, M. T. R., & Kaleem, Z. (1990–1992). Khan, M. T. R., & Kaleem, Z. (2016). Uplink optimal power allocation for heterogeneous multiuser SIMO SC-FDMA networks. *Electronics Letters*, *52*(24), 1990–1992. doi:10.1049/el.2016.2872

Ahmad, A., & Khan, R. (2017). Resource allocation for SC-FDMA based cognitive radio systems. *International Journal of Communication Systems*, *30*(5), e3046. doi:10.1002/dac.3046

Gao, D. Y. (2000). Canonical dual transformation method and generalized triality theory in nonsmooth global optimization. *Journal of Global Optimization*, *17*(1–4), 127–160. doi:10.1023/A:1026537630859

Hwang, J.-K., Tsai, Y.-T., & Li, J.-D. (2012). Multiuser channel estimation for SC-FDMA system with CAZAC sounding reference sequence and Instruments-in-MATLAB verification. In *2012 IEEE 16th International Symposium on Consumer Electronics* (pp. 1–5). IEEE. 10.1109/ISCE.2012.6241750

Jia, Z., & Shen, L. (2014). Differential coding and detection methods in SC-FDMA systems. In *The 7th IEEE/International Conference on Advanced Infocomm Technology* (pp. 206–211). IEEE. 10.1109/ICAIT.2014.7019554

Kim, D., Kim, H.-M., & Im, G.-H. (2012). Iterative channel estimation with frequency replacement for SC-FDMA systems. *IEEE Transactions on Communications*, *60*(7), 1877–1888. doi:10.1109/TCOMM.2012.050812.100436

Kim, K., Kim, H., Han, Y., & Kim, S.-L. (2004). Iterative and greedy resource allocation in an uplink OFDMA system. In *2004 IEEE 15th International Symposium on Personal, Indoor and Mobile Radio Communications (IEEE Cat. No. 04TH8754)* (Vol. 4, pp. 2377–2381). IEEE.

Kwak, Y.-J., Lee, J.-H., Cho, J.-Y., & Cho, Y.-O. (2007). Method and apparatus for time multiplexing uplink data and uplink signaling information in an SC-FDMA system.

Orakzai, F. A., Ahmad, A., Khan, M. T. R., & Iqbal, M. (2017). Optimal energy-efficient resource allocation in uplink SC-FDMA networks. *Transactions on Emerging Telecommunications Technologies*, *28*(8), e3153. doi:10.1002/ett.3153

Ruder, M. A., Heinrichs, S., & Gerstacker, W. H. (2012). *Codebook aided user pairing and resource allocation for SC-FDMA. In 2012 IEEE Globecom Workshops.* IEEE.

Tsai, Y.-T., Li, J.-D., & Hwang, J. K. (2013). Joint CFO-multiuser channel estimation for SC-FDMA uplink system with CAZAC sounding reference sequence. In *2013 36th International Conference on Telecommunications and Signal Processing (TSP)* (pp. 175–180). IEEE. 10.1109/TSP.2013.6613914

Xiong, X., & Luo, Z. (2011). SC-FDMA-IDMA: A hybrid multiple access scheme for LTE uplink. In *2011 7th International conference on wireless communications, networking and mobile computing* (pp. 1–5). IEEE. 10.1109/wicom.2011.6040400

Yune, T.-W., Choi, C.-H., Im, G.-H., Lim, J.-B., Kim, E.-S., Cheong, Y.-C., & Kim, K.-H. (2010). SC-FDMA with iterative multiuser detection: Improvements on power/spectral efficiency. *IEEE Communications Magazine, 48*(3), 164–171. doi:10.1109/MCOM.2010.5434389

Zhang, C., Wang, Z., Yang, Z., Wang, J., & Song, J. (2010). Frequency domain decision feedback equalization for uplink SC-FDMA. *IEEE Transactions on Broadcasting, 56*(2), 253–257. doi:10.1109/TBC.2010.2046972

KEY TERMS AND DEFINITIONS

Centralized Resource Allocation: In centralized resource allocation, first, a user reports its channel station information to the BS and then BS sends the resource allocation information to the user.

CSI: In wireless communications, channel state information refers to known channel properties of a communication link. This information describes how a signal propagates from the transmitter to the receiver and represents the combined effect of signal.

Distributed Resource Allocation: In distributed resource allocation, resource allocation decision are taken at user side.

OFDMA: OFDMA stands for orthogonal frequency division multiple access technique. In this technique, multiple users orthogonally access the channel simultaneously to avoid interference.

PAPR: PAPR stands for Peak Average to Power Ratio. It is defined as the ratio of peak power to average power. It is a relationship between peak power and average power.

QoS: Quality of service is the description of the overall performance of a service. It refers to any technology that manages data traffic to reduce packet loss, latency, and jitter on the network. QoS controls and manages network resources by setting priorities for specific types of data on the network.

SC-FDMA: SC-FDMA is a single carrier frequency division multiple access technique. It is also known as linear pre-coded OFDMA because an extra block of DTFT is present in SC-FDMA.

Chapter 7
Analysis of Vulnerabilities in IoT and Its Solutions

Puspanjali Mallik

https://orcid.org/0000-0002-3896-3457
Shailabala Women's Autonomous College, India

ABSTRACT

The internet of things (IoT) fulfils abundant demands of present society by facilitating the services of cutting-edge technology in terms of smart home, smart healthcare, smart city, smart vehicles, and many more, which enables present day objects in our environment to have network communication and the capability to exchange data. These wide range of applications are collected, computed, and provided by thousands of IoT elements placed in open spaces. The highly interconnected heterogeneous structure faces new types of challenges from a security and privacy concern. Previously, security platforms were not so capable of handling these complex platforms due to different communication stacks and protocols. It seems to be of the utmost importance to keep concern about security issues relating to several attacks and vulnerabilities. The main motive of this chapter is to analyze the broad overview of security vulnerabilities and its counteractions. Generally, it discusses the major security techniques and protocols adopted by the IoT and analyzes the attacks against IoT devices.

INTRODUCTION

Everyday information security brings new challenges because of it's wide availability in each field starting from personal to commercial lives. Data must be protected from interception, theft and attack caused by unauthorized persons or hackers (F.Meneghello et.al, 2019). Network security is one part of information security

DOI: 10.4018/978-1-5225-9493-2.ch007

which may be challenged by Denial-of-Service attack and Cloning attack . Several countermeasures have been introduced to reduce the vulnerability risk and to strengthen the network security (M.Irshad et.al, 2016) Among the counter measures the best form is prevention action that includes monitoring control to detect threats. This network security expanding in dimension becomes complex every day as per as the varying mode of uses and applications. Traditionally, only the edge devices (J.W.Jones et.al, 2018) were included in the formation of network but towards the last part of twentieth century all devices within a proposed area became enable to communicate with each other. Fig 1. shows the types of devices integrated within a common area satisfying the communication criteria to form the IoT.

This newly formed network linked to Internet by following cloud technology

Figure 1. IoT Devices

and facilitates the automation of applications such as smart agriculture, smart healthcare, smart home, smart city, smart car, and smart transportation system etc. By the year 1999, this idea was first introduced by Kevin Asthon, founder of one automated organization at MIT to describe a system where the Internet is connected to the physical world via ubiquitous sensors. He successfully implemented the data connectivity and after this communication started between any two devices where as previously it was only limited between any two routing devices. Fig.2 includes the list of number of connected devices used in IoT . According to this, in the year 2015, there were only 3.8 billion IoT devices were connected, the number will reach to 9.9 billion in this year 2020 and the number is expected to reach by 21.5 billion by the year 2025.

Figure 2. Size of IoT according to the number of connected devices

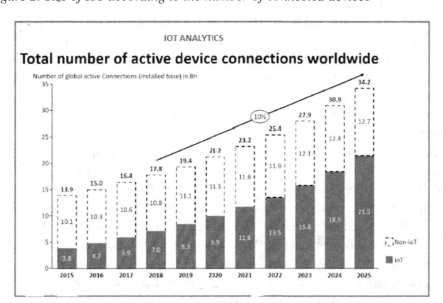

By the study of Fig. 2 it can be observed that a varying number of IoT devices are merging into it and form a complex structure. This complex form, the IoT in it's architecture includes three tiers where each tier is having a coherent set of similar type of elements. Tier one includes sensors and actuators, tier two includes the edge gateway and tier three includes the cloud. The function of sensor is to read and collect information, convert it to electrical signal and forwards it to other processed part. These wide range of sensors are classified as: inductive, capacitive, photoelectric, ultrasonic, and magnetic field sensors depending upon their functions. Actuators work in the reverse way to sensors. They receive electrical signals and convert it into physical actions. The cloud or more appropriately the cloud server is used to aggregate data and process data. So in IoT, cloud acts as a main component and thus it is called as the Head or Brain of IoT.

With the realization that billions of devices are included in the representation of IoT, a deep insight also focuses towards the origin of data sources. The different types of data sources are summarized as follows (M.A. Iqbal et.al, 2016):

Passive sources: These kind of sensors are low-energy and low-opearational ability particularly applicable for remote and rough locations such as ground water testing. Mainly they produce current data when application program interface (API) is called.

Active sources: This category of sensors are active and placed in fast application of information, such as continuously streaming data from jet engine.

Dynamic sources: Here the sources are in the form of physical, mechanical and electrical systems attached with the sensors. Here the sensors have the capacity to carry out communication with the organization, and web based applications.

In case of IoT, the security challenges become more complex because it embodies a hierarchical set of heterogeneous devices which are resource constraints. With any variation to it's software or hardware construction is challenged by attacker. So vulnerabilities need to be addressed and counter measured to avail the secure communication.

A typical variation of vulnerability countermeasure is the frequently faced cyber security(N.Neshenko et.al, 2019). A cybercrime can caused a major loss in billions of dollars. For finding the frauds, the Internet Crime Complaint Center (ICC) was established with the purpose to receive and prepare report based on cyber crimes. In present scenario the online media are suffering from cyber crime which involves the use of tools and techniques of World Wide Web. Challenges are created by the confidentiality, integrity, authenticity and availability of data in cyberspace. Social networking and online communication have created new vulnerabilities that are more cyber crime in nature(C.Vorakulpipat et al, 2018).

There are some factors which are vital for causing cyber crime in most cases:

1. Data access and sharing policies.
2. Data leakage
3. Monitoring of data
4. Sentencing of cyber criminals.

Following figure delineates the different types of cyber crime happened at present. In order to combat from such crimes a clear vision must be set to encourage device authentication prior to data exchange.

So, countermeasures are needed in order to refrain from security risks.

Device identification credentials also puts importance to ensure secure transmission As the devices are deployed in open places it becomes crucial to know the intended device and the intruder. So, device authentication is also required in addition to message integrity where connected devices can ensure trust prior to their communication. Authentication certifies user identity which is an evidence of whether data received to correct user or not (Z.Ling et.al, 2017). There are various authentication methods that can be used to provide user's identity ranging from a password to higher levels of security such as multi-factor authentication keys.

This book chapter focuses on the practical problems of IoT domain, IoT Generic Framework with technologies, security mechanisms along with general discussion of the concerned protocols with their comparisons and authentication technique as a major security essential.

Figure 3. Types of Cyber Crime

Security Requirements in Information System

The main objective lying within this is to protect the resources of automated information system such as hardware, software, firmware, data and telecommunication from misuses so that it can ensures confidentiality, integrity and availability (CIA) triad (K.Zhao et.al, 2013).

Considering the possibility of vulnerability in real scenarios, the security measures are adopted. Any information system must follow these three fundamental security requirements: confidentiality, integrity and availability, popularly known as security triad.

Confidentiality in Security

It states that only authorised devices have the access rights to the information system which can be classified into two parts, such as: data confidentiality and privacy(H. Djeloual et.al, 2018). Confidentiality in data connectivity assures that private or confidential information is not available to unauthorized users. Privacy in data connectivity states that information must be given to the user for whom it is.

Figure 4. CIA triad

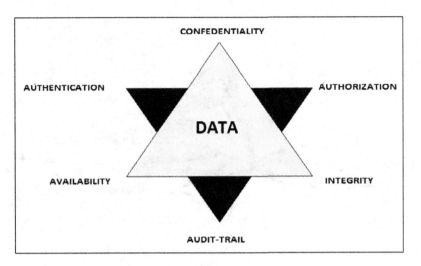

Integrity in Security

It states that only authorized users are allowed to modify the information content such as write, delete and update modifications. Integrity assures two concepts such as: integrity in data and integrity in system. Data integrity conforms that information are modified by authorized way and system integrity assures that a system perform it's function in ann unimpaired manner.

Availability in Security

It states that only authorized people have access to the information system as and when it is needed. Fault Tolerance can be avoided by ensuring the backup components such as servers, storage and networks (V.Hassija et.al, 2019). Fault tolerance in case of storage can be avoided by adopting highly scalable RAID architecture. Network back up can be ensured by providing multiple switches, ports, multiple cables between the end points.

Additional security requirements are:

Authentication in Security

It checks to ensure that the user's credentials are valid. The simplest way to use authentication technique is by putting user's name and password, called as the first factor to the authentication process. Sophisticated authentication can be achieved by applying multifactor authentication. Most of the information systems use two-factor authentication scheme where a secret key is generated by a random number generator which is known only to the user. A third factor authentication scheme seems still more secure where biometric parameter is considered as a third factor.

Authorization in Security

It confirms that a particular user has a certain access permission to an application. This process grants different levels of authorization to different categories of users. All sorts of permissions are mapped and stored in a table, called as Access Control List (ACL).

Security Requirements in IoT

IoT is an extended and integrity form of traditional Internet. The peculiarities of this concept is that everyday new issues are coming, so it considers as a separate issue.
The identified properties are:

- Uncontrolled Environment: The devices are placed in public places and many such places which are vulnerable to attack.
- Mobility: As some devices are not stable in their state so they are vulnerable.
- Physical Accessibility: Some sensors are physically accessible such as Environmental sensors.
- Trust: It is difficult to maintain trust because there are a large set of devices.
- Heterogeneity: With the combination of a varying class of devices IoT seems to be heterogeneous in nature.
- Constrained Resources: IoT devices have energy and computational limitations. IoT is the integration of RFID nodes and WSN nodes where resources are limited in energy and operation. In order to sustain from limitation, lightweight and energy aware algorithms are used. Communication between any two nodes is less secure in IoT than Intenet which results easy data leakage and easily node compromising.

So it is confirmed that IoT system exists with less secure network guards.

IoT Generic Framework

Fig. 4 describes the architectural details of all layers, protocols and technologies in respect to each layer with their assigned work.
The task of perception layer is to collect information, It forward the data to Network layer for further transmission . A node in perception layer is used for data

acquisition and data control. So, it forwards the data whatever it collected to the gateway or sends control instruction to the controller. The assigned technologies meant for perception layer are: RFID, WSNs, RSN, NFC, etc.

Generally Radio-frequency identification (RFID) identifies and tracks the tags attached to objects where the tags contain electronically enabled information. In case of bar code, the tags are placed in the line-of-sight to be readable but here tags don't need to be within the line of sight of the reader, so it may be embedded in the tracked object or placed anywhere within the object. There are two types of tags: Passive tags and active tags. Passive tags collect energy from a nearby RFID reader's interrogating radio waves where as active tags have a local power source (such as a battery) and it can operate hundreds of meters from the RFID reader. WirelessHart is a communication protocol that uses 2.4 GHz band for its radio technologies such as Wireless Local Area Network (WLAN), Bluetooth and ZigBee. It is structurally represented as a flat mesh network where all wireless stations work as a network. Z-Wave used as a communication protocol to connect wireless devices and particularly useful in smart home applications.

Figure 5. IOT architecture with its technologies and protocol

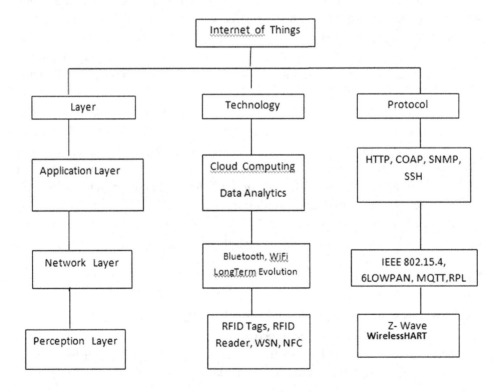

There are five communication technologies: Bluetooth, RFID, NFC, WSN and Cloud Computing.

Network layer security can be categorized as, core network security and local network security according to its access . In this typical IoT concern, for adopting wireless communication, there are technologies, such as 3G, Ad-Hoc, are placed at this layer.. Different network transmission has different technology. In addition to communication further security strategy is to consider the resource constraint devices. With this regard the most widely accepted protocol placed here is 6LoWPAN (IPv6 over Low-Power Wireless Personal Area Networks), is a low power wireless mesh network where every node has its own IPv6 address. This allows the node to connect directly with the Internet using open standards. MQTT (Message Queuing Telemetry Transport) is an open ISO standard for lightweight, network protocol that transports messages between devices. The protocol usually runs over TCP/IT. However, any network protocol that provides ordered, lossless, bi-directional connections can support MQTT. It is designed for connections with remote locations where a "small code footprint" is required or the network bandwidth is limited.

Application layer includes special security in form of resource constraint considerations. The security provided in this layer are: middleware technology security, cloud technology platform security and so on. At present, IoT applications in different industries have different requirements.

Security Challenges and Issues

Internet of Things safety issues is mainly observed in the following points:

- To ensure security in physical devices such as sensors, actuators and processors.
- To ensure security in the operation of the assigned devices
- To ensure security for the information network

Above mentioned remarks focuses to extend the security challenges and measures in all layers.

Perception Layer Challenges

A). Unauthorized Access to Tags

The devices assigned at this layer are RFIDs. which bear tags for reading data and as the devices are openly placed, they might be vulnerable to several physical attacks by adversaries [7]. Adversaries may change the data, delete the data and deny the communication.

B). Unauthorized Access to Nodes

Due to the open space deployment, adversaries easily capture the node, seize it, and change it accordingly. The hardware is tampered or replicated and it gives a serious impact on the entire network.

C). Tag Cloning

Cybercriminals capture the device and not only tags are accessed unauthorized manner but also the tags are reproduced to create confusion and to make the device compromise for cyber attack so that reader can not distinguish between the original and the compromising node.

D). False Data Injection Attacks

A compromise node always suffers from the illegal operation by the adversaries. The mishandling like the fake data transmission, false facts and erroneous instructions are always fed by the adversaries which degrades the performance of the communication network.

Network Layer Challenges

Network layer consists of the devices capable of proving Bluetooth, WiFi, and Long Term Evolution by which transmission of data happens from one point to another point. Underlying security problems resulted at this layer are:

A) Spoofing Attack

Adversary saatisfies the right to enter into the information system, controls the entire IoT system and feeds malicious data into the network system

B). Attacks doe to Sinkhole

Here the opponent makes the compromised device so attractive that the nearby devices divert their mode of communication and blend towards this compromised device. After a pick of data transmission the path towards this compromised device becomes so heavy that it drops the packets causing data loss to the transmission. The device seems itself fool assuming as if data transmission is continuing as it is. This attack resulted extra energy and invites denial-of-service (DoS) attack.

C). Attack due to Sleep Deprivation

Sensor devices are powered by batteries assembled within their body with the provision for a limited period of battery life. In order to maintain network life time the batteries need to be used properly. It is wise to use the communication in case of need otherwise it draws a heavy loss in life time of network. The solution is to apply sleep routines to the devices whenever there is no communication. Particularly the devices which are functioning dynamically are preferring this form of communication.

D). Attack due to Denial of Service (DoS)

Adversary creates a lot of traffic by which network is suffered with a lot of traffic and drains a lot of useful resources. It creates unavailability of resource to the target system.

Application Layer Challenges

The primary reason of the Application layer is to help offerings requested by customers.

A) Phishing Attack

Performing tricky ways adversary wants to collect useful information from the user such as identity, and password by means of spoofing the authentication credentials and phishing data from websites..

B) Malicious Virus/worm

IoT applications are infected by adversaries by injecting malicious self-propagation attacks, suh as: worms, malicious program, and so on. This causes private data lost

C) Sniffing Attack

Attacker can add sniffer utility to the device by which data in the network can be corrupted.

Security Solutions at Different Layers

A) Perception Layer Security Solutions

Unlawful activities like brute-force attack, collision attack, and side channel attack are caused by adversaries. Most of the devices acting at the perception layer are affected by this, so it becomes essential that the communicating devices must have to identify each other with their credentials so that security can be maintained. Various algorithms required for authentication are RSA, DSA, Hash based algorithms, AES, and DES.

Unauthorized means of accessing the data violates privacy and adversaries seeks interest to enter into the data. By adopting the tricky interactions it collects the vital information and try to compromise the node. This results vulnerability to the whole network. To ensure privacy symmetric and asymmetric encryption is followed according to the need of the security.

Anonymous techniques are followed to achieve data hiding. Most of the identification information such as place, type and uses of information are needed to be protected from the adversaries. K-anonymity [23,29] technique is a popular anonymous technique to ensure information hiding.

In order to ensure node cloning, RFID tag attack, and injecting unlawful data to the node, physical layer applies data encryption, tag frequency modification, hash-chain, hash-lock protocol hash-chain, and anti-jamming, Hash-chain is easy to maintain as it requires less computational overhead, and fast. Hash functions are mapping one entity to another entity uniquely. RFID safety protocols are Hash lock protocol, LCAP, and Hash-Chain.

B) Network Layer Security Solutions

The main function of network layer is to transmit data from one host to another host and encapsulate data together to form packets and this data transmission happens after the completion of authentication. With data transmission the additional objective is to ensure security in transmission. This safety is achieved by assigning more than one path and adopting resource constrained protocols suitable for the network.

The operations are defined by IEEE 802.15.4 standard which includes low-cost, and computationally lower rate communication devices such as RFID, sensors, and actuators. The protocols assigned for network layer are RPL, MQTT, and 6LoWPAN. In general the Routing Protocol for Low-Power and Lossy Network (RPL) is a distance vector network that supports for a variety of link protocols. In case of routing it builds a Destination Oriented Directed Acyclic Graph (DODAG) that has only one route from leaf node to root. There is one added protocol of RPL such as Cognitive RPL (CORPL) that is designed for cognitive networks and uses the DODAG topology for routing.

Other protocols are IPV6 over Low power wireless personal area networks (6LoWPAN), Datagram transport Layer security (DTLS), and Constrained Application Protocol (COAP) .

Message Queuing Telemetry Transport (MQTT) protocol designed for this layer to provide easy transmission of lightweight communication by using publish/ subscribe operations. It uses two types of devices: server and client. The servers are treated as message broker that receives all messages from the client and routes the message to appropriate destinations called as clients. Both the server and client receive a two-level handshake to ensure that only one copy received.

C) Middleware and Application Layer

There are several mechanisms applied for this layer. It acts as a service provider to facilitate service to devices and integrates the middleware benefits and security requirements to the devices. Various security provisions are: symmetric key and public key cryptosystem, certification transfer technology to gain authentication, Anti-DoS, firewalls, updated spyware, and intrusion detection techniques. Here the Data integrity is executed through message integrity codes (MIC). It ensures availability through the use of intrusion detections (IDS) and firewalls. It uses time stamps for maintaing integrity, sequence numbers which are used for replay protection.

Security Techniques for IoT

This section presents main security mechanisms to satisfy the requirements described previously.

A) Encryption

It ensures the confidentiality which was mentioned earlier and well known as one security triad. Data confidentiality is the main motive of encryption . By this the actual message known as plaintext changes to another form known as ciphertext and this ciphertext is transferred. If message became leaked by eavesdropper then only ciphertext can be accessible. This encryption mechanism can be classified as symmetric encryption and asymmetric encryption. In case of symmetric encryption only one key is used between any two users to revert plaintext to ciphertext but in case of asymmetric form two keys are used, The key which can be sharable is known as public key while the private key should kept secret. The principle behind the design of public and private key is that the message with one is decrypted by other. To ensure confidentiality, generally sender encrypts its data by using the public key of receiver and receiver recovers the original data by using its own private key.

Encryption Mechanisms

This is applicable to encrypt a long message either bit by bit the complete block and is treated at a time as a single plaintext to encrypt a long message. There are four types of operating modes to be used for encrypting long messages: Electronic

CodeBook (ECB), Cipher Block Chaining (CBC), Cipher feedback (CFB) and Output FeedBack (OFB).

National Institute of Standards and Technology (NIST) published one of the most useful block cipher in the year 2001, called as AES.

Figure 6. AES Structure

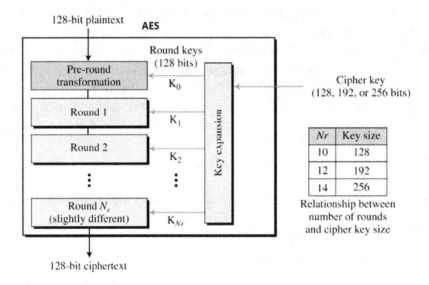

The most usable key sizes are 128, 192 or 256 bits. The sequence of operations performed are: add key, shift rows, sub bytes and mix columns and each set of these four series of operations constitute as one round. AES can have 10 to 14 numbers of such operations. About 128 bit key, there are 2^{128} number of attempts can be made to break the key which ensures that it is a safe protocol.

From its drawbacks it is treated as a protocol which is very hard to implement in practically, here every block is encrypted in the same way, and it uses very simple algebraic structure.

Asymmetric Key Encryption

It ensures confidentiality and authentication. The process uses two different keys as it was stated earlier. It differs from traditional symmetric algorithms because the former uses mathematical functions in its algorithm where as the later uses only substitution and permutation. Public key encryption seems more robust than symmetric encryption. In order to obtain public key, a Public key certificate is needed. The public key certificate is a digital document is issued and digitally signed

Figure 7. Public Key cryptosystem

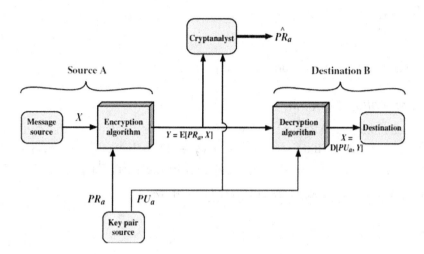

by the concerned Certification authority that binds the name of the subscriber to the public key.

The encrypted message is $Y=E(PR_A,X)$
And the original message is $X=D(PU_A,Y)$
PR_A is A's private key
PU_A is A's public key

In this communication, source A starts the communication with destination B. A initiates the communication and encrypts the message by using its own private key and B decrypts the message by using A's public key. As the message was encrypted using A's private key, in this case only A can prepare the message. Here, the total form of encrypted message is called as a digital signature. It is not possible to edit any part of the message without A' s permission which justifies the validity of the credentials of user A. The process is called as authentication. So, in context to asymmetric encryption where only message confidentiality can be preserved other security constraints such as digital signature and authentication also maintained.

B) The RSA Algorithm

This algorithm was named by three scientist Rivest, Shamir and Adleman. It uses mathematical exponentiation function where plaintext is encrypted in blocks and the size of the block is less than or equal to $\log_2(n)+1$, where n is any positive iteger. There are three steps:

- Key Generation Algorithm
- Encryption
- Decryption

The above steps are elaborated as follows:

Key Generation Algorithm

First user named as Alice selects two prime numbers, calculates the number 'n' where $n=p \times q$. Claculate $\varphi(n)=(p-1)(q-1)$, selects integer e, such that $\gcd(\varnothing(n),e)=1; 1<e<\varnothing(n)$, calculate d, where $d \equiv e^{-1}(mod\varnothing(n))$. Then public key and private keys are computed as follows:

PU=[e,n]

PR=[d,n]

Where *PU* and *PR* are public key and private key of Alice.

Suppose, in this case, user Bob wants to send the message 'M' to Alice. Bob calculates $C=M^e$ mod n sends to Alice. Alice decrypts the original message by calculating $M=C^d$ mod n.

C) Elliptic Curve Digital Signature Algorithm (ECDSA)

Another robust asymmetric pattern IoT security algorithm but unlike RSA which is computationally complex this is lightweight and facilitates key agreement digital signature, pseudo-random number generation in addition to confidentiality and authentication by using smaller key size in a faster and more efficient way.. For providing security it uses elliptic curve cryptography. The concept of elliptic curve is that it is a plane algebraic curve satisfying the non-singular equation $y^2=x^3+ax+b$; where y^2 is a curve with no cusps and no self-intersections. Overall, it is infeasible to find the discrete logarithm of a random elliptic curve element with respect to a publicly base point. In this case, the elliptic curve group provides the same level of security to that of RSA-based system with a smaller key size, reducing storage and transmission requirements. A cryptosystem in RSA uses 3072 bits key size where as in case of ECDSA, it uses only 256 bits of key size. National Institute of Standards and Technology (NIST) of U.S. accepts elliptic curve ECDSA in its Suite A of cryptographic algorithms. It uses hash functions that maps messages of any arbitrary length into its fixed length equivalent hash values. These form of hash functions are constructed through Merkle-Damgard functions meant to build collision resistant hash functions from its collision resistant one-way compression functions. This construction was followed by other popular hash algorithms such as MD5, SHA1 and SHA2. The MD5 message- digest algorithm is a widely used hash function produces 128-bit hash value, The Secure Hash Algorithm, SHA1 produces 160-bit hash value and SHA2 family consists of six hash functions with hash values of 224,256, 384 or 512 bits. Disadvantages of ECDSA are still needs alternate solutions because it does not provide a direct of method encryption Its implementation requires a secure random number generator, so if the same random value is reused then the private key can not be calculated

D) Constrained Application Protocols (COAP)

Resource constrained devices are not able to use the traditional wireless network (web based) protocols such as HTTP and FTP. CoAP has a service layer that supports Internet Application Protocol to resource constrained devices such as RFID and sensors in case of low-power and lossy wireless networks. Fig. 7 shows the structural representational representation of CoAP.

Figure 8. CoAP Architecure

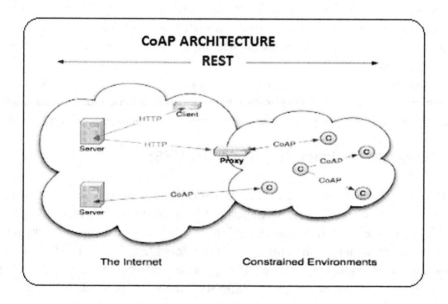

At present CoAP DTLS (Datagram Transport Layer Security) at the transport layer for transparently accessing the application layer.. DTLS provides end-to-end security.

E) Lightweight Encryption

At present it seems as most preferable IoT cryptography designed for security implementation in resource constrained devices having low energy and low computation. It focuses on block and stream ciphers in its encryption, message authentication codes and hash functions in its authentication. It is standardised by the International Organization for Standardization (ISO). There are a number of lightweight functions specified by International Electrical Commission (IEC). These lightweight cryptosystem must uses two process: The Random Number Generator process and the Secure Hardware process.

While Random number generator ensures robustness in security it suffers from computational time because the sequence of operations are not predictable in polynomial time. It ensure randomness and require the generation of pseudorandom numbers for several purposes as, e.g., to create a communication start up statement called as, nonce during authentication phase, to generate public keys, and to avoid replay attack . These random number generators are of two types: (1) True Random Number Generator (TRNG) and (2) Pseudo Random Number Generator (PRNG). Usually TRNG uses noise source and Pseudo Random Number Generator (PRNG) expands a short key onto a long sequence of bits. Most of the real applications in IoT use PRNG.

Hardware security can be maintained by applying Physically Uncloneable Functions (PUFs) to the authentication techniques. Authentication is needed when devices are placed in open spaces and vulnerable to physical attacks by adversaries.

F) Physically Uncloneable Functions (PUFS)

PUF stands as an identifying feature of a specific instance of a class of objects, alternatively called as the object's fingerprint as similar to the fingerprint of a person [18]. PUF creates random physical variations into the micro-structure of an integrated circuit, which is unique in nature [28]. It is infeasible to break the structure and thus uncloneable property satisfied. PUF provides one-way function to its internal structure that can not be duplicated. Most well known property of PUF is that it is hard to predict but it is construct that makes it a good chiice for use as a security primitive for IoT devices. PUFs are used for designing authentication techniques in wireless networks and radio frequency identifications (RFIDs).

The operation of PUF is very complex because of the variations in the o\ of PUF outcome. PUFs are represented with the following notations:

- PUF Class
- PUF Instance
- PUF Evaluation

A PUF class, denoted as *P*, is a set of a particular PUF construction type and represent the structural design of a PUF[23][28]. A PUF instance, denoted as puf stands as a discrete instantiation of a PUF class *P*.

Mathematically, it is represented as:

$$P \equiv \left\{ puf_i \leftarrow P.Create\left(r_i^c\right) : \forall i, r_i^c \overset{\$}{\leftarrow} \{0,1\}^* \right\}$$

Here, a PUF instance is treated as a state of its PUF class *P* that allows part of a PUF instance's state is configurable which is represented as *puf(x)*. PUF instance is not fixed but can be easily modified by an external input. A PUF evaluation procedure called as *puf.Eval* produces an outcome of PUF instance. When PUF instance is challengeable, it can be represented as *puf(x).Eval* which can be represented mathematically as:

$$puf(x).\text{Eval}\left(r^E \overset{\$}{\leftarrow} \{0,1\}^* \right).$$

Here a measurement is generally stated as the response of the PUF instance . The class of all possible response values, is denoted as γP

PUF response values can be ordered in a number of different ways. There are three important PUF responses such as: different PUF instances, different PUF instances, same PUF instance but from distinct evaluations.

A PUF embedded in an integrated circuit (IC) serves as an identifying feature in authentication process due to its uniqueness and reproducibility.

CONCLUSION

The IoT expands dramatically to avail information and it is virtually transforming the industries, the organizations, and the daily lives with new benefits in the form of automation and availability of information. IoT is going to replace Internet where the later is considered as a single large network but the former is a heterogeneous cluster of network of all devices. This book chapter reviews the possibilities of classical and recent threats, challenges and technologies in context to IoT vulnerability analysis.

Possible Future Actions

To leverage IoT robustness possible actions will be enforced which are feasible and easy to maintain, e.g., data analytics, deep learning methodologies, data driven approaches, and machine learning that can successfully adopts the currently available approaches. IoT will be more easier by integrating it with cloud technologies and machine learning. In order to effectively address the security, a collaboration is needed between the security provider and security user to successfully aware about the recent upgradations and security provisions. There is a need to trace out week programming practices and vendors, platforms, device time, and deployment time. In overall, stringent IoT programming standards, technologies, user awareness, deployment in environmental and risk management strategies are considered as per the security provision to get services effectively from an IoT.

REFERENCES

Djeloual, Amira, & Bensaali. (2018). Compressive Sensing-Based IoT Applications: A Review. *Journal of Sensor and Actuator Network.*

Hassija, V., Chamola, V., Saxena, V., Jain, D., Goyal, P., & Sikdar, B. (2019). A Survey on IoT security: Application Areas, Security Threats, and Solution Architectures (vol. 7). IEEE Access.

Iqbal, Olaleye, & Bayoumi. (2016). A Review on Internet of Things (IoT): Security and Privacy Requirements and the Solution Approaches. *Global Journal of Computer Science and Technology: E Network, Web & Security, 16.*

Irshad, M. (2016). A Systematic Review of Information Security Frameworks in the Internet of things. In *18th International Conference on High Performance Computing and Communications,* (pp. 1270-1275). IEEE. 10.1109/HPCC-SmartCity-DSS.2016.0180

Jones. (2018). Security Review On The Internet of Things. In *Third International Conference on Fog and Mobile Edge Computing (FMEC).* IEEE.

Ling, Z., Luo, J., Xu, Y., & Gao, C. (2017). *Security Vulnerabilities of Internet of Things: A case study of the Smart Plug System*. IEEE Internet of Things Journal.

Meneghello, F., Calore, M., Zucchetto, D., Polese, M., & Zanella, A. (2019). *IoT: Internet of Threats. In A survey of practical security vulnerabilities in real IoT devices*. IEEE Internet of Things Journal.

Neshenko, N., Bou-Harb, E., Crichigno, J., Kaddoum, G., & Ghani, N. (2019). *Demystifying IoT Security: An Exhaustive Survey on IoT Vulnerabilities and a First Empirical Look on Internet-scale IoT Exploitations*. IEEE.

Vorakulpipat, R. Thaenkaew, & Hai. (2018). Recent Challenges, Trends and Concerns Related to IoT Security: An Evolutionary Study. IEEE.

Zhao & Ge. (2013). A Survey on the Internet of Things Security. *9th International Conference on Computational Intelligence and Security*.

Chapter 8
Taxonomy of Computer Network Congestion Control/ Avoidance Methods

Mirza Waseem Hussain
Baba Ghulam Shah Badshah University, India

Sanjay Jamwal
Baba Ghulam Shah Badshah University, India

Tabasum Mirza
Baba Ghulam Shah Badshah University, India

Malik Mubasher Hassan
Baba Ghulam Shah Badshah University, India

ABSTRACT

The communication platform in the computing field is increasing at a rapid pace. Technology is constantly budding with the materialization of new technological devices, specifically in the communication industry. The internet is expanding exponentially. Internet-enabled devices are becoming part and parcel of our daily lives. It has turned out to be almost impossible to think about the world without the internet. The internet structures might be reinforced to meet coming prerequisites in mobile communication. Congestion plays a vital role in regulating the flow of data to accelerate the exchange of data in between the wired and wireless devices. In this chapter, the authors try to highlight various network congestion techniques with their limitations proposed from time to time by various researchers. This chapter plays a vital role in highlighting the history of networking congestion detection/ avoidance techniques starting from the early days of networking.

DOI: 10.4018/978-1-5225-9493-2.ch008

1. INTRODUCTION to network congestion

Congestion can be defined as the state of the network where total demand for resources is more than availability of the resource. Mathematically it can be defined as: If $\sum$ Total demand $> \sum$ Total available resources, then it is in the state of congestion. In other words, the congestion can be defined as the situation where the total number of packets are more than it can accumulate thus resulting in deterioration of performance. Usually congestion occurs when availability of resources is insufficient to meet the demand of the network. Often increasing the number of resources i.e. larger buffers, reducing processing time, using high speed links does not alienate the problem of congestion (C.-Q. Yang & Reddy, 1995). Although there is no universal accepted definition for network congestion nevertheless one of commonly used definition not in technical perspective but in users perspective is defined as: "A network is said to be congested from the perspective of a user if the service quality noticed by the user decreases because of an increase in network load"(Keshav, 1991).

1.1 Reasons for Network Congestion

Congestion in the network can be caused due to various reasons, which are defined as under:

1.1.1 Inadequate Memory to Store Incoming Packets

Sometime, all of the sudden a large chunk of data arrives simultaneously from several input sources and this data needs to be forwarded through a common output link, then a long queue of packets will be created at the output link. If the size of the queue is small, the packets will be dropped. Adding more buffers also may not help even. If the router has infinite memory the situation may even get worse, for example if there is a long queue of packets, the packets a tail end will take some time to reached the head of queue to be processed but in the meantime, time out for majority packets occurs resulting in useless processing and duplication of packets (Raj Jain, n.d.). Thus, increasing the load needlessly and making the situation worse. It was Nagel who proved that a network with infinite buffer is also susceptible to congestion as a network with normal buffer (Nagle, 1987). The size of the buffer should not be too large as it will create more delay and waste processing time as in Figure 1(a), nor should it be too small resulting in packet loss as shown in Figure 1(b).

Figure 1. Effect on packets with large and small buffer: (a) Queue with infinitely large buffer; (b) Queue with small buffer

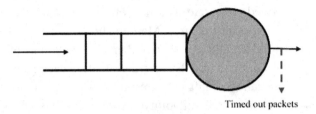

Timed out packets

1.1.2 Arrival Rate of Packets Exceeds the Outgoing Link Capacity/Size

Providing a high-speed link seems to control the congestion but it may worsen the condition in some situations. Since high speed balanced links and network processing need to be configured. In reality it's almost impossible. If in case the balanced links with same speed are configured with processing, still bottleneck possess problem. Bottleneck causes significant degradation of performance due to the limited amount of resources available. This can be perceived in the Figure 2. In spite of the fact that all the links are of same speed 30 Mbps, but at the bottleneck link the data is thrice the capacity of link causing congestion and plunging the quality of the entire network.

1.1.3 Laggy Processor

The Laggy processor of routers can also cause congestion. If the processor is slow in performing various tasks like processing queues, processing routing table etc. The large queue will be still built, although having links with excess bandwidth. Increasing the bandwidth of the line will not increase the speed of the CPU and vice versa. Therefore, there is only the shift of bottleneck from one part of the system to another. With the real cause as the mismatch between the various parts of the system.

1.1.4 Bursty Traffic

This is one of the major causes of congestion. If the source host can send data at a uniform rate, then the problem of congestion will be very less as all the above causes just act as the catalyst for congestion for bursty traffic.

Figure 2. Representing the bottleneck in a balanced network

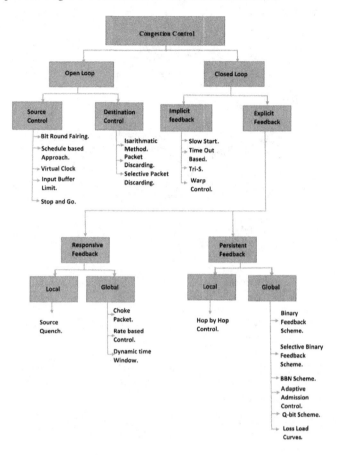

1.2 Consequence of Network Congestion

The congestion has an adverse effect on network throughput and delay. The effect of congestion on throughput can be shown in Figure 3. Initially throughput increases linearly with against the offered load, but when the load of the network increases beyond some capacity, the throughput drops. Further increase in the load beyond this point results in deadlock where not even a single packet is delivered. The Figure 3. shows the curves with three situations. The dark yellow curve shows the ideal one, where each and every packet is delivered with its maximum capacity. The blue curve in the uncontrolled one, where no technique is employed. The green curve shows the controlled one, where the congestion technique was used to prevent congestion collapse but with less throughput than ideal, due to overhead employed in congestion technique.

Figure 3. Packet delivery vs load (Congestion Control Video Lecture by Prof Ajit Pal of IIT Kharagpur, n.d.)

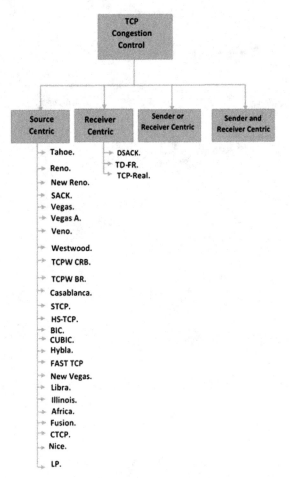

The delay in the network increases as the load in the network approaches the capacity of the network, irrespective of technique used. Although, initially the delay is more in control as shown in Figure 4 by green curve, due to overhead of congestion controlling technique but it saturates a higher capacity. The uncontrolled network is shown in Figure 5 by blue curve that saturates at very low load.

The Figure 5 shows the effect of congestion on packet delivery. Initially the packets that are generated at source are delivered to destination in linear fashion. However, on the occurrence of congestion the packet delivery rate gradually begins to fall, finally ending when none of the packets are delivered.

Figure 4. Packet delay vs load (Congestion Control Video Lecture by Prof Ajit Pal of IIT Kharagpur, n.d.)

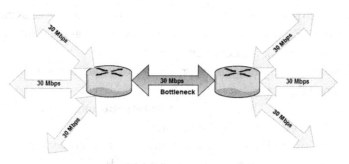

Figure 5. Packet sent vs packet delivery (Congestion Control Video Lecture by Prof Ajit Pal of IIT Kharagpur, n.d.)

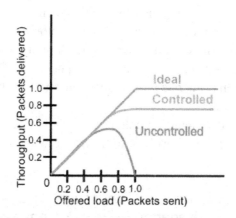

1.3 Congestion Collapse

Congestion collapse is the condition in the network where the announcement of useful information is either limited or totally hampered (J. Singh & Lambay, n.d.).

The first occurrence of congestion collapse was reported in 1984, where the capacity of the network was dropped from 32 Kbit/s to 40bit/s and the situation remained continuous until Jacobson's congestion control was implemented on end nodes (Nagle, 1984) (Jacobson, 1988).

The congestion collapse can occur due to various reasons which are as (Floyd & Fall, 1999):

1.3.1 Unresponsive Flows

The flows which do not employ end-end congestion control are termed as unresponsive flows. Since there is no end-end congestion control, it cannot adjust the flow of packets in the network on the occurrence of packet loss.

The congestion collapse in unresponsive flows is due to retransmission of the packets that are either in transit or already have been received by the receiver and need to be discarded later. The congestion collapse due to unnecessary retransmission of packets is called classical congestion collapse.

Classical collapse can be solved by implementing better timers and using modern congestion control mechanisms of TCP (Jacobson, 1988).

1.3.2 Network Congestion Collapse Due to Non-Deliverable Packets

This type of congestion collapse occurs when the bandwidth is wasted in transporting packets in the network but are eventually dropped before reaching the concluding destination. The condition gets worse when open loop applications are deployed as an alternative of end-end congestion control applications. The condition is even more degenerated when using rate base application where packet rate is intensified in response to packet drop.

This type of network congestion collapse can be prevented by using end-end network congestion control algorithms to all traffic without taking into account type of flows (UDP, TCP).

1.3.3 Fragment/Cell Depended Congestion Collapse (Kent & Mogul, 1987) (Kadangode Ramakrishnan & Floyd, 1999) (K Ramakrishnan & Floyd, 1999)

In this type of network congestion collapse, the cells/fragments no longer can be reassembled into a legit packet and are eventually dropped at the receiving end. Typically, this type of collapse occurs due to inconsistency of transmission units of data link layer and higher layer (datagram/fragment). Fragment/cell-based collapse can be resolved by providing information amongst the layers. Some examples of such mechanisms are Early Packet Discard (Romanow & Floyd, 1995), Path MTU discovery (Mogul et al., 1988).

The other variant of fragment/cell-based congestion collapse is where the packet is successfully delivered to the transport layer but is discarded by the end node before being utilized to the user (Varghese, 1996).

Figure 6. Growth of internet users (Internet Top 20 Countries - Internet Users 2020, n.d.)

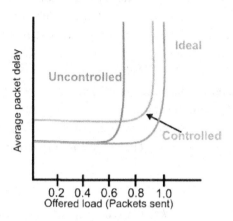

1.3.4 Network Congestion Collapse Due to Increased Control Traffic

This type of network congestion collapse can arise when load on the network increases, consequently resulting in congestion. As such, much of the bandwidth is consumed by control data (DNS message, packet header message, routing update etc.) with hardly anything for actual data.

1.3.5 Network Congestion Collapse Due to Stale Packets

In this scenario the congested link is occupied in transporting the packets that are stale or useless. For example, the user inquires some content from the web, due to the lengthy queue he had to wait, now the user is not anymore interested but the packet arrived.

2. INSPIRATION TO DEVELOP NETWORK CONGESTION CONTROL ALGORITHMS/TECHNIQUES

In communication networks the Internet provides a global infrastructure to exchange information. With the early motive of the network to link peripherals with the central computer, at the beginning computers were pretty big and uneconomical. The DARPA couldn't bear a large number of computers for their researchers and came up with a resolution to share resources. This led to the existence of the Internet.

Internet has experienced marvelous growth with 562 users in 1983 to 4.5 billion in 2020 (*Internet Top 20 Countries - Internet Users 2020*, n.d.)(*Number of Internet Users Worldwide | Statista*, n.d.)(*Internet Growth Statistics 1995 to 2019 - the Global Village Online*, n.d.) as shown in Figure 6. The number websites climbed from 623 in 1993 (Gray, 1996) to 1.7 billion websites in 2020 (*Total Number of Websites - Internet Live Stats*, n.d.). With the growth in size, the Internet has also witnessed growth in diversity of applications like audio-video streaming, Emails, P2P file sharing, VOIP, FTP, gaming and E-commerce. In fact, researchers are working to provide interplanetary Internet (Farrell & Jensen, 2004). The heterogeneity in applications also lead to improvement in latency, Bit error rate, and channel capacity as shown in Table 1.

Table 1. Various types of networks with Latency, Bit error rate and Capacity.

Network	Latency	Bit-Error Rate	Capacity
Datacentre(Alizadeh et al., 2010)	100 μsec to 1 millisecond	10-12	1Gbps to 1Tbps
Satellite Network (M Allman et al., 1999)	250 milliseconds to 1 second	10-10	100 Kbps to 155 Mbps
Wired Wide Area Network(Optical Carrier Transmission Rates - Wikipedia, n.d.) (Hui et al., 2001) (Paxson, 1999 (News Release 060929a, n.d.)	100 milliseconds to 300 milliseconds	10-2	50 Mbps to 14Tbps
Wired Local Area Network(10 Gigabit Ethernet - Wikipedia, n.d.)	Less than 1 millisecond	10-12	10 Mbps to 10 Gbps
802.11 Wireless Local Area Network and Mesh Network(Khorov et al., 2019) (Bicket et al., 2005)	1 millisecond to 200 milliseconds	10-5	1 Mbps to 600 Mbps
Cellular networks (e.g., 5G)(5G, Gear Up - Huawei, n.d.)	1 millisecond	-	10 Gbps

With the tremendous explosion of data (Zaman et al., 2012), every day 2.5×10^{18} bytes of data is being generated (*39+ Big Data Statistics for 2020*, n.d.)(*30 Eye-Opening Big Data Statistics for 2020: Patterns Are Everywhere*, n.d.)(*Infographic: How Much Data Is Generated Each Day?*, n.d.). The data arises not only from single source but from numerous sources like: Posts of social websites, sensors used in weather forecasting, audio-video, cell phones, GPS, business transactions etc. as shown in Figure 7.

Figure 7. Various types of data generated in Exabytes (Cisco Visual Networking Index Predicts Annual Internet Traffic to Grow More Than 20 Percent (Reaching 1.6 Zettabytes) by 2018 | The Network, n.d.)

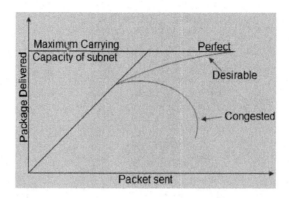

With the revolution in computing and software engineering, the trend is approaching in the direction of the devices with smart applications. These smart applications furthermore have their own bandwidth requirement but also are bandwidth hungry. As the bandwidth is limited, with smart applications driving most of the bandwidth. According to Cisco's Group report (2013), on regular basis the common user uses 15 Gb/month and in future usage to grow to 75 Gb/month (*Cisco Visual Networking Index Predicts Annual Internet Traffic to Grow More Than 20 Percent (Reaching 1.6 Zettabytes) by 2018 | The Network, n.d.*). In the same report 36 Gb/month of bandwidth is utilized by average household and this bandwidth to grow to 150 Gb/month (*Cisco Visual Networking Index Predicts Annual Internet Traffic to Grow More Than 20 Percent (Reaching 1.6 Zettabytes) by 2018 | The Network, n.d.*) as shown in Figure 9. When the bandwidth requirements upsurge it is very important to ensure that each and every user gets a fair share of bandwidth while guaranteeing quality of service.

With billions of devices connected, generating enormous amount of data therefore, there is a requirement for an efficient and reliable mechanism to route data. In the 21th century, the priorities changed, now the focus is on management and performance. Therefore, network congestion control plays an indispensable role in routing the data in a proficient manner.

The necessity of a network congestion control mechanism was first realized, when network congestion collapse occurred in 1984 were the speed of the network dropped from 32 Kbps to 40 bps. Later the congestion control mechanisms were modified and added to TCP.

Figure 8. Normal global Internet bandwidth usage (Cisco Visual Networking Index Predicts Annual Internet Traffic to Grow More Than 20 Percent (Reaching 1.6 Zettabytes) by 2018 | The Network, n.d.).

#	Country or Region	Internet Users 2020 Q1	Internet Users 2000 Q4	Population, 2020 Est.	Population 2000 Est.	Internet Growth 2000 - 2020
	TOP 20 COUNTRIES WITH HIGHEST NUMBER OF INTERNET USERS - 2020 Q1					
1	China	854,000,000	22,500,000	1,439,062,022	1,283,198,970	3,796 %
2	India	560,000,000	5,000,000	1,368,737,513	1,053,050,912	11,200 %
3	United States	313,322,868	95,354,000	331,002,651	281,982,778	328 %
4	Indonesia	171,260,000	2,000,000	273,523,615	211,540,429	8,560 %
5	Brazil	149,057,635	5,000,000	212,392,717	175,287,587	2,980 %
6	Nigeria	126,078,999	200,000	206,139,589	123,486,615	63,000 %
7	Japan	118,626,672	47,080,000	126,854,745	127,533,934	252 %
8	Russia	116,353,942	3,100,000	145,934,462	146,396,514	3,751 %
9	Bangladesh	94,199,000	100,000	164,689,383	131,581,243	94,199 %
10	Mexico	88,000,000	2,712,400	132,328,035	2,712,400	3,144 %
11	Germany	79,127,551	24,000,000	83,783,942	81,487,757	329 %
12	Philippines	79,000,000	2,000,000	109,581,078	77,991,569	3,950 %
13	Turkey	69,107,183	2,000,000	84,339,067	63,240,121	3,455 %
14	Vietnam	68,541,344	200,000	68,541,344	200,000	34,250 %
15	United Kingdom	63,544,106	15,400,000	67,886,011	58,950,848	413 %
16	Iran	67,602,731	250,000	83,992,949	66,131,854	27,040 %
17	France	60,421,689	8,500,000	65,273,511	59,608,201	710 %
18	Thailand	57,000,000	2,300,000	69,799,978	62,958,021	2,478 %
19	Italy	54,798,299	13,200,000	60,461,826	57,293,721	415 %
20	Egypt	49,231,493	450,000	102,334,404	69,905,988	10,940 %
	TOP 20 Countries	3,241,273,512	251,346,400	5,233,377,837	4,312,497,691	1,289 %
	Rest of the World	1,332,876,622	109,639,092	2,563,237,873	1,832,509,298	1,216 %
	Total World	4,574,150,134	360,985,492	7,796,615,710	6,145,006,989	1,267 %

NOTES: (1) Top 20 Internet Countries Statistics were updated for Dec 31, 2019. (2) Growth percentage represents the increase in the number of Internet users between the years 2000 and 2020. (3) The most recent user information comes from data published by Facebook, International Telecommunications Union, official country telecom reports, and other trustworthy research sources. (4) Data from this site may be cited, giving the due credit and establishing a link back to www.internetworldstats.com. Copyright © 2020, Miniwatts Marketing Group. All rights reserved worldwide.

TCP is the leading protocol as long as most of the traffic is still driven by it (Postel, 1981b)(Jacobson, 1988)(Mark Allman et al., 1999)(Partridge & Shepard, 1997) . However, TCP showing significant performance limitations after the introduction of multimedia (Nguyen & Hwang, 2002) and wireless technology therefore there is the need to incorporate new protocols(M Allman et al., 1999)(Katabi et al., 2002) (Li et al., 2007)(Dukkipati et al., 2005)(Alizadeh et al., 2010)(Qazi et al., 2009) (Vasudevan et al., 2009) .

The TCP works well for applications like FTP and Web (Wu et al., 2000), where reliable delivery is given priority over latency, but the multimedia applications requires stricter and smaller latency, hence gives more priority to latency over reliable delivery. Therefore, multimedia applications don't use pure TCP(Loguinov & Radha, 2002)(Dukkipati et al., 2010)(Chilamkurti & Soh, 2002)(Q. Liu & Hwang, 2003) .

Additional challenge that influences the TCP's performance improvement is the heterogeneity of the Internet. Different bandwidths are available to TCP connections due to multiplexing, mobility and access control (D. Katabi, 2002). Most of the times different flows share the same bottleneck bandwidth affecting the other flows (Mark

Allman & Paxson, 2001). In addition to this, in the shared medium, the channel utilization and access control protocols also affect the availability of bandwidth to various flows.

For wireless networks, the TCP possesses another serious challenge. The TCP is incapable to identify why the packet was dropped. The packet can be dropped due to several reasons other than congestion; it can be due to; fading channel, noisy channel, interference, mobility, limited availability of bandwidth.

Since the TCP is incapable to differentiate why the packet was dropped. When the packet is dropped the TCP unnecessarily decreases the congestion window and back offs its retransmission timer to reduce load without realizing that the packet might be dropped due to various reasons additional to congestion like; fading, mobility etc. Therefore, causing severe performance degradation resulting in wastage of resources.

3. TAXONOMY OF NETWORK CONGESTION CONTROL SCHEMES

The taxonomy of congestion control scheme is based on decision making capability and how useful information is extracted from congestion control algorithms for making various decisions. The main two subcategories of congestion schemes are open loop and closed loop congestion control algorithm (C.-Q. Yang & Reddy, 1995) as shown in Figure 10.

3.1 Open Loop Network Congestion Control Scheme

In an open loop congestion scheme, there is no feedback information from the congested node. The decision is purely based on local knowledge of nodes such as local bandwidth and local buffer capacity with no dynamic monitoring of the network. The open loop is based on an admission control mechanism and is able to stabilize the arrival traffic rate but is inefficient for varying/all traffic patterns. Based on the location of decision making or control, the open loop can be further sub categorized as source and destination control.

3.1.1 Source-Based Control

In open loop source-based control, the rate of traffic is controlled at the source. The various types of this algorithm are: Bit Round Fairing (Demers et al., 1989), Schedule based Approach (Mukherji, 1986), Virtual Clock (Zhang, 1990), Input Buffer Limit (Lam & Reiser, 1979), Stop and Go (Golestani, 1991) with detail shown in Table 2.

3.1.2 Destination-Based Network Congestion Control

In open loop destination-based control, the rate of traffic is either controlled at destination or at intermediate nodes without using any feedback. The various algorithms of this type are: Isarithmatic Method (Davies, 1972), Packet Discarding (Tanenbaum 1944-, 1981), and Selective Packet Discarding (N Yin, S.-Q Li, 1990). The detailed summary of this scheme is shown in Table 2.

Table 2. Detail of open congestion schemes controlled at source and destination.

S.no.	Type of Algorithm	Control at	Method
1	Bit Round Fairing (Demers et al., 1989)	Source	Each Switch maintains different queues for every different sources
2	Schedule based Approach (Mukherji, 1986)	Source	Dividing the bandwidth of channel into equal slots and assign each slot to the user
3	Virtual Clock (Zhang, 1990)	Source	For each data a virtual clock is assigned with each tick equal to mean inter packet interval.
4	Input Buffer Limit (Lam & Reiser, 1979)	Source	The input and transit data is differentiated at each node and imposing restriction on input buffer.
5	Stop and Go (Golestani, 1991)	Source	Time frame is used at the source end for each and every connection for admission control. The Switching node is forced for multiplexing.
6	Isarithmatic Method (Davies, 1972)	Destination	The limit of packets is imposed on each and every node thus limiting the overall traffic on the network
7	Packet Discarding (Tanenbaum 1944-, 1981)	Destination	No advance reservation is done and the amount of buffer is limited at switching node.
8	Selective Packet Discarding (N Yin, S.-Q Li, 1990)	Destination	Same as packet discard but with selective discard policy.

3.1.2.1 Closed-Loop Network Congestion Control

The closed-loop congestion scheme uses feedback information for decision making. The feedback can be either globally generated by destination node or locally generated by immediate node. The closed loop can monitor the network dynamically owing to feedback information that can be either global or local. The closed loop scheme has been further branched into implicit feedback and explicit feedback. The detailed summary is shown in Table 3.

Table 3. Detailed summary of closed loop congestion scheme

S.no	Type of algorithm	Method	Type of feedback
1	Slow Start(Jacobson, 1988)	Window based control where initially the window size is increased as long as no congestion occurs. On the occasion of congestion window size is decreased in multiplicative order.	Implicit
2	Time Out Based (R. Jain, 1986)	Uses window based policing that acts on source to specify minimum value, maximum value, increasing and decreasing size of source window	Implicit
3	Tri-S(Z. Wang & Crowcroft, 1991)	In this window-based policy, source uses a quick and fair approach that tries to optimize the network as quick as possible instead of slow start.	Implicit
4	Warp Control (Park, 1993)	Same as timeout based, with additional features of using admission control on the source based on rate and time stamp(warp) to observer utilization of network	Implicit
5	Source Quench (Postel, 1981a)	The source receives a message known as quench message indicating the decrease of data rate, the rate is then gradually increased after a certain period of time if no other quench message is received in that particular amount of time.	Explicit Responsive Local
6	Choke Packet (Varakulsiripunth et al., 1986)	The source on receiving the choke packet decreases the rate of data by some percentage. The source for a fixed amount of time ignores the receiving of repeated choke packet. After the completion of a fixed amount of time, if no choke packet is received, then data rate at source is increased.	Explicit Responsive Global
7	Rate based Control (Comer & Yavatkar, 1990)	The monitoring of the data to destination is done by the source node. When a new packet is sent from source to destination, if the rate of sending increases the available capacity, the packet is discarded. Based on the capacity the inter packet gap between the packets is adjusted by the source end to adjust the rate of data.	Explicit Responsive Global
8	Dynamic time Window (Faber et al., 1992)	The time window is used by the user to reserve the rate and to calculate the source average rate. The variance of traffic in the time window is used to control the throughput.	Explicit Responsive Global
9	Hop by Hop Control(Mishra & Kanakia, 1992)	The feedback information of neighboring switches are used to adjust the rate dynamically	Explicit Persistent Local
10	Binary Feedback Scheme (K. K. Ramakrishnan & Jain, 1990)	When the congestion occurs, the congestion bit is set to 1 by the switch for each packet. The destination on receiving the packet sends back the congestion bit to the source indicating the source to adjust the window accordingly.	Explicit Persistent Global
11	Selective Binary Feedback Scheme (K. Ramakrishnan, R. Jain, 1991)	For each user the count on the number of packets within the average queue interval is set at the switching node. The users sending more than fair share count are informed to slow the rate.	Explicit Persistent Global
12	BBN Scheme (Rose, 1992)	The feedback information is used to know the resource utilization. On the basis of resource utilization, the ratio for each flow is calculated at source end. This flow ratio is used as a throttle point to limit the data rate.	Explicit Persistent Global

continues on following page

Table 3. Continued

S.no	Type of algorithm	Method	Type of feedback
13	Adaptive Admission Control(Haas, 1991)	For each flow the round-trip time (RTT) between the source and destination is used as a parameter to indicate the congestion and to put into effect the adaptive admission control rate.	
14	Q-bit Scheme (Rose, 1992)	In this scheme in addition to the C bit used in binary feedback scheme the Q-bit scheme uses the Q-bit to indicate the status of the queue. This scheme also makes uses of rate-based load control.	Explicit Persistent Global
15	Loss Load Curves (Williamson, 1993)	The local load at each switching node is monitored and the loss load curve is sent to source as feedback. Based on local load the rough throughput curve is calculated. This scheme is less dependent on RTT and hop count therefore achieves more rapid response time as compared to slow start.	Explicit Persistent Global

3.1.2.1.1 Implicit Use of Feedback

In this scheme the feedback is not sent as a separate packet. Irrespective of traffic conditions neither the feedback message is specific nor explicit action oriented. Various algorithms of this scheme are: Slow Start (Jacobson, 1988), Time Out Based (R. Jain, 1986), Tri-S (Z. Wang & Crowcroft, 1991), and Warp Control (Park, 1993).

3.1.2.2 Explicit Use of Feedback

In this scheme the feedback is sent explicitly as a separate packet (piggybacked). This can be further categorized into Responsive and persistent.

3.1.2.2.1 Responsive Use of Feedback

The feedback that responses on the occurrence of certain events is called as responsive (threshold), which has been further classified as local and global.

3.1.2.2.1.1 Local Use of Feedback

The feedback is generated only when congestion occurs and the message is only sent to upstream neighboring nodes. The only algorithm of this scheme is Source Quench (Postel, 1981a).

3.1.2.2.1.2 Global Use of Feedback

The feedback is generated by the destination node based on network traffic conditions such that the threshold value of the queue has reached. The feedback is directed way back to source from destination. The various algorithms of this scheme are: Choke Packet (Varakulsiripunth et al., 1986), Rate based Control (Comer & Yavatkar, 1990), Dynamic time Window (Faber et al., 1992).

3.1.2.2.2 Persistent Use of Feedback

The feedback that remains present all of the time constantly or periodically is called persistent feedback, which is further classified as local and global.

3.1.2.2.2.1 Local Use of Feedback

In this category information is sent between the neighboring nodes only. The lone algorithm in this category is Hop by Hop Control (Mishra & Kanakia, 1992).

3.1.2.2.2.2 Global Use of Feedback

Most of the algorithms falls in this class. In global the network information is sent continuously or periodically from destination to source. The various algorithms in this class are: Binary Feedback Scheme (K. K. Ramakrishnan & Jain, 1990), Selective Binary Feedback Scheme (K. Ramakrishnan, R. Jain, 1991), BBN Scheme (Robinson et al., 1989), Adaptive Admission Control (Haas, 1991), Q-bit Scheme (Rose, 1992), and Loss Load Curves(Williamson, 1993) .

Figure 9. Taxonomy Network congestion control schemes

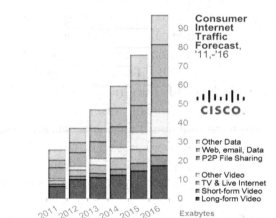

3.2 Network Congestion Control Schemes in TCP

TCP (Transmission Control Protocol) is one of the oldest and most extensively used end-end protocol for providing reliable data transfer. It was Cerf and Kahn who first gave the concept of TCP and used variable length segments for sending and receiving data (Cerf & Icahn, 2005).To end the network congestion collapse Jacobson modified the TCP in 1988 to provide first end-end control algorithm (Jacobson, 1988).TCP comprises of four fundamental algorithms; slow start, congestion avoidance, fast retransmission and fast recovery for controlling flows and congestion by adjusting congestion window, retransmission time out, slow start threshold and round trip time(Jacobson, 1988)(Kho et al., 2013).

The basic principle of TCP is based on additive increase multiplicative decrease (AIMD). Once the connection is established, the connection enters into a slow start phase, where to begin with only one segment is sent. The sending value for a number of segments is increased exponentially upon the recipient of successful acknowledgement. If the time expires or the sender receives three duplicate acknowledgements from the receiver, the network assumes congestion and enters into congestion avoidance phase. The slow start threshold value is set at half of the congestion window and congestion window size is increased linearly with every round trip time (RTT)(Cerf & Icahn, 2005)(Stevens, 1997). To minimize the effect of congestion, the sender reduces the transmission rate by reducing the growth rate of the congestion window. The congestion window is reset to one, if the time out occurs. For congestion avoidance algorithm, if the congestion window size is less or equal to the slow start threshold value, then TCP enters again into the slow start phase, but if the congestion window size is greater than the slow start threshold, then it executes the congestion avoidance algorithm.

For fast transmission, when the packet is lost, the packet is directly retransmitted without waiting for transmission time out (RTO) to expire. Slow start threshold is set to half of congestion window with congestion window set half of original value (Jacobson, 1988) (Mascolo et al., 2001). The retransmission time out (RTO) is estimated and is much higher than RTT. The efficiency of sending data is increased as the waiting for RTO is reduced. Typically, fast recovery is implemented with fast retransmission. In fast retransmission, if three duplicate acknowledgements are received by the sender then a fast recovery process is started and uses the received acknowledgements to control the sending rate. If no duplicate packet is received, the fast recovery process is exited and fast retransmission is restarted.

3.2.1 Taxonomy of TCP Congestion Techniques/Algorithms

TCP algorithms can be either classified on basis of location control or on basis of metric used to indicate congestion (Kho et al., 2013) [306].

3.2.1.1 Classification on The Basis of Location of Control

Based on the entity to control the congestion, the TCP algorithms are categorized into four groups; sender centric, receiver centric, sender or receiver centric, sender and receiver centric. The Figure 10 shows classification of TCP algorithms based on location.

3.2.1.1.1 Sender-Centric Control

In sender-centric, the congestion control, flow control and reliable data transfer is performed by the sender unit. The receiver is only responsible for sending acknowledgement for feedback purpose. The data transfer is done on the basis of DATA-ACK message conversation (Hsieh et al., 2003) and is one of most widely deployed techniques. The various algorithms of this type are: Tahoe(Cerf & Icahn, 2005), Reno (Z. Wang & Crowcroft, 1992), New Reno (Sally Floyd et al., 2004)[309], SACK (Mathis et al., 1996), Vegas(Brakmo & Peterson, 1995), Vegas A (Srijith et al., 2005), Veno (C. P. Fu & Liew, 2003), Westwood (Mascolo et al., 2001), TCPW CRB (R. Wang et al., 2002), TCPW BR(G. Yang et al., 2003), Casablanca (Biaz & Vaidya, 2003), STCP (Kelly, 2003), HS-TCP (Sally Floyd, 2003), BIC(L. Xu et al., 2004), CUBIC (Ha et al., 2008), Hybla (Caini & Firrincieli, 2004), FAST TCP (Jin et al., 2005), New Vegas(Sing & Soh, 2005), Libra (Marfia et al., 2010), Illinois (S. Liu et al., 2008), Africa (King et al., 2005), Fusion(Kaneko et al., 2007), CTCP(Tan et al., 2006), Nice(Venkataramani et al., 2002), LP (Kuzmanovic & Knightly, 2006), Light-Weight Congestion Control for the DCCP(Chodorek & Chodorek, 2020)

3.2.1.1.2 Receiver-Centric Control

Receiver-centric TCP congestion control algorithms were first introduced in 1997 and are the second most used algorithms after sender centric. In this type, the receiver is responsible for flow control, congestion control and reliable data transfer using REQUEST-DATA exchange. The receiver sends a message to the sender requesting to send the data at a rate specified by the receiver. The various algorithms in this subclass are: DSACK (S Floyd et al., 2000), TD-FR (Paxson, 1997), and TCP-Real(Tsaoussidis & Zhang, 2002), Cctcp (Saino et al., 2013),CCN (Iwamoto et al., 2018), Receiver-side TCP countermeasure in cellular networks (Dong et al., 2019), Towards Efficient Parallel Multipathing: A Receiver-Centric Cross-Layer Solution to Aid Multipath TCP (Cao et al., 2019).

3.2.1.1.3 Sender- Centric or Receiver Centric Control

In this type either sender or receiver can employ congestion control, flow control and reliable data transfer, depending upon where the mobile is sending or receiving. The only example of this type is MCP (Shu et al., 2008) .

3.2.1.1.4 Sender-Centric and Receiver-Centric Control

Also known as hybrid-centric and can perform congestion control, flow control, and reliable data transfer. For example, the receiver performs flow control and calculates congestion window. The sender uses the information of the receiver to

Figure 10. Classification of TCP congestion algorithms based on location

adjust the congestion window. The algorithms in this category are: TFRC (S Floyd et al., 2008), DOOR (F. Wang & Zhang, 2002), PBE-CC (Xie et al., 2020).

3.2.1.2 Classification on the Basis of Congestion Metrics

The congestion metrics have been divided into:

3.2.1.2.1 Loss-Based Congestion Metric Scheme

Loss-based schemes are one of the oldest schemes. The packet loss is used to calculate packet arrival rate, packet loss rate, packet probability. The various algorithms in this category include: Reno (Z. Wang & Crowcroft, 1992), New Reno (Sally Floyd et al., 2004), Casablanca(Biaz & Vaidya, 2003), STCP (Kelly, 2003), HS-TCP (Sally Floyd, 2003), BIC(L. Xu et al., 2004), CUBIC (Ha et al., 2008) Tahoe (Cerf & Icahn, 2005), Hybla (Caini & Firrincieli, 2004), Libra (Marfia et al., 2010), Illinois (S. Liu et al., 2008), and Africa(King et al., 2005), TD-FR, TCP-Real (Tsaoussidis & Zhang, 2002), MCP (Paxson, 1997), DOOR(F. Wang & Zhang, 2002) .

3.2.1.2.2 Delay-Based Congestion Metric Scheme

The delay-based scheme uses Round trip time (RTT) for calculating, length of queue, queueing delay, instant queue length etc. The various algorithms in this scheme are: Vegas(Brakmo & Peterson, 1995), Vegas A(Srijith et al., 2005), FAST TCP (Jin et al., 2005), New Vegas(Sing & Soh, 2005), Nice (Venkataramani et al., 2002).

3.2.1.2.3 Hybrid-Based Congestion Metric Scheme

The scheme combines delay based, Loss based scheme and bandwidth as the congestion metric. Some of the algorithms of this category are: Veno (C. P. Fu & Liew, 2003), Westwood(Mascolo et al., 2001), TCPW CRB(R. Wang et al., 2002), TCPW BR (G. Yang et al., 2003), Fusion(Kaneko et al., 2007), and DSACK(S Floyd et al., 2000).

3.3 Techniques/Schemes to Control Congestion AD HOC Network (MANET)

The TCP mechanism to control the congestion is not suitable for MANETS due to their inherent properties like mobility, shared wireless media, Multihop nature, nature of media. Therefore, requires a different/special approach to control the congestion. For this purpose, either existing protocols are modified or new techniques are developed by taking into the account the nature of MANET.

3.3.1 TCP With Feedback (Chandran et al., 2001)

The TCP with feedback (TCP-F) is the modified approach of TCP to tackle the problem of congestion in MANET. In TCP-F two messages Route Failure Notification (RFN) and Route Re Establishment Notification (RRN) are used to notify route failure and establishment of new route respectively. As soon as the intermediate node detects the failure in the route, it immediately sends an RFN message to the source and tries to find the new route. In case if no route is found the sender is put in a sleep state by freezing the retransmission timer and window size. When at an intermediate node, a new route is found, it immediately sends RRN to the sender to notify availability of a new route to destination. The sender after receiving the RRN restarts the frozen values and uses additional timer to keep the count on RRN messages.

3.3.2 TCP- Explicit Link Failure Notification (Holland & Vaidya, 2002)

(TCP-ELFN) is similar to TCP-F except it uses ELFN (Explicit Link Failure Notification) to report link failure and uses probe packets for route reestablishment, The ELFN is send by intermediate node upon the discovery of no route and can be implemented via ICMP (Internet Control message Protocol) or by Piggybacking the information.

The sender state is frozen upon the receipt of ELFN and a probe packet is generated periodically to enquire about a new route. Once the acknowledgement is received for the probe packet, indicating a new route, the frozen state at sender end is restored to continue its normal operation.

3.3.3 TCP with Buffering and Sequencing (Kim et al., 2001)

TCP-BUS (TCP with Buffering and Sequencing) is similar to TCP-F and TCP-ELFN with added buffering capabilities at intermediate nodes. This approach is more routing oriented with associativity-based routing (ABR)(Toh, 1997) by using local query and reply to find partial path. When the intermediate node detects the route failure, it sends an explicit route disconnection notification (ERDN) message to the sender. Upon the recipient of ERDN the sender freezes the state as in TCP-F and TCP-ELFN. Instead of dropping the packet's intermediate node when route failure occurs, the packets are buffered and retransmission timeout is extended. The new route discovery information at intermediate node is conveyed to the sender via Explicit Route Successful Notification (ERSN) packets and the sender is retained to normal state. The local query and reply message are modified to include TCP segment information. The local query contains the segment number of packets in the head of the queue at the intermediate node and the reply contains the sequence number of the last successful received segment. This allows it to distinguish between the packets that are buffered at intermediate node and those lost. Thus, lost packets are transmitted using selectivity acknowledgement policy.

3.3.4 Ad Hoc TCP (J. Liu & Singh, 2001)

The ATCP (Ad Hoc TCP) uses the network layer feedback to get information of the path by periodically probing the status of the network. Based on the information the state of the TCP sender is changed to persist state, congestion control state or retransmission state. This protocol is implemented as a thin layer between TCP and IP with additional features of distinguishing loss the packets due to congestion and error. If the intermediate node detects the partition in the network, the ATCP puts TCP on persistent state and changes the value of congestion window to one to ensure that it does not use old value of congestion window, thus forcing the TCP to probe the correct value of congestion window to be used in new route. In case the intermediate node detects the loss of packets due to error then it will immediately retransmit the packets at sender end without invoking congestion algorithm. To maintain the state of the network ATCP uses ECN (Explicit Congestion Notification) and can be in any of the four states; (a) Normal (b) Congested (c) Loss (d) Disconn.

Initially when the connection is established, the sender is put on normal state, during this state ATCP is kept in invisible state and does not interfere with the normal operation of TCP. When the packet loss occurs or packets are received in out of order fashion duplicate acknowledgement packets are generated. In traditional TCP when three duplicate acknowledgement are received, it retransmits the packets and reduces the contention window size, but in this approach the sender counts the number duplicate acknowledgements, if the counts reaches three, it changes the state of TCP to persist and ATCP to Loss state, thus avoiding unnecessary invocation of congestion algorithm. In Loss state ATCP resends unacknowledged packets from the buffer. Upon the receipt of new acknowledgment, the state of TCP is removed from persistence and ATCP is changed to Normal. If the state of ATCP is in Loss and ECN (Explicit Congestion Notification) message is received, its state is changed to congested state and TCP is removed from persistent state. During the congestion state the ECN flag of both data and acknowledgement is set. If ATCP is in normal state and receives ECN, then it changes the state to congested state and remains invisible, thus allowing TCP to invoke normal congestion control algorithm. In case of partition in the network and destination an unreachable message is received by the sender, then the state of TCP is changed to persist and that of ATCP to Disconn state until the network is connected or new acknowledgement is received.

3.4.5 Split TCP(S Kopparty, SV Krishnamurthy, M Faloutsos, 2002)

In this approach longer, paths are split into shorter TCP connections known as segments or zones with the help of intermediate nodes known as proxy node, thus acting as end point of segment or zone. In addition to this congestion control and reliability are separated with the help of local and end-end acknowledgement. When the proxy node receives the packets, it stores them and generates a local acknowledgement that is sent to the source or predecessor segment. The proxy node empties the buffer only when it receives the local acknowledgement from its immediate proxy successor. To maintain the reliability end-end acknowledgement is used and the source only clears the buffer only after receiving end-end acknowledgement. The sender maintains two windows, one for congestion that changes according to local acknowledgement from proxy nodes and an end-end window that changes according to end-end acknowledgement.

3.5.6 Enhanced Interlayer Communication and Control (Sun & Man, 2001)

In ENIC (Enhanced interlayer communication and control) the delay acknowledgement is combined with route failure. The intermediate node has less functionalities, therefore requires less support of the intermediate node. The message is reused and no explicit message is used in case of route failure or congestion. This technique has no buffering capabilities at intermediate node as the route failure occurs, the packets are simply discarded. Here both the sender and receiver are informed about the route failure, therefore helping them to recalculate new timeout and retransmission time etc.

3.5.7 TCP Re-Computation (Zhou et al., 2003)

TCP-RC (TCP Re-Computation) is the extension of TCP-ELFN and uses the characteristics like round trip time, route length of new route to calculate slow start threshold, congestion window etc. This technique concentrates mainly on new routes rather than old one.

3.5.8 LRED and Adaptive Pacing (Z. Fu et al., 2003)

The LRED and Adaptive pacing are used in combination to counter the congestion in wireless networks. This approach provides the best possible way of predicting and adjusting the TCP window before the overload occurs. The transmission at MAC layer is being monitored to count the possibility of packets to be dropped and the adaptive pacing is called only when LRED value exceeds the threshold value.

3.5.9 NRED (K. Xu et al., 2005)

The NRED works on stage. In the first stage link utilization is used to estimate the neighborhood queue size. It is assumed that if link utilization crosses some threshold value then there is the possibility of congestion. The second stage calculates the packet drop probability and this information is made available to neighboring nodes. In the last stage each and every node calculates the drop probability based on previous information. Finally, packets are dropped based on drop probability.

3.5.10 Non-Work Conservative Scheduling(L. Yang et al., 2003)

It was Yang et al. who observed that when MANET are connected to the wired backbone by reducing the congestion window, the performance is decreased many folds. So, they devised a technique called Non-Work Conservative Scheduling where the timer is set for each and every node for sending a particular packet. The node is refrained from sending the data until the timer expires, thus reducing the packet forwarding rate at intermediate node.

Other Newer Ad Hoc Congestion control techniques include: TCP-NCE(Sreekumari & Chung, 2011),F-ECN(X. Yang et al., 2009),Adaptive load distribution approach based on congestion control scheme in ad-hoc networks(Sharma & Kumar, 2019), Standbyme(Muchtar et al., 2020), modified on-demand routing protocol with local route establishment process and fuzzy-based reliable node detection (Pillai & Singhai, 2020)

4.CONCLUSION

The congestion is said to the state of the network when the availability of the resources to serve the request is less than demand. The congestion can be caused due to various reasons like insufficient memory to store arriving packets, packet arrival rate exceeding the outgoing link capacity, low speed processor and bursty traffic. Often it is misinterpreted that overprovisioning the resources will eliminate the congestion, in some cases it may make the situation even worse. The congestion can have adverse effects on network throughput and delay. If no proper congestion control technique is selected, the congestion collapse may occur, where none of the packets is delivered. So, there is the need to implement the congestion control mechanism. The congestion control schemes can be broadly classified in open loop and closed loop based on the feedback which can be further classified into various subclasses. Most of the traffic driven on the Internet is based on TCP as it provides end-end reliable support. Various algorithms have been proposed based on TCP that can be bifurcated based on location of control and congestion control metric used. Since TCP is unable to perform to its fullest for Adhoc networks due various characteristics possessed by Ad Hoc networks. Therefore, either already existing TCP algorithms are modified or new algorithms are proposed to control the congestion in the Ad Hoc network.

REFERENCES

Alizadeh, M., Greenberg, A., Maltz, D., Padhye, J., Patel, P., Prabhakar, B., Sengupta, S., & Sridharan, M. (2010). *DCTCP: Efficient packet transport for the commoditized data center*. Academic Press.

Allman, M., Paxson, V., & Stevens, W. (1999). *TCP congestion control*. Academic Press.

Allman, M, Glover, D., & Sanchez, L. (1999). *RFC2488: Enhancing TCP Over Satellite Channels using Standard Mechanisms*. RFC Editor.

Allman, M., & Paxson, V. (2001). On estimating end-to-end network path properties. *Computer Communication Review*, *31*(2, supplement), 124–151. doi:10.1145/844193.844203

Biaz, S., & Vaidya, N. H. (2003). Differentiated services: A new direction for distinguishing congestion losses from wireless losses. Technical report, University of Auburn.

Bicket, J., Aguayo, D., Biswas, S., & Morris, R. (2005). Architecture and evaluation of an unplanned 802.11 b mesh network. *Proceedings of the 11th Annual International Conference on Mobile Computing and Networking*, 31–42.

39. Big Data Statistics for 2020. (n.d.). Retrieved July 10, 2020, from https://leftronic.com/big-data-statistics/

Brakmo, L. S., & Peterson, L. L. (1995). TCP Vegas: End to end congestion avoidance on a global Internet. *IEEE Journal on Selected Areas in Communications, 13*(8), 1465–1480. doi:10.1109/49.464716

Caini, C., & Firrincieli, R. (2004). TCP Hybla: A TCP enhancement for heterogeneous networks. *International Journal of Satellite Communications and Networking, 22*(5), 547–566. doi:10.1002at.799

Cao, Y., Yu, D., Zeng, L., Liu, Q., Wu, F., Gui, X., & Huang, M. (2019). Towards Efficient Parallel Multipathing: A Receiver-Centric Cross-Layer Solution to Aid Multipath TCP. *2019 IEEE 25th International Conference on Parallel and Distributed Systems (ICPADS)*, 790–797.

Cerf, V. G., & Icahn, R. E. (2005). A protocol for packet network intercommunication. *Computer Communication Review, 35*(2), 71–82. doi:10.1145/1064413.1064423

Chandran, K., Raghunathan, S., Venkatesan, S., & Prakash, R. (2001). A feedback-based scheme for improving TCP performance in ad hoc wireless networks. *IEEE Personal Communications, 8*(1), 34–39. doi:10.1109/98.904897

Chilamkurti, N. K., & Soh, B. (2002). A simulation study on multicast congestion control algorithms with multimedia traffic. *Global Telecommunications Conference, 2002. GLOBECOM'02. IEEE, 2*, 1779–1783. 10.1109/GLOCOM.2002.1188504

Chodorek, A., & Chodorek, R. R. (2020). Light-Weight Congestion Control for the DCCP: Implementation in the Linux Kernel. In Data-Centric Business and Applications (pp. 245–267). Springer.

Cisco Visual Networking Index Predicts Annual Internet Traffic to Grow More Than 20 Percent (reaching 1.6 Zettabytes) by 2018 | The Network. (n.d.). Retrieved July 10, 2020, from https://newsroom.cisco.com/press-release-content?articleId=1426270

Comer, D. E., & Yavatkar, R. S. (1990). A rate-based congestion avoidance and control scheme for packet switched networks. *Proceedings 10th International Conference on Distributed Computing Systems*, 390–397. 10.1109/ICDCS.1990.89307

Congestion Control video lecture by Prof Ajit Pal of IIT Kharagpur. (n.d.). Retrieved July 10, 2020, from https://freevideolectures.com/course/2278/data-communication/23

Davies, D. (1972). The Control of Congestion in Packet Switch Networks. *IEEE Transactions on Communications*, *20*(3), 546–550. doi:10.1109/TCOM.1972.1091198

Demers, A., Keshav, S., & Shenker, S. (1989). Analysis and simulation of a fair queueing algorithm. *Computer Communication Review*, *19*(4), 1–12. doi:10.1145/75247.75248

Dong, P., Gao, K., Xie, J., Tang, W., Xiong, N., & Vasilakos, A. V. (2019). Receiver-side TCP countermeasure in cellular networks. *Sensors (Basel)*, *19*(12), 2791. doi:10.339019122791 PMID:31234375

Dukkipati, N., Kobayashi, M., Zhang-Shen, R., & McKeown, N. (2005). Processor sharing flows in the internet. *International Workshop on Quality of Service*, 271–285.

Dukkipati, N., Refice, T., Cheng, Y., Chu, J., Herbert, T., Agarwal, A., Jain, A., & Sutin, N. (2010). An argument for increasing TCP's initial congestion window. *Computer Communication Review*, *40*(3), 26–33. doi:10.1145/1823844.1823848

30 . Eye-Opening Big Data Statistics for 2020: Patterns Are Everywhere. (n.d.). Retrieved July 10, 2020, from https://kommandotech.com/statistics/big-data-statistics/

Faber, T., Landweber, L. H., & Mukherjee, A. (1992). Dynamic Time Windows: packet admission control with feedback. *Conference Proceedings on Communications Architectures & Protocols*, 124–135. 10.1145/144179.144255

Farrell, S., & Jensen, C. (2004). A Flexible Interplanetary Internet. *Tools and Technologies for Future Planetary Exploration*, *543*, 87–94.

Floyd, S. (2003). *RFC3649: HighSpeed TCP for large congestion windows*. RFC Editor.

Floyd, S., Henderson, T., & Gurtov, A. (2004). *RFC3782: The newreno modification to TCP's fast recovery algorithm*. RFC Editor.

Floyd, S., Handley, M., Padhye, J., & Widmer, J. (2008). TCP friendly rate control (TFRC): Protocol Specification. IETF RFC 5348.

Floyd, S, Mahdavi, J., Mathis, M., & Podolsky, M. (2000). *RFC2883: An Extension to the Selective Acknowledgement (SACK) Option for TCP*. RFC Editor.

Fu, C. P., & Liew, S. C. (2003). TCP Veno: TCP enhancement for transmission over wireless access networks. *IEEE Journal on Selected Areas in Communications*, *21*(2), 216–228. doi:10.1109/JSAC.2002.807336

Fu, Z., Zerfos, P., Luo, H., Lu, S., Zhang, L., & Gerla, M. (2003). The impact of multihop wireless channel on TCP throughput and loss. *IEEE INFOCOM 2003. Twenty-Second Annual Joint Conference of the IEEE Computer and Communications Societies (IEEE Cat. No. 03CH37428), 3,* 1744–1753.

5G . Gear Up - Huawei. (n.d.). Retrieved July 10, 2020, from https://carrier.huawei.com/en/spotlight/5g

10 . Gigabit Ethernet - Wikipedia. (n.d.). Retrieved July 10, 2020, from https://en.wikipedia.org/wiki/10_Gigabit_Ethernet

Golestani, S. J. (1991). Congestion-free communication in high-speed packet networks. *IEEE Transactions on Communications, 39*(12), 1802–1812. doi:10.1109/26.120166

Gray, M. (1996). *Web Growth Summary.* https://stuff.mit.edu/people/mkgray/net/web-growth-summary.html

Ha, S., Rhee, I., & Xu, L. (2008). CUBIC: A new TCP-friendly high-speed TCP variant. *Operating Systems Review, 42*(5), 64–74. doi:10.1145/1400097.1400105

Haas, Z. (1991). Adaptive admission congestion control. *Computer Communication Review, 21*(5), 58–76. doi:10.1145/122431.122436

Holland, G., & Vaidya, N. (2002). Analysis of TCP performance over mobile ad hoc networks. *Wireless Networks, 8*(2–3), 275–288. doi:10.1023/A:1013798127590

Hsieh, H.-Y., Kim, K.-H., Zhu, Y., & Sivakumar, R. (2003). A receiver-centric transport protocol for mobile hosts with heterogeneous wireless interfaces. *Proceedings of the 9th Annual International Conference on Mobile Computing and Networking,* 1–15. 10.1145/938985.938987

Hui, R., Zhu, B., Huang, R., Allen, C., Demarest, K., & Richards, D. (2001). 10-Gb/s SCM fiber system using optical SSB modulation. *IEEE Photonics Technology Letters, 13*(8), 896–898. doi:10.1109/68.935840

Infographic: How Much Data is Generated Each Day? (n.d.). Retrieved July 10, 2020, from https://www.visualcapitalist.com/how-much-data-is-generated-each-day/

Internet Growth Statistics 1995 to 2019 - the Global Village Online. (n.d.). Retrieved July 10, 2020, from https://www.internetworldstats.com/emarketing.htm

Internet Top 20 Countries - Internet Users 2020. (n.d.). Retrieved July 10, 2020, from https://internetworldstats.com/top20.htm

Iwamoto, D., Sugahara, D., Bandai, M., & Yamamoto, M. (2018). Adaptive Congestion Control for Handover in Heterogeneous Mobile Content-Centric Networking. *2018 Eleventh International Conference on Mobile Computing and Ubiquitous Network (ICMU)*, 1–6. 10.23919/ICMU.2018.8653625

Jacobson, V. (1988). Congestion avoidance and control. *Computer Communication Review, 18*(4), 314–329. doi:10.1145/52325.52356

Jain, R. (1986). A timeout-based congestion control scheme for window flow-controlled networks. *IEEE Journal on Selected Areas in Communications, 4*(7), 1162–1167. doi:10.1109/JSAC.1986.1146431

Jain, R. (n.d.). Congestion Control in Computer Networks. *Issues (Chicago, Ill.)*.

Jin, C., Wei, D., Low, S. H., Bunn, J., Choe, H. D., Doylle, J. C., Newman, H., Ravot, S., Singh, S., & Paganini, F. (2005). FAST TCP: From theory to experiments. *IEEE Network, 19*(1), 4–11. doi:10.1109/MNET.2005.1383434

Kaneko, K., Fujikawa, T., Su, Z., & Katto, J. (2007). TCP-Fusion: A hybrid congestion control algorithm for high-speed networks. *Proc. PFLDnet, 7*, 31–36.

Katabi, M. H. C. R. (2002). Internet Congestion Control for Future High Bandwidth-Delay Product Environments. *ACM SIGCOMM 2002*.

Katabi, D., Handley, M., & Rohrs, C. (2002). Congestion control for high bandwidth-delay product networks. *Proceedings of the 2002 Conference on Applications, Technologies, Architectures, and Protocols for Computer Communications*, 89–102. 10.1145/633025.633035

Kelly, T. (2003). Scalable TCP: Improving performance in highspeed wide area networks. *Computer Communication Review, 33*(2), 83–91. doi:10.1145/956981.956989

Kent, C. A., & Mogul, J. C. (1987). *Fragmentation considered harmful* (Vol. 17). Academic Press.

Keshav, S. (1991). *Congestion Control in Computer Networks PhD Thesis*. UC Berkeley TR-654.

Kho, L. C., Défago, X., Lim, A. O., & Tan, Y. (2013). A taxonomy of congestion control techniques for tcp in wired and wireless networks. *2013 IEEE Symposium on Wireless Technology & Applications (ISWTA)*, 147–152. 10.1109/ISWTA.2013.6688758

Khorov, E., Kiryanov, A., Lyakhov, A., & Bianchi, G. (2019). A tutorial on IEEE 802.11ax high efficiency WLANs. *IEEE Communications Surveys and Tutorials*, *21*(1), 197–216. doi:10.1109/COMST.2018.2871099

Kim, D., Toh, C.-K., & Choi, Y. (2001). TCP-BuS: Improving TCP performance in wireless ad hoc networks. *Journal of Communications and Networks (Seoul)*, *3*(2), 1–12. doi:10.1109/JCN.2001.6596860

King, R., Baraniuk, R., & Riedi, R. (2005). TCP-Africa: An adaptive and fair rapid increase rule for scalable TCP. *Proceedings IEEE 24th Annual Joint Conference of the IEEE Computer and Communications Societies*, *3*, 1838–1848. 10.1109/INFCOM.2005.1498463

Kopparty, S., Krishnamurthy, S. V., & Faloutsos, M., S. T. (2002). Split TCP for mobile ad hoc networks. *IEEE Global Telecommunications Conference GLOBECOM'02*, 138–142.

Kuzmanovic, A., & Knightly, E. W. (2006). TCP-LP: Low-priority service via end-point congestion control. *IEEE/ACM Transactions on Networking*, *14*(4), 739–752. doi:10.1109/TNET.2006.879702

Lam, S., & Reiser, M. (1979). Congestion Control of Store-and-Forward Networks by Input Buffer Limits - An Analysis. *IEEE Transactions on Communications*, *27*(1), 127–134. doi:10.1109/TCOM.1979.1094280

Li, Y.-T., Leith, D., & Shorten, R. N. (2007). Experimental evaluation of TCP protocols for high-speed networks. *IEEE/ACM Transactions on Networking*, *15*(5), 1109–1122. doi:10.1109/TNET.2007.896240

Liu, J., & Singh, S. (2001). ATCP: TCP for mobile ad hoc networks. *IEEE Journal on Selected Areas in Communications*, *19*(7), 1300–1315. doi:10.1109/49.932698

Liu, Q., & Hwang, J.-N. (2003). End-to-end available bandwidth estimation and time measurement adjustment for multimedia QoS. *2003 International Conference on Multimedia and Expo. ICME'03. Proceedings (Cat. No. 03TH8698)*, *3*.

Liu, S., Başar, T., & Srikant, R. (2008). TCP-Illinois: A loss-and delay-based congestion control algorithm for high-speed networks. *Performance Evaluation*, *65*(6–7), 417–440. doi:10.1016/j.peva.2007.12.007

Loguinov, D., & Radha, H. (2002). Increase-decrease congestion control for real-time streaming: Scalability. *Proceedings. Twenty-First Annual Joint Conference of the IEEE Computer and Communications Societies*, *2*, 525–534. 10.1109/INFCOM.2002.1019297

Marfia, G., Palazzi, C. E., Pau, G., Gerla, M., & Roccetti, M. (2010). TCP Libra: Derivation, analysis, and comparison with other RTT-fair TCPs. *Computer Networks, 54*(14), 2327–2344. doi:10.1016/j.comnet.2010.02.014

Mascolo, S., Casetti, C., Gerla, M., Sanadidi, M. Y., & Wang, R. (2001). TCP westwood: Bandwidth estimation for enhanced transport over wireless links. *Proceedings of the 7th Annual International Conference on Mobile Computing and Networking*, 287–297. 10.1145/381677.381704

Mathis, M., Mahdavi, J., Floyd, S., & Romanow, A. (1996). *RFC2018: TCP selective acknowledgement options*. RFC Editor.

Mishra, P. P., & Kanakia, H. (1992). A hop by hop rate-based congestion control scheme. *Conference Proceedings on Communications Architectures & Protocols*, 112–123. 10.1145/144179.144254

Mogul, J. C., Kent, C. A., Partridge, C., & McCloghrie, K. (1988). *RFC1063: IP MTU discovery options*. RFC Editor.

Muchtar, F., Al-Adhaileh, M. H., Alubady, R., Singh, P. K., Ambar, R., & Stiawan, D. (2020). Congestion Control for Named Data Networking-Based Wireless Ad Hoc Network BT. In *Proceedings of First International Conference on Computing, Communications, and Cyber-Security (IC4S 2019)*. Springer Singapore.

Mukherji, U. (1986). *A schedule-based approach for flow-control in data communication networks*. Massachusetts Inst of Tech Cambridge Lab for Information And Decision Systems.

Nagle, J. (1984). *Congestion control in IP/TCP internetworks. Request For Comment 896*. Network Working Group.

Nagle, J. (1987). On packet switches with infinite storage. *IEEE Transactions on Communications, 35*(4), 435–438. doi:10.1109/TCOM.1987.1096782

News Release 060929a. (n.d.). Retrieved July 10, 2020, from https://www.ntt.co.jp/news/news06e/0609/060929a.html

Nguyen, A. G., & Hwang, J.-N. (2002). SPEM online rate control for realtime streaming video. *Proceedings. International Conference on Information Technology: Coding and Computing*, 65–70. 10.1109/ITCC.2002.1000361

Number of internet users worldwide | Statista. (n.d.). Retrieved July 10, 2020, from https://www.statista.com/statistics/273018/number-of-internet-users-worldwide/

Optical Carrier transmission rates - Wikipedia. (n.d.). Retrieved July 10, 2020, from https://en.wikipedia.org/wiki/Optical_Carrier_transmission_rates

Park, K. (1993). Warp control: A dynamically stable congestion protocol and its analysis. *Journal of High Speed Networks*, *2*(4), 373–404. doi:10.3233/JHS-1993-2404

Partridge, C., & Shepard, T. J. (1997). TCP/IP performance over satellite links. *IEEE Network*, *11*(5), 44–49. doi:10.1109/65.620521

Paxson, V. (1997). End-to-End Internet Packet Dynamics. *Proceedings of the ACM SIGCOMM '97 Conference on Applications, Technologies, Architectures, and Protocols for Computer Communication*, 139–152. 10.1145/263105.263155

Paxson, V. (1999). End-to-end internet packet dynamics. *IEEE/ACM Transactions on Networking*, *7*(3), 277–292. doi:10.1109/90.779192

Pillai, B., & Singhai, R. (2020). *Congestion Control Using Fuzzy-Based Node Reliability and Rate Control BT*. In K. N. Das, J. C. Bansal, K. Deep, A. K. Nagar, P. Pathipooranam, & R. C. Naidu (Eds.), *Soft Computing for Problem Solving* (pp. 67–75). Springer Singapore.

Postel, J. (1981a). *Internet control message protocol*. RFC Editor.

Postel, J. (1981b). *Rfc0793: Transmission control protocol*. RFC Editor.

Qazi, I. A., Andrew, L. L. H., & Znati, T. (2009). Congestion control using efficient explicit feedback. *IEEE INFOCOM*, *2009*, 10–18.

Ramakrishnan, K., & Floyd, S. (1999). *A proposal to add explicit congestion notification (ECN) to IP*. RFC 2481.

Ramakrishnan, K, & Floyd, S. (1999). *RFC2481: A Proposal to add Explicit Congestion Notification (ECN) to IP*. RFC Editor.

Ramakrishnan, K., & Jain, R. (1991). *A Selective Binary Feedback Scheme far General Topologies*.

Ramakrishnan, K. K., & Jain, R. (1990). A binary feedback scheme for congestion avoidance in computer networks. *ACM Transactions on Computer Systems*, *8*(2), 158–181. doi:10.1145/78952.78955

Robinson, J., Friedman, D., & Steenstrup, M. (1989). Congestion control in BBN packet-switched networks. *Computer Communication Review*, *20*(1), 76–90. doi:10.1145/86587.86592

Romanow, A., & Floyd, S. (1995). Dynamics of TCP traffic over ATM networks. *IEEE Journal on Selected Areas in Communications, 13*(4), 633–641. doi:10.1109/49.382154

Rose, O. (1992). The Q-bit scheme: Congestion avoidance using rate-adaptation. *Computer Communication Review, 22*(2), 29–42. doi:10.1145/141800.141803

Saino, L., Cocora, C., & Pavlou, G. (2013). Cctcp: A scalable receiver-driven congestion control protocol for content centric networking. *2013 IEEE International Conference on Communications (ICC)*, 3775–3780. 10.1109/ICC.2013.6655143

Sharma, V. K., & Kumar, M. (2019). Adaptive load distribution approach based on congestion control scheme in ad-hoc networks. *International Journal of Electronics, 106*(1), 48–68. doi:10.1080/00207217.2018.1501613

Shu, Y., Ge, W., Jiang, N., Kang, Y., & Luo, J. (2008). Mobile-Host-Centric Transport Protocol forEAST Experiment. *IEEE Transactions on Nuclear Science, 55*(1), 209–216. doi:10.1109/TNS.2007.914319

Sing, J., & Soh, B. (2005). TCP New Vegas: improving the performance of TCP Vegas over high latency links. *Fourth IEEE International Symposium on Network Computing and Applications*, 73–82. 10.1109/NCA.2005.52

Singh, J., & Lambay, M. A. (n.d.). *Network Border Patrol: Preventing Congestion Collapse and Promoting Fairness in the Network*. Academic Press.

Sreekumari, P., & Chung, S.-H. (2011). TCP NCE: A unified solution for non-congestion events to improve the performance of TCP over wireless networks. *EURASIP Journal on Wireless Communications and Networking, 2011*(1), 23. doi:10.1186/1687-1499-2011-23

Srijith, K. N., Jacob, L., & Ananda, A. L. (2005). TCP Vegas-A: Improving the performance of TCP Vegas. *Computer Communications, 28*(4), 429–440. doi:10.1016/j.comcom.2004.08.016

Stevens, W. (1997). *RFC2001: TCP slow start, congestion avoidance, fast retransmit, and fast recovery algorithms*. RFC Editor.

Sun, D., & Man, H. (2001). ENIC-an improved reliable transport scheme for mobile ad hoc networks. *GLOBECOM'01. IEEE Global Telecommunications Conference (Cat. No. 01CH37270), 5*, 2852–2856.

Tan, K., Song, J., Zhang, Q., & Sridharan, M. (2006). A compound TCP approach for high-speed and long distance networks. *Proceedings - IEEE INFOCOM*, 1–12. doi:10.1109/INFOCOM.2006.188

Tanenbaum, A. S. (1981). *Computer networks*. Prentice-Hall.

Toh, C.-K. (1997). Associativity-based routing for ad hoc mobile networks. *Wireless Personal Communications*, 4(2), 103–139. doi:10.1023/A:1008812928561

Total number of Websites - Internet Live Stats. (n.d.). Retrieved July 10, 2020, from https://www.internetlivestats.com/total-number-of-websites/

Tsaoussidis, V., & Zhang, C. (2002). TCP-Real: Receiver-oriented congestion control. *Computer Networks*, 40(4), 477–497. doi:10.1016/S1389-1286(02)00291-8

Varakulsiripunth, R., Shiratori, N., & Noguchi, S. (1986). A congestion-control policy on the internetwork gateway. *Computer Networks and ISDN Systems*, 11(1), 43–58. doi:10.1016/0169-7552(86)90028-0

Varghese, G. (1996). On avoiding congestion collapse. *Viewgraphs, Washington University Workshop on the Integration of IP and ATM*.

Vasudevan, V., Phanishayee, A., Shah, H., Krevat, E., Andersen, D. G., Ganger, G. R., Gibson, G. A., & Mueller, B. (2009). Safe and effective fine-grained TCP retransmissions for datacenter communication. *Computer Communication Review*, 39(4), 303–314. doi:10.1145/1594977.1592604

Venkataramani, A., Kokku, R., & Dahlin, M. (2002). TCP Nice: A mechanism for background transfers. *ACM SIGOPS Operating Systems Review, 36*(SI), 329–343.

Wang, F., & Zhang, Y. (2002). Improving TCP performance over mobile ad-hoc networks with out-of-order detection and response. *Proceedings of the 3rd ACM International Symposium on Mobile Ad Hoc Networking & Computing*, 217–225. 10.1145/513800.513827

Wang, R., Valla, M., Sanadidi, M. Y., Ng, B. K. F., & Gerla, M. (2002). Efficiency/friendliness tradeoffs in TCP Westwood. *Proceedings ISCC 2002 Seventh International Symposium on Computers and Communications*, 304–311. 10.1109/ISCC.2002.1021694

Wang, Z., & Crowcroft, J. (1991). A new congestion control scheme: Slow start and search (Tri-S). *Computer Communication Review*, 21(1), 32–43. doi:10.1145/116030.116033

Wang, Z., & Crowcroft, J. (1992). Eliminating periodic packet losses in the 4.3-Tahoe BSD TCP congestion control algorithm. *Computer Communication Review*, *22*(2), 9–16. doi:10.1145/141800.141801

Williamson, C. L. (1993). Optimizing file transfer response time using the loss-load curve congestion control mechanism. *Computer Communication Review*, *23*(4), 117–126. doi:10.1145/167954.166249

Wu, D., Hou, Y. T., & Zhang, Y.-Q. (2000). Transporting real-time video over the Internet: Challenges and approaches. *Proceedings of the IEEE*, *88*(12), 1855–1877. doi:10.1109/5.899055

Xie, Y., Yi, F., & Jamieson, K. (2020). *PBE-CC: Congestion Control via Endpoint-Centric, Physical-Layer Bandwidth Measurements*. ArXiv Preprint ArXiv:2002.03475

Xu, K., Gerla, M., Qi, L., & Shu, Y. (2005). TCP unfairness in ad hoc wireless networks and a neighborhood RED solution. *Wireless Networks*, *11*(4), 383–399. doi:10.100711276-005-1764-1

Xu, L., Harfoush, K., & Rhee, I. (2004). Binary increase congestion control (BIC) for fast long-distance networks. *IEEE INFOCOM*, *2004*(4), 2514–2524.

Yang, C.-Q., & Reddy, A. V. S. (1995). A taxonomy for congestion control algorithms in packet switching networks. *IEEE Network*, *9*(4), 34–45. doi:10.1109/65.397042

Yang, G., Wang, R., Sanadidi, M. Y., & Gerla, M. (2003). TCPW with bulk repeat in next generation wireless networks. *IEEE International Conference on Communications, 2003. ICC'03.*, *1*, 674–678. 10.1109/ICC.2003.1204260

Yang, L., Seah, W. K. G., & Yin, Q. (2003). Improving fairness among TCP flows crossing wireless ad hoc and wired networks. *Proceedings of the 4th ACM International Symposium on Mobile Ad Hoc Networking & Computing*, 57–63. 10.1145/778415.778423

Yang, X., Ge, L., & Wang, Z. (2009). F-ECN: A Loss Discrimination Based on Fuzzy Logic Control. *2009 Sixth International Conference on Fuzzy Systems and Knowledge Discovery*, *7*, 546–549. 10.1109/FSKD.2009.140

Yin & Li. (1990). Stern Congestion control for packet voice by selective packet discarding. *IEEE Trans. Comm*, *38*(5), 674–683.

Zaman, M., Quadri, S. M. K., & Butt, M. A. (2012). Information Translation: A Practitioners Approach. *Proceedings of the World Congress on Engineering and Computer Science*, *1*.

Zhang, L. (1990). Virtual clock: A new traffic control algorithm for packet switching networks. *Proceedings of the ACM Symposium on Communications Architectures & Protocols*, 19–29. 10.1145/99508.99525

Zhou, J., Shi, B., & Zou, L. (2003). Improve TCP performance in Ad hoc network by TCP-RC. *14th IEEE Proceedings on Personal, Indoor and Mobile Radio Communications, 2003. PIMRC 2003, 1*, 216–220.

Chapter 9

M–Commerce Location– Based Services:
Security and Adoptability Issues in M–Commerce

Archana Sharma
Institute of Management Studies Noida, India

ABSTRACT

Truthful authentication with secure communication is necessary in location-based services to protect from various risks. The purpose of this research is to identify security risks in mobile transactions especially in location-based services like mobile banking. The factors need to be identified the reasons of customer distrust in mobile banking. In addition, the security issues with mobile banking systems and mobile devices are highlighted. The chapter finds which approach is more suitable and secure for mobile banking transaction between customer and bank. The research predominantly focuses upon customer trust, security issues, and transaction costs owing to different technology standards of mobile commerce. The first phase highlights the various location-based services in m-commerce, various technology standards, customer trust, and perceived risk, and further, at next level, it highlights the various problems associated mobile database and a comparative study of various replication protocols, transaction security issues, and LBS security challenges.

DOI: 10.4018/978-1-5225-9493-2.ch009

INTRODUCTION

Mobile communication has become the basic need of people and society in today's world as it improved the lifestyle as also the business processes with its innovative applications. The information availability unsurpassed and everyplace is the mobilizing force for the growth of internet, portable computing devices and wireless communication. Communication is 2-way 'transmission' and 'reception' of data streams wherein voice, data or multimedia streams are transmitted as signals and received by a receiver. Mobile communication requires transmission of data to and fro from handheld devices whereas out of the two or more communicating devices at least, one is handheld or mobile. The location of the device varies, locally or globally and communication takes place through wireless network. In the beginning stages of wireless communication, the mobility range was defined by type of used antenna, transmitter power and the frequency of operation. The transmission and receiving of various allotted channels is conducted from the antenna atop tower. Any vehicle within range could try to seize one of those channels and complete the call. However, the number of channels made available never came even close to satisfying the requirement. The solution of this problem was given by cellular radio. In this, the area coverage were divided into cells and average cell was 2-10 miles across and depended actually on the number of users in the cell and reduce the cell size as more mobile users add up to increases because of nearby the transmitter. In addition to this, the upcoming wireless and mobile networks have added another dimension of mobility and moved the E-Commerce to M-Commerce. New business opportunities have been opened due to Mobile Commerce with addition of location-based services (LBS) which has not been provided by stationary Internet. For example, the nearest petrol pump location, traffic route, nearby good restaurant etc. could only be provided by determining the current geographic position of the mobile user. There are basically three Mobile Commerce applications which need location support and these location-intensive applications are mobile financial applications, mobile advertising and location-based services.

These services utilize location of mobile user to provide location-aware content like information about nearby petrol pumps, ATM machines and products. 'Pull mode' and 'Push mode', both type of services are offered in Location-based services. For instance, in case of pull mode a mobile user may want to know the availability and waiting time on one or more theatres close to his/her current location. And in the case of push mode the mobile user, as an instance, might like to be informed as to if one of his relatives is located in the same area. In general, these services need location tracking of fixed, portable, and of mobile entities. Location information of all fixed entities are kept in a separate database area-wise while location tracking of portable and mobile entities could be performed on-demand. Upon entering in that location by a mobile user, the services list and location information is provided based on current preferences and/or the history of choices. Authentication process identifies a specific device which is allowed to use services. Thus it is a significant factor for accessing Mobile Location Based Services as secure and trustworthy whereas security and privacy are important factors in a mobile device.

Difference Between M-Commerce and E-Commerce

The retailing of goods and services over the Web is described as E-Commerce (Kalakota & Robinson,2002). There are various definitions of E-Commerce emphasizing its different aspects. One of the definition states: 'A transaction would be considered as E-Commerce, when monetary payment has been involved in the purchase or the actual receiving of service or product with the help of electronic mean over network.(Fischer,2003). It includes various types of customer activities in the retail organization, service organization and business to business activities such as net banking, inventory management, SMS banking, fund transfer electonically, secure electronic transaction(SET), E-Advertising, automated data collection system, E-Marketing, E-Business and EDI (Electronic Data Exchange)System.

M-Commerce is an extensive appearance of E-Commerce. The services offered in E-Commerce with the support of electronic mean over computer network. Within Mobile- Commerce the transaction of products or other services are accessible by means of telecommunication networks or wireless network through mobile devices. Either of these approaches are essentially correct which could be derived from the following facts:

a)   There are several services offered by Mobile Commerce which may as well be availed without using mobile services with the help of stationary Internet like purchasing movie ticket etc.
b)   Mobile Commerce unlocks the novel commerce prospects by enabling location aware services which the stationary Internet does not provide. For example, the location of the nearby petrol pump, movie halls, restaurant etc. in real time could only be provided by determining the current geographic position of the mobile user.

Mobile phones are generally used for the fiscal transaction of commodities and merchandise . This is a platform where a mobile user could avail various transaction facilities which are related to commercial facilities through mobile phone. This also provides information and services which could trigger a future transaction. Mobile

Commerce is significant in SMS or the text messaging, mobile payments, financial & banking services, logistics, location based services like location tracking, ticket booking, mobile advertising etc.

Electronic Commerce Architecture

Electronic Commerce applications follow client-server architecture. Clients are the devices with software which request information from servers and servers are the computers that serve information upon the request made by the clients. Client devices manage the user interface. The server holds application tasks, storage management and various security issues with the scalability to add more clients as are needed for serving more number of customers. The client-server architecture links Personal Computers to a database server or storage, where most computing is made on the client. The client-server representation permits the client to act together with the server via a request-reply sequence managed by a paradigm called as message passing. Profit-making users, now widely have started client –server network to run and economize their applications and this trend in accelerating the E-Commerce. The technology at the back of the interaction of server and client are World Wide Web and Internet.

To carry out transactions successfully and without compromising security and trust, business communities, financial institutions and companies offering technological solutions wanted a protocol that works very similar to the way how a credit card transactions work (Daniel et al,2007).

Secure Electronic Transaction: The payment process in secure electronic transaction ensures that the information about order and payment of the customers remain confidential. Further check the authenticity of the customer concerning the credit account. The payment process is easy and simple. When the customer makes a purchase, the SET authenticates the credit card as per the details provided by the customer, and thereafter the person which could be at the online Store sends the order details to the bank. Thereafter transaction occurs between the two after the approval about purchase. In addition, bank approve the digitally sign and authorize the merchant to process the order.

Mobile Commerce Architecture

The Mobile Commerce architecture is based upon the facility provided through GPRS, GSM, CDMA and 3G enabled mobile devices. These services are provided to mobile users directly by the bank or through a 3rd party vendor.

Case-1: In first case the set up consists of mobile network, application server and the bank's database server at the bank premises. The dissimilar sort of services which are to be provided to the mobile user is the responsibility of the application server whereas the security of infrastructure depends upon the banking services provided to the customer.

Case-2: In this case, the bank outsources the same facility to 3rd party vendor and the database may possibly be similar as the Core Banking database while maintaining the additional table for the access the services of mobile banking by mobile users. The mobile transactions now does via the mobile network and for authentication verification, autherization, transaction processing etc. the mobile banking server communicate with the database server of the intended user's bank.

Mobile User Requirements

The features and characteristics of Mobile Commerce are different from Electronic Commerce which are considered during the development and design of Mobile Commerce applications and services. Few of them has been described below:

Ubiquity

Mobile users be required to be proficient to receive information and execute transactions in real time regardless of location.

Personalization

The enormous information, applications and services that is given on Internet is of great importance. However as per the preferences of mobile users, different services and applications should be personalized.

Flexibility

The activities such as receiving information, and conducting transactions should be flexible and comfortable for the mobile user.

Localization

The access to local information and services should be provided to mobile users by the service providers. Therefore Service Providers need know the location of mobile users in order to promote their services and products directly on to their customers in a local environment. The value added services could be more demanding by mobile customer that can be unachievable or achievable depending on the existing technologies. There are certain constraints over Mobile Commerce like reliability, easiness, performance, security, bandwidth, disconnection, etc. The services may be following:

a.   Easy and timely access to information e.g. to know the latest availability of flights etc.
b.   Immediate purchase opportunity information e.g. for immediate purchase of movie tickets etc.
c.   Withdrawal of money from an account through SMS banking etc.
d.   Location management like locating a person etc.

Location Based Mobile Commerce Services

There are basically three Mobile Commerce applications which need location support and these location-intensive applications are following:

a.   Mobile financial applications
b.   Mobile advertising
c.   Location-based services

Mobile Financial Applications

Mobile transaction, SMS banking, mobile payments etc. as financial application could convert a mobile device into a trade instrument replacing banks, credit cards and ATMs by allowing the financial transactions for mobile use with mobile money. As a mobile user does the any transaction of item or services from a trader or service provider further connect with reliable third party - a financial institution for the authenticity of the mobile user and the amount of purchase. The mobile payments is done in next step after the purchase completion approval. The corresponding amounts are then withdrawn from 'm-wallet' of mobile user charged on user's 'phone bill' or detected in his 'bank account'. Otherwise the mobile user could pay by making use of mobile money supplied to her or him from another user or a 3rd party mobile money provider. The mobile money may be floated among mobile users without restraint either by, use of a local area wireless network or wide area wireless service provider's network.

Other two requirements of financial applications through mobile are support for following:

a.   Mobile payments
b.   Secure transactions

Various associations are already operational on mobile payments including Pay Circle that is established by HP, Lucent, Oracle, Sun, and Siemens (Varshney, Vetter & Kalakota,2002). Macro payments and micro-payments are two categories of mobile payment. For mobile payment providers issues like the transaction costs of mobile micro-payments and the ways to make profit on mobile micro-payments may exist.

Mobile Advertising

Mobile advertising is most likely to become a significant revenue for telecommunication. As compared to web-based advertising, there are many benefits of mobile advertising which includes high penetration rate, personal communication device, multimedia capability, interactive and individually addressable. Thus, advertisers could associate with each user for full personalized advertisements to improve large value of mobile advertisements. The existing mobile advertising methods could be divided into 3 categories viz. SMS, Applets, and Browser.

Location-Based Services

These services utilize location of mobile user to provide location-aware content like information about nearby petrol pumps, ATM machines and products. Pull' and 'Push' modes are two types mode are offered by Location-based services. For instance, in case of pull mode a mobile user may want to know the availability and waiting time on one or more theatres close to his/her current location. And in the case of push mode the mobile user, as an instance, might like to be informed as to if one of his relatives is located in the same area. In general, these services need location tracking of fixed, portable, and of mobile entities. Location information of all fixed entities are kept in a separate database area-wise while location tracking of portable and mobile entities could be performed on-demand. Upon entering in that location by a mobile user, the services list and location information is provided based on current preferences and/or the history of choices.

The present research focus on the location based services for mobile users with privacy and perceived risk issues using these service and problems associated with mobile database .

BACKGROUND

Adoptability of Mobile Services

The present research work has been aimed to cover the fissure of the familiar practice of mobile services and, in particular, effect of demographic characteristics on usage. The concepts of innovation and dissemination of innovation are even more complicated as technology and service aspects have an effect on the characteristics of mobile banking services (Thornton & White,2001). The innovation research within financial services (Black et al,2001) have applied Rogers' model to Internet banking. The precise infrastructure is required in location based services for positioning the mobile workstation which determines the location of the object in a reference system that can be cell division or path, a coordinate or address system (Ashban & Burney,2001). Location Based Applications are classified as device oriented or user-oriented.

Mobile User-Oriented Applications

These are applications where a service is based on mobile user. Examples of such applications are social networking where the aim is to locate friends or family in agreement of the mobile user.

Device-Oriented Applications

These are applications which are external to the mobile user and focus on the position of a mobile user. In lieu of only a mobile user, an object as a vehicle, or a group of persons as a fleet could be located. In device-oriented applications, the object or person located does not usually control the service. Car tracking is one of the example of this kind of applications for theft recovery where the car is sending information without human intervention (Renuka,2012).

Mobile Banking as Location Based Service

For mobile users, the mobile banking could be considered as the banking services facility for their mobile device as the operation of bank's 'savings' or 'current' accounts. The research findings and conclusion of various streams like academic world, different industries as media has shown their interest that the mobile devices, which was expected firmly recognized as an other option of payment in most technologically developed societies (Taga, Karlsson & Arthur,2004). Despite consistent efforts of mobile network operators, core bank and mobile payment service providers in promoting and offering various mobile payment services, the mobile customer acceptance of this innovation had shown the interest quite less in the adoptability payments through mobile as another option of payment mechanism (Karnouskos,2004).

While each of these players approach the market with different expectations, several studies have shown that merchant/consumer adoption is key to the success of mobile payments (Karnouskos,2004). For the smooth progress of micropayments in M-Commerce transactions and to encourage reduced use of cash at point-of-sales terminals, mobile payments as solution has been proposed(Mallat,2007). If efforts, which are being made in the promotion of mobile payments adoptability succeed, may be the destroyer service in 2G, 3G etc and will motivate M-Commerce to adopt the same (Jayawardhena & Foley,2000). The early development of mobile payment has jumped owing to the far above the ground dissemination rate of mobile phones and handheld devices. Mobile phones outnumber every type of mobile device today. In 2004, the Gartner Group anticipated that by 2008 more mobile phones are expected across the globe as compare to televisions and fixed land line phones and personal computers.

Indeed, M-Commerce provides several types of location based services such as steering, directory exploration and ticketing or permission for entry where the location for mobile consumer has main significance. Hence, location and personalized information are the few supplementary dimensions of M-Commerce value creation (Anckar & D'Incau,2002) (D'Roza & Bilchev,2003). These M-Commerce services are divided mainly two types:

a)   Those which are requested by mobile users as the their location is get awared.
b)   Auto generated once a specified condition is satisfied .

Technical Challenges in Mobile Uses

The technical requirement in mobile commerce is a satisfactory transmission network of payment systems between the sellers and buyers (Sharma,2011). The important area of mobile commerce is the security in mobile payment and network technology. Making commercial transactions through mobile phone requires high level of security (Zhang, Yuan & Archer,2002).

Different Technology Standards

A technology development and creation has started in Mobile wireless industry since 1970's. Mobile communication has grown up quite much in past few years in matching with the present day need. It has become an essential part for personal growth as also for the business and economy of the organizations. GSM has improved its services generation wise despite of wireless transmission is more complicated. Mobile networks could be differentiated from each other generation-wise from 1G to 4G.

The Future Roadmap: Fifth Generation (5G): A few new additions for the mobile users are ultra low latency, ultra high data speeds, new devices and form factors. The serious processing power has been added to the base stations for mobile edge computing. Unlike 4G which is driven using video, 5G would be driven by the internet while 5G network would be intelligent enough to understand as to how to allocate its capability and resources which are not currently available. That means the location based services, content aware services and user-aware services are expected in 5G. The lack of these three mean the wastage of precious bandwidth as the service would allocate more bandwidth where this least requires it. A regulatory framework would also be required to develop and to speed up this entire change (Tomar & Sharma,2014).

Adoptability Scopes of Mobile Commerce in 4G and 5G

Regardless of 3G services was launched in 2010 across the world, its penetration in India is hardly 10%. Like 3G, 4G has also not been quite meaningful so far. This has been over 5 years now since operators won the spectrum to launch 4G services but progress is yet to become visible fully. Merely a few have managed so far to launch 4G services significantly in big cities. India is lagging behind for the 4G adoptability. Challenges faced in launching 4G services are following:

1. Low cost 4G enabled handsets
2. Lack of technology ecosystem
3. Relevant content and services
4. Delayed government processes like permissions, etc.

The launch of 4G services has been delayed due to above mentioned challenges and penetration of 3G become more deeper. Besides this, there are some other challenges for operators like low tariff, elevated venture and a spirited market. Like India, most of the countries across the world are still struggling even with 4G which would be major hurdle the 5G may be years far away. However few countries like Japan which have 4G networks and ecosystems at grown-up stage, it would be easier to implement the same. Because the 5G environment is not possible to be created by any one company or group of companies it needs concerted efforts from all tech companies, service providers and governments. One of the basic prime challenge is in setting up a universal standard in order to ensure that everything is communicated in same language.

Application Implementation Threats (Khan & Barman,2015):

1. Since all the network operators and service providers would share a common core network infrastructure which may lead to collapse due to compromise of a single operator, if not carefully guided.
2. The masquerade attack is possible by third parties as the billing frauds can easily arise while legitimate users accessing the information.
3. Since 5G is a secure IP based solution it will be vulnerable to all the security threats as the current Internet world.

Mobile Cloud Computing and Its Impact on M-Commerce

"Mobile Cloud Computing" term was introduced after "Cloud Computing" in 2007. It is a novel paradigm for mobile applications in which the cloud take care of the most of the processing and data storage related to the applications of that mobile device. The centralized mobile applications over cloud further accessed over mobile internet with the support of mobile device web browser. However, the commercialization and mobile network communication capabilities are not fully leverage the powerful context in Mobile Cloud Computing model.

The Mobile Cloud Computing concept is based upon the principles of cloud computing, which brings attributes such as no on premise software, on demand access, and"XaaS i.e "Everything as a Service" within the mobile domain, adding further "Network as a Service" i.e."NaaS" and also Payment as a Service to the maximum of on demand capabilities and permit applications to leverage the full power in mobile networking. Mobile commerce is a growing trend among tablet users and smartphones. Finance and shopping apps are usually used because these enable easy access to financial information and accelerate payment processes. As more bank transactions from mobile devices are required thus the need of secure online gateways is a significant issue in cloud computing. Still, the most popular applications of mobile cloud is mobile commerce (Parsad et al,2012).

Mobile Database

Portability and mobility are latest challenges in the mobile database management and distributed environment. The support for mobile computing by the centralized database is still in the developing stage. Mobile computing is essential to design specifications to manage long periods of disconnection and also to manage other constrained resources in mobile computing like variable bandwidth, limited battery life etc. In mobile computing, shared data have big competition owing to its ability to access services and information through wireless connections that could be reserved even while the mobile user is moving. The shared data consistency assurance become more difficult due to constraints and limitations of wireless communication channel and data sharing by mobile users with others. The location is not the constraint for mobile users while accessing via wireless connection. The mobility, low bandwidth, disconnection, security risks and heterogeneous networks in mobile database systems make conventional database processing schemes no longer fully matched.

Disconnected Operation and Data Inconsistency

Disconnected operations support is necessary for mobile transactions as the moving host could frequently disconnect from the network and should not be preserved as a failure as traditional database. Replication has significant importance in distributed mobile computing environment due to mobile nature of a mobile host which do not remain connected to the remaining system all the time. Moreover a mobile host may move out of the cell range which causes disconnection while the data made available may result in conflicting partitioned sharing during disconnected operation. In case of a transaction system, the conflict operation is defined if both of them are in the same object and one of them at least is a write operation. The table presents the operations that do not conflict as 'No' while a 'Yes' entry indicates that operations do conflict.

The data made available in disconnected operation may result in conflicting partitioned sharing during disconnected operation. These conflicts are of following two types:

a. write/write, and
b. read/write

With *write/write* conflict, the data (object) is updated on a disconnected client and on the corresponding servers both. With *read/write* conflicts, the data (object) is updated in one partition (mobile host or corresponding server) and it is read on other. The conflict resolution process begins with successful detection of these conflicts. If partitioned sharing conflicts are not detected, then these cause data inconsistencies in different ways. The apparent disadvantage of data replication is about the overhead which is needed to maintain data consistency across multiple sites. In order to execute a transaction on a replicated data, the transaction manager must first translate every operation on a data item into an operation on one i.e. in the case of a *read* or, several i.e. in the case of a *write* operations of copies of that data item.

Data item availability in a mobile environment has been solved by replication, however the problem is the motivation of various replicas consistency maintenance of a data item. Under mobile computing, the mobility could be possible for a replica of a data item which is maintained by the mobile host. The availability of the replica on the mobile host is dependent upon whether or not the mobile host is disconnected or connected. Similarly in disconnected mode, the fixed host data maintained is not available to a mobile host.

Replication Protocols in Mobile Environment

Replication improves database availability and reliability in distributed database which means that certain data objects are redundantly and intentionally stored at multiple sites. In mobile environment, replication solves the problem about availability of a data item but data item consistency between different replicas of a data item is overstated. The mobile host dis-connectivity or connectivity is the basic reason for availability of replica on the mobile host. Similarly, data maintained at the fixed hosts during dis-connectivity mode is not available for a mobile host. The study extends further the evaluation framework of different replication models from various perspectives to stimulate a viable base or foundation of mobile data-sharing systems.

Two intense replication models for replica management exist: synchronous and asynchronous (Coulouris et al, 1994). Synchronous replication (Cavalleri et al,2000) updates the operations performed to all replicas at the same time. while in asynchronous replication instead of the updates at same time, the operations at one site performed and updated to be disseminate to other sites replica managers. The synchronous replication ensures the transmission bandwidth and the maximum data integrity with consistent availability of contributing sites in the transaction. (Salman & Lakshmi, 2010). Due to variation of time interval of database synchronization with in different service provider applications, the asynchronous replication is more flexible as compare to synchronous replication. Apart from that, mobile disconnections is very frequent as compare to failure and a transaction could work if remote server is not connected or down in asynchronous replication. The same is not applicable in synchronous replication (Salman & Lakshmi,2008). In ideal replicated mobile database system all three features adaptability, availability, and reliability are expected to be employed effectively.

VARIOUS PROBLEMS ASSOCIATED WITH M-COMMERCE ADOPTABILITY

Customer Trust and Associated Issues for low Adoptability of M-Commerce

Various problems associated with the wireless network are following:

1. Limitations in bandwidth
2. Connection stability
3. Function predictability

These networks lack a standard protocol and have comparatively high operation costs besides the data transmitted wirelessly is more prone to losing the confidentiality in a private conversation and therefore, there is short of trust in adoptability of Mobile Commerce. Some of such major issues are described below:

Reliability and Security Issues

To build online trust as significant for M- Commerce particularly in early stages, the reliability of data and security are required. As the unsatisfactory performance of the wireless communication system makes customer distrustful, the mobile technology focus is likely to shift from producing customer trust in vendors.

The mobile technology trust and vendor trust are equally important to secure the customer trust (Sharma, 2011). Building customer's trust on M- Commerce is a unremitting practice which widen when the trust built up at the initial step. Mobile technology and mobile vendors are necessary framework elements to generate the customer trust. In process to get better trust in mobile technology, technical obstacles should be overcome. Generally, Location Based Services integrate a mobile device's location with other information in order to provide added value to the user (Bouwman et al,2007) (Barbará,1999). Location Based Services have become a worldwide fact due to progress in Wireless Communication Technology and increasing numbers of mobile users(Chen et al,2002). The service provider facilitate mobile users by location awareness with certain degree of accuracy (Constantinides,2002) in Location aware Services of M-Commerce. And due to their location disclosure, mobile users compromises their privacy of personal information and increase the security risk. (Fox,2000) which reflects the potential losses associated with the release of personal location information to the Location Based Services provider.

Perceived Risk, Privacy and Trust

Perceived risk in Mobile Commerce transactions has been observed in several studies due to possibilities of frauds and product quality dis- satisfaction by mobile customer. Unauthorized access and privacy of the mobile user personal information are the major factors of the perceived risk in Mobile Commerce. Thus, perceived risk correspond to a negative consequence on mobile customers to use M – Commerce deliberately (Tomar & Sharma,2014). In Location Based M-Commerce services especially in Mobile Banking, the service provider provides services to mobile users like their locations with a certain level of granularity to maintain a degree of secrecy (Tomar & Sharma,2014).

In banking services, particularly the perceived risk associated with the financial product itself as is high and with consumer goods electronic delivery channel which has increased the innovation significance (Harrison,2000). Confidentiality and authenticity various security aspects are essential and considered as the basic fundamental requirements to access the sensitive information from bank(Jayawardhena & Foley,2000).

Mobile Transaction Security

One of Mobile Commerce services is considered as mobile payment. It is defined as the financial transaction process of goods or services through a mobile device between two parties. Mobile payment security is a prime consideration which might be challenged during the sensitive payment information transmission or handling. Authentication of intended payment request, confidentiality of transaction details, integrity and non-repudiation are the essential security aspects for secure transactions (Park & Song,2001):

a. **Authentication:** It is about the verification of the identities of parties in a communication.
b. **Confidentiality:** It is about the privacy and secrecy of the transaction over network to ensure that a message or contents of transaction transmitted over network has been read by only intended receiver.
c. **Integrity:** Integrity means to ensure that the no alteration has been in contents of the message during transaction over network and same contents has been received by the intended receiver.
d. **Non-Repudiation:** It is a mechanism to ensure that alleged mobile user involved in a transaction cannot deny later that he was not the intended user in that transaction.

Security Issues of Location Based Mobile Financial Transaction

A mobile user often needs the location information while it roams under another service provider whose network might or might not have the same location performance. Other additional issues like privacy and security issues, service interoperability and ownership of location information are considered in such scenario. Owing to this reason, some mobile users are vulnerable to privacy and security issues. A robust transaction security is necessarily required in Mobile financial transaction. WAP and SMS based banking services are the common services made use of by the mobile user in location based mobile financial transactions.

SMS Banking

Mobile user finds out the details about their account balance and gets desired data in SMS banking. Despite its few features like ease to use, suitability, cost efficiency and fast exchange of messages, the challenges are to be overcome for wide acceptance of SMS based payment transaction (Basudeo & Jasmine,2012) as there is no encryption technique which can be applied for sending and receiving SMS. Therefore, the data are not secure while transmitted through the SMS which is a store and forward service where message/s is/are not directly sent to the recipient but through a network SMS centre(Basudeo & Jasmine, 2012).

Security Challenges in SMS Banking Transactions

In general, initially in GSM network architecture the security aspects like authenticity of mobile user, confidentiality of data, end point to end point security were ignored and mobile user were intended for SMS usage only. (Manoj & Bramhe,2011). The spoofing attack is major issues with SMS banking under which replay of SMS message sends out by malicious mobile u which appears to be sent by the intended sender. Although in recent trends SMS architecture secures the original address of sender by altering respective fields in original SMS header whereas the SMS has encryption only during path originating from base trans-receiver station and upto mobile station. Despite the complex cryptography, a hacker may get the password from the device stolen.

WAP Based Mobile Transactions

Using WAP, mobile users access internet and other networking services via mobile WAP, which gives access to an indefinite amount of information for all time. This could be expanded but has uncertainty of use while connected. This enables wireless communication and Mobile Commerce where Internet data moves to and fro from wireless devices like smart phones, mobile phones, etc. Presently encryption process is made use of to secure data transmission between bank and users, however, the problem faced with is that this encryption process is not effective enough to secure the protection of sensitive data between the customer and bank. Internet banking is much more powerful in computer systems and have well defined complex encryption process to make sure security.

Indian Bank offers WAP based Mobile Banking services to mobile users who have GPRS enabled facility in their mobile phones. The account details could be accessed through WAP based mobile banking while accessing their phone. Various account related enquiries, transfer of funds - both inter-bank and intra-bank are supported by WAP based mobile banking for customers' convenience at 24 * 7 round the clock. This is secured through login password i.e. MPIN and OTT - two-factor authentication meant for funds transfers. The present mobile banking implementations which use WAP have proven to be very secure but there exist certain ambiguity which could lead towards insecure communications, which are as follows:

a.   No end point -to-end point encryption security between client and bank server.
b.   No end point -to-end point encryption security between the client and the Gateway.
c.   No end point -to-end point encryption security between the Gateway and the Bank Server.

Security Challenges using WAP in Mobile Banking Transactions

For mobile transactions the essential requirement is end point to end point security. It means that the mobile device which is used for mobile banking by mobile user and the data transacted through it are secured only on the core Bank and not at the mobile User end and therefore, more probability of attacks due to data vulnerability. Using wireless application protocol, end point –to-end point security is difficult most of the time. The main reason for the probability of attack is due to no encryption of data for data confidentiality at gateway during the protocol switching process (Narendiran et al,2009). The vulnerability to hacker's attacks over WAP during protocol switching process besides the compression of contents.

The Wireless Applicational Security moel comprises the Web server, internet, WAP gateway, wireless networkand WAP enabled handset where the SSL works only between Web Sever and WAP gateway with Public Key Infrastructure and WTLS over wireless network between WAP gateway to WAP-enabled handset which provides secure authentication and End-to End Transport Layer Security. It provides functional similarity to the internet transport layer security systems:

a. Transport Layer Security (TLS)
b. Secure Sockets Layer (SSL)

In case of WAP model, all the applications and contents are based on world wide web format which implements the wireless application security. Data transportation is made by using some standards of the protocols of communication of world wide web. In spite of security support added at middle ware such as WAP end to end security is still a problem. In case of financial application, wireless PKI is a system meant to manage key and certificate, which is used to authenticate and get digital signature from the mobile users. But failure of a wireless infrastructure affects greatly the M-transaction failure of HLR/VLR that stores approximately the location of mobile user (Sharma, Kansal & Tomar,2015). Various security risks at WAP and server are following:

Security Risks at WAP

a. Not encrypted data over gateway while switching of protocol process
b. Attacker can access unencrypted access
c. Eavesdropping attack
d. Malicious software
e. Man in middle attack

Security Risks at Server

a. Server failure
b. Virus attack
c. System crash

It is critical for a mobile user with wireless data services to access Location Based Services (LBS).

SOLUTIONS AND RECOMMENDATIONS

Customer Trust in Mobile Commerce

The mobile users get the services like about their locations with the support of service provider provides with some intensity of granularity to preserve a degree at level of secrecy in location based Mobile Commerce services. This level of granularity is dependent upon their perceived risk A in addition to the incentives the service providers receive as monetary benefits for improved Mobile Commerce services. Factors of perceived risks include unauthorized access besides the unconscious computing derived from mobile applications. Therefore, a negative effect of perceived risk has generated on mobile users about Mobile Commerce. Different perceptions for mobile users about LBS services exist as these are 'value added services' than these are 'must have services' considering their accessing cost. Mobile user may consider LBS services to be more attractive if mobile network operators permit the wireless data services more affordable or free for mobile users. Hence, the location aware services success is largely dependent upon the cost of wireless data services. In general, with the change of mobile user's location a previous query is not valid. As a result, the user may have to avoid doing the broadcast channel repetitively (Lee et al,2002). It, therefore, significantly increases not only the access time but also the access cost involved.

Following three factors have been considered which reflect the cost of Mobile Commerce due to cell change of mobile user and how Mobile Commerce could be better utilized by a mobile user:

a.   Location,
b.   Volume and
c.   Time

If these three parameters are maintained then the cost of Mobile Commerce could be reduced further as also the services could be made much attractive. Caching and pre-fetching have been adopted (Sharma & Tomar,2013) to increase the performance of information transformation in mobile requirement and the quality of LBS could be further updated to reduce the congestion.

Evaluation Framework of Replication Protocols

In order to determine the imperfections and boundaries of different replication approaches, the framework compares the quality of different significant features from a mobile perspective which includes: availability, adaptability and reliability. The sustainability of semantics of information is also as equally important as the availability is a considered as main goal (Sharma & Kansal,2013).

1.   Availability: This ascertains the probability for requested data to be ready for access which also includes responsivity and performance of data.
2.   Reliability: This determines the consistency level and fault–tolerance of the data access.

3.   Adaptability: This provides reasonable use of shared resources and connectivity in dynamically evolving environment. In case of high adaptability the manual interference and preconfigured information to have balanced control of data flow and shared resources is not needed.

Evaluation Criteria

The main goal is to analyze the replication in mobile environment for data access highly available. In numerical evaluation and comparison, it is necessary to make categorizations and discrete boundaries so that it may allow grading. The grading criteria for availability adaptability and reliability has been used as scale of One to five.

1.   **Availability**: Data access availability is divided into two transactions viz. the update requests and the read requests. Also different amount of network resources are utilized by these type of requests which lead to specific effects on the managing of the replica system. These are enforced usually in wide varying methods. Generally, update requests cannot adversely read requests at the level of availability. Therefore, in this evaluation framework, update transactions has only been considered as they determine the lowest possible availability. The five scales which help in the evaluation of availability of data access are following:
   a.   Poor grade: It is given to the weakest possible availability of replica model and is necessary to contact each replica manager before it is possible to update data.
   b.   Low availability: Data access is possible without be in touch with every replica manager. Certain fixed groups or group have the authority for granting data update access.
   c.   Average grade: It needs still more complexity and availability and it gives access to data even if certain changes are made into the environment. Dynamic group membership for authoring of data are included into this category.
   d.   Good availability: It is given in case it is possible to utilize data on dynamic network partitioning except in the most hostile situations.
   e.   Excellent availability: It is the highest assigned ranking and describes that data is available for access always, if any copy could be found.

2.   **Adaptability:** Under mobile environment where displacement in host and unreliable connections play main role the network is in an evolving stage consistently and intervention manually could be used to modify configurations

to provide temporal operability, however, they are slow and expensive to use. Henceforth, automatic adaptability to the changes in an environment is crucial. The five scales which help in evaluation of adaptability are following:

a.   Poor adaptability: It describes a system which is completely preconfigured and does not adapt to any changes in the environment.

b.   Low adaptability: It is given for minor development in the environment and does not hinder the functionality of a system while the control model is still as preconfigured.

c.   Average adaptability: Despite of self-configuration features some manual configuration also required.

d.   Good adaptability: It ensures that it is possible to make a system able to adapt to environments without manual intervention. Such environments could not be too complex and heterogeneous.

e.   Excellent adaptability: It does not need manual configuration and could adapt to any environment.

3.   **Reliability:** This is necessary to maintain the semantics of data whereas the methods utilized by the application has been the criteria for judging the reliability and it may be possible to give the essential level of reliability for appropriate operation for lower consistency guarantee by limiting the freedom of applications. The five scales which help the evaluation of reliability are following:

a.   **Poor reliability:** It implies that consistency during data access is not guaranteed while update on any data item is possible either locally or in the network.

b.   **Low reliability:** It ensures some amount of consistency in data access and most but not every data item is available for mobile users according to the replica model.

c.   **Average reliability:** It must forward in terms of availability for the mobile users that are in a consistent state. Nevertheless this does not support inconsistency as generally, inconsistency is not permitted in a data sharing system.

d.   **Good reliability:** It states that at times some inconsistent information access is possible by the mobile user.

e.   **Excellent reliability**: It ensures that all the available data items are consistent with the latest of versions. In general, only excellent reliability is acceptable in database. However alternatively it might be possible to give data sharing for mobile users without strict consistency need for data access.

Replica Control Protocols

Certain present replication protocols deficiencies and advantages from the perspective of mobile computing have been highlighted in this research work. These protocols have been categorized as asynchronous and synchronous protocols.

Primary Copy Replication

This technique forwards and executes all transactions by the primary copy while all rest replicas are merely its backups or, are secondary and apply updates in writing transactions before are committed in primary.

Table 1. Evaluation of Primary Copy (Sharma & Kansal,2013)

Criteria	Implementation method	Success
Availability	To increase availability of data, read access is available if a connection to any slaves exists.	Poor
Adaptability	In case of failure of primary server, one among the slaves has a possibility to claim status of the master.	Low
Reliability	The primary replica manager controls data integrity and updates on slaves. A possibility for conflicting updates does not exist.	Excellent

Dynamic Voting

Availability, reliability and fault tolerance are the main features of dynamic voting. (Dolev, Keidar & Lotem,1997).

The tables 2 clearly present that dynamic voting protocol are better than primary copy and static voting algorithms in terms of adaptability and availability. The availability and adaptability are much higher with no negative impact on reliability.

Table 2. Evaluation of dynamic voting (Sharma & Kansal,2013)

Criteria	Implementation Method	Success
Availability	Allow the system to adopt its quorum requirements according to state of the system.	Average
Adaptability	Adjust proper weights, group selections and observers change in the network environment.	Average
Reliability	It offers similar level of reliability as primary copy.	Excellent

Lazy Replication

A replica does not always connected consistently to the rest of the system in mobile applications. Thus waiting for updation and dissemination to all replicas is not a better solution (Golding,1997). The asynchronous replication is used when the performance is the primary goal. Lazy replication propagates the updates at one replica and then further replicas lazily exchanges new information via gossip messages.

Table 3. Evaluation of Lazy Replication (Sharma & Kansal,2013)

Criteria	Implementation Methods	Success
Availability	Permits transmission swiftly.	Excellent
Adaptability	Weak consistency and serializability due to dynamic change in network.	Average
Reliability	Reliability is low due to dependency on central node which generates the sequence order support for periodic updation.	Low

Grapevine

Grapevine is an asynchronous replication protocol to fulfil eventual consistency. While a node executes the update, a timestamp is related to each updated item and then the node makes use of an unreliable multicast in propagating that update to rest other nodes.

Table 4. Evaluation of Grapevine (Sharma & Kansal,2013)

Criteria	Implementation Methods	Success
Availability	Anti-entropy ensures all the updates until they are mutually consistent.	Good
Adaptability	Update propagation depends on group especially on anti–entropy.	Low
Reliability	Ensure reliability by spanning all copies at each comparison.	Excellent

Time Stamped Anti-Entropy

An asynchronous protocol, time stamped anti-entropy avoids synchronous communication, (Golding,1992). As a site is partitioned from the rest of the network, it continues to give service and receive updated information once it reconnects.

Therefore a time stamped anti-entropy is a group communication protocol which provides reliable and final delivery and which delivers messages to each process into the group even when processes fail temporarily or are disconnected from network.

This asynchronous protocol gives high availability of data owing to group communication .

Table 5. Evaluation of timestamped anti-entropy (Sharma & Kansal,2013)

Criteria	Implementation Method	Success
Availability	Uninterrupted service in network partition.	Good
Adaptability	*Adaptability* is achieved by restricting the number of strict transactions.	Good
Reliability	Group communication, stable storage.	Excellent

Secure Mobile Commerce and Adoptability of M-Commerce recommendations

In Mobile Commerce, the information is transmitted through mobile internet which may include the customers' personal detail, the order information, payment detail and so on. These details are to be kept private from unauthorized access. Therefore, security requirements in Mobile Commerce essentially should include the following aspects:

Network Access Security: For Secure access of the 3G services by mobile users with appropriate protection from the attacks in the (radio) access link, the group of security features needs to be added.

Network Domain Security: To exchange secure signals of data and safeguard from attacks over connected network in the service provider domain, various security options are needed.

Configurability and Visibility of security: Features are required which facilitate the user with security feature to check whether it is operational or not.

Application domain security: To exchange messages as secure, the group of security features are needed which are enabled in applications of the user along with service provider domain.

User domain security: Mobile station access security features are needed.

For this all stakeholders like Regulators, Government, Telecom Service Providers and Mobile device manufactures need make all out efforts to encourage penetration of Mobile Commerce widely in metros, towns and rural areas as also from high-end to low-end users

FUTURE RESEARCH DIRECTIONS

To strengthen the security further during mobile transactions a security model may be developed to include the risks of various possible threats and attacks on the mobile devices and, the vulnerable points as in communication channels, web browser, mobile operating system thereby which may verify the Mobile Commerce security based authentication methods. The latest positioning technologies like GPS or Global Positioning System permit companies to offer services and goods to the mobile users based upon the current location of the mobile user which could as well be included further. There exist further scopes to analyze the adoptability of GPS in Location based Mobile Commerce services.

CONCLUSION

It is widely recognized that mobile phones have huge potential to conduct financial transactions and lead the financial growth with much reduced costs and lots of convenience. It is also an established fact that technology advancement plays an essential role in Mobile Commerce adoption. Taking advantage of technology and further improving upon shortcomings in present technical conditions the efficiency of Mobile Commerce model need be improved considerably. Considering this, the research was undertaken to study different technology standards for Mobile Commerce and accordingly the factors that affect Mobile Commerce adoption, in turn, which were concerned with the service quality and communication costs of mobile transactions and the trust of mobile users in Mobile Commerce were explored, evaluated and analyzed.

The basic factors that affect Mobile Commerce adoption primarily are Trust and its Service quality. Trust among players as for authorization and payment are key aspects in the Mobile Commerce infrastructure while quality of service ensures that the services offered are able to meet the consumers' expectations. Mobile customers need security before accessing mobile transactions. The research has thus focused upon the security issues related to the distrust of customer for Mobile Transactions. Mobile Transactions are based upon Location Based Services, especially related to mobile financial transactions as SMS banking transactions and WAP based transactions which include the authentication process and security aspects in mobile transactions.

Generally the mobile user is not fully knowledgeable about the technology being provided, henceforth, the research has strived to explain different mobile technologies and in-depth analyses of such technologies. It has been concluded that mobile and related security technology could be provided the much needed security capability in order to protect mobile transactions and services. However the end-to-end security threats and attacks are still possible. Besides various technology standards, interoperability among different applications and the mobile devices also affect the convenience to adopt Mobile Commerce by mobile users. Therefore, ease of use should be taken into account while analysing the applicability of Mobile Commerce as also about secure technologies in mobile transactions. Therefore, Mobile Service Providers need make aware mobile users in respect of the secure technology.

Further as a mobile user with wireless data services, it is critical for the mobile user to access location aware services. However in location based Mobile Commerce services, the service provider gives service to mobile users as their locations with some level of granularity so as to maintain a degree of secrecy. This level of granularity depends upon the perceived risk as also on the incentives received by the service providers in terms of monetary benefits or improved Mobile Commerce services. Thus, perceived risk has a negative effect on intention of mobile users to use Mobile Commerce. Therefore in order to build trust in the mobile users as also to reduce the access cost of the Mobile Commerce, all service providers come forward to get some concrete solution. Further, the communication cost depends on wired, wireless cost of mobility of mobile clients and frequent disconnection which increases the battery consumption and other maintenance costs. Frequent mobility from one cell to other, Hit ratio, probability of accessing the data from Local data server and central server also increase the communication cost.

REFERENCES

Al-Ashban, A., & Burney, M. A. (2001). Customer adoption of tele-banking technology: The case of Saudi Arabia. *International Journal of Bank Marketing*, *19*(5), 191–204. doi:10.1108/02652320110399683

Anckar, B., & D'Incau, D. (2002). Value creation in mobile commerce: Findings from a consumer survey. *Journal of Information Technology Theory and Application*, *4*(1), 41–62.

Barbará, D. (1999). Mobile Computing and Databases - A Survey. *IEEE Transactions on Knowledge and Data Engineering*, *11*(1), 108–117. doi:10.1109/69.755619

Basudeo, S., & Jasmine, K. (2012). Comparative Study on Various Methods and types of mobile payment system. *Proc. International Conference on Advances in Mobile Network, Communication and Its Applications*, 10.

Black, N. J., Lockett, A., Winklhofer, H., & Ennew, C. (2001). The adoption of Internet financial services: A qualitative study. *International Journal of Retail & Distribution Management, 29*(8), 390–398. doi:10.1108/09590550110397033

Bouwman, H., Carlsson, C., Molina-Castillo, F. J., & Walden, P. (2007). Barriers and drivers in the adoption of current and future mobile services in Finland. *Telematics and Informatics, 24*(2), 145–160. doi:10.1016/j.tele.2006.08.001

Cavalleri, M., Prudentino, R., Pozzoli, U., & Veni, G. (2000). A set of tools for building PostgreSQL distributed database in biomedical environment. *Proc.* 22ⁿᵈ Annual *International conference on Engineering in Medicine & Biology*, 540-544. 10.1109/IEMBS.2000.900796

Chen, P. Y., & Ht, L. M. (2002). Measuring switching costs and the determinants of customer retention in Internet-enabled businesses: A study of the online brokerage industry. *Information Systems Research, 13*(3), 255–274. doi:10.1287/isre.13.3.255.78

Constantinides, E. (2002). The 4S Web-marketing mix model, Electronic Commerce Research and Applications. *Elsevier Science, 1*(1), 57–76.

Fox, J. (2000). A river of money will flow through the Wireless Web in coming years. All the big players want is a piece of the action. *Fortune, 142*(8), 140–146.

Harrison, T. (2000). *Financial Services Marketing*. Prentice Hall.

Jayawardhena, C., & Foley, P. (2000). Changes in the Banking sector - the case of Internet Banking in the UK'. *Internet Research: Electronic Networking Applications and Policy, 10*(1), 19–30. doi:10.1108/10662240010312048

Karnouskos. (2004). Mobile Payment: A Journey through Existing Procedures & Standardization Initiatives. *IEEE Communications Surveys & Tutorials*, 44-66.

Khan, M. H., & Barman, P. C. (2015). 5G - Future Generation Technologies of Wireless Communication. Revolution 2020. *American Journal of Engineering Research, 4*(5), 206-215.

Lee, D. L., Lee, W.-C., Xu, J., & Zheng, B. (2002). Data management in location-dependent information services: Challenges and issues. *IEEE Pervasive Computing, 1*(3), 65–72. doi:10.1109/MPRV.2002.1037724

Mallat, N. (2007). Exploring Consumer Adoption of Mobile Payments - A Qualitative Study. *The Journal of Strategic Information Systems, 16*(4), 413–432. doi:10.1016/j.jsis.2007.08.001

Manoj, V. & Bramhe. (2011). SMS Based Secure Mobile Technology. *International Journal of Engineering and Technology, 3*(6), 472-479.

Narendiran, C., Albert Rabara, S., & Rajendran, N. (2009). Public Key Infrastructure for Mobile Banking Security. *Global Mobile Congress*, 1-6. 10.1109/GMC.2009.5295898

Park, N. J., & Song, Y. J. (2001). M-Commerce security platform based on WTLS and J2ME. *International Symposium on Industrial Electronics (ISIE).*

Prasad, M. R., Gyani, J., & Murti, P. R. K. (2012). Mobile Cloud Computing: Implications and Challenges. *Journal of Information Engineering and Applications, 2*(7), 7 - 15.

Renuka, B. (2012). Location Based Services on Mobile E-Commerce. *International Journal of Computer Science and Information Technologies, 3*(1), 3147–3315.

Salman, A. M., & Lakshmi, R. (2008). Disconnected Modes of Operations in Mobile Environments. *Proc. INDIACom-2008, 2nd National Conference on Computing for Nation Development*, 253-256.

Salman, A.M. & Lakshmi, R. (2010). Replication Strategies in Mobile Environments. *BVICAM'S International Journal of Information Technology, 2*(1).

Sharma, A. (2011). Adoption Analysis of Mobile banking in India. In *Proc. International Conference SPIN-2011 (Speech Processing and Integrated Networks).* Amity College of Engineering.

Sharma, A. (2012). M-Commerce Technology Adoption and Trust Challenges. *International Journal of Scientific and Research Publications, 2*(2), 1-5.

Sharma, A., & Kansal, V. (2013). An Evaluation Framework of Replication Protocols in Mobile Environment. *International Journal of Database Management Systems, 5*(1), 45-51.

Sharma, A., Kansal, V., & Tomar, R.P.S. (2015). Location Based Services in M-Commerce: Customer Trust and Transaction Security Issue. *International Journal of Computer Science and Security, 9*(2), 11–21.

Taga, K., Karlsson, J., & Arthur, D. (2004). Little Global M-Payment Report. Academic Press.

Thornton, J., & White, L. (2001). Customer Orientations and Usage of Financial Distribution Channels. *Journal of Services Marketing, 15*(3), 168–185. doi:10.1108/08876040110392461

Tomar, R.P.S., & Sharma, A. (2014). Mathematical Model to Study the Cost Effects and Mobile-Users Trust on Location based Data Access in Mobile-Commerce Transactions. *International Journal of Computer Applications, 105*(12), 22-26.

Zhang, J., Yuan, Y. F., & Archer, N. (2002). Driving Forces for M-Commerce success. Michael G. De Groote School of Business, McMaster University. doi:10.1007/0-306-47548-0_4

KEY TERMS AND DEFINITIONS

EDI: Electronic data interchange is process to exchange the business data through electronic means or inter communication of official documents through computer instead of paper.

GSM: Global system for mobile is digital communication technology used for data transmission and mobile voice transmission service.

GPRS: General packet radio service is cellular network service over packet-switching communication with the progress of 2G GSM evolution.

PKI: Public key infrastructure is set of roles or system of processes to support hardware, software, revoke and distribution of digital certificates, keys and cryptographic algorithms, etc.

Replication: It is a process of data storing at more than one database server to maintain the consistency across all database servers.

SSL: Secure socket layer is security protocol in a TCP/IP Model for encrypted communication between web client and web server by enabling the cryptographic algorithm at transport layer.

TLS: Transport layer security is a security protocol at transport layer in TCP/IP model to provide the end-end communication security over internet.

WAP: Wireless application protocol is a specification designed to access information via wireless devices like mobile phone over internet. WAP uses micro browsers to access files over internet on handheld device with low bandwidth constraints.

Chapter 10
Software–Defined Networking (SDN):
An Emerging Technology

L. Naga Durgaprasad Reddy
Sri Yerramilli Narayanamurthy College, India

ABSTRACT

This chapter researches in the area of software-defined networking. Software-defined networking was developed in an attempt to simplify networking and make it more secure. By separating the control plane (the controller)—which decides where packets are sent—from the data plane (the physical network)—which forwards traffic to its destination—the creators of SDN hoped to achieve scalability and agility in network management. The application layer (virtual services) is also separate. SDN increasingly uses elastic cloud architectures and dynamic resource allocation to achieve its infrastructure goals.

I. INTRODUCTION

In the early days of basic Internet protocols development no native support for access control was provided at the network level. It was expected that applications would connect to each other in the global network without any restrictions. Along with the growth of commercial use of Internet mechanisms for L3 (and higher) network access control became necessary for normal operations, and packet filtering solutions were developed (including software implementations in operating systems) — firewalls, Intrusion Prevention Systems (IPS), network anti viruses (V. Yazici, M. O. et.al, 2014).

DOI: 10.4018/978-1-5225-9493-2.ch010

In terms of client devices mobility, network configuration is changing rapidly and the information about network topology changes could not be used directly for access control. That is why the problem of network access control based on the information about the expected behavior (flows) of network applications is becoming more and more important.

SDN is a step in the evolution towards programmable and active networking and allows network administrators to have programmable central control over the entire network.

II. SOFTWARE DEFINED NETWORKING

Software-defined networking (SDN) is a new emerging technology for networking in which control is Decoupled a hardware and given to software part called a controller[1]. When a packet arrives at a switch in a foreseeable network rule built into the switch patented firmware tell the switch where to forward the packet. The switch sends every packet going to the same destination along the same path and treats all the packets the exact same way. In the campus network, smart switches designed with application-specific integrated circuits (ASICs) are Refined enough to recognize different types of packets and treat them differently, but such switches can be quite expensive (H. Kim and N. Feamster,2013).

The aim of SDN is to allow network administrators respond quickly to changing to the requirements. In a software-defined network, a network administrator can shape traffic from a centralized control software without having to touch individual switch. The administrator can change any network switch rules when necessary ordering, de-ordering or even blocking specific types of packets with a very level gritty of control (M, Algarni, 2013). Currently, the most popular specification for creating a software-defined network is an open standard called Open Flow. Open Flow lets network administrators remotely control routing tables.

A . *Open Flow Switches*

Open Flow provides an open protocol to program the flow table in different switches and routers. A network administrator can partition traffic into production and research fellows. Researchers can control their own flows by choosing the routes their packets follow and the processing they receive. In this way, researchers can try new routing protocols, security models, addressing schemes, and even alternatives to IP. On the same network, the production traffic is isolated and processed in the same way as today. The data path of an Open Flow Switch consists of a Flow Table, and an action associated with each flow entry. The set of actions supported by an Open

Flow Switch is extensible, but below we describe a minimum requirement for all switches. For high-performance and low-cost the data-path must have a carefully prescribed degree of flexibility. A Flow Table, with an action associated with each flow entry, to tell the switch how to process the flow.

By specifying a standard interface the Open Flow Protocol through which entries in the Flow Table can be defined externally, the Open Flow Switch avoids the need for researchers to pro-gram the switch. It is useful to categorize switches into dedicated Open Flow switches that do not support normal Layer 2 and Layer 3 processing, and Open Flow-enabled general purpose commercial Ethernet switches and routers, to which the Open Flow Protocol and interfaces have been added as a new feature.

B. Open Flow Controllers

The open Flow controller is a bid that manages flow mechanism in a software-defined networking (SDN) environment. The present SDN controllers are based on the Open Flow protocol where SDN controller serves as a sort of operating system (OS) for the network. Entirely communications between applications and devices have to go through the controller. The Open Flow protocol connects controller software to network devices so that server software can tell switches where to send packets. The controller uses the Open Flow protocol to configure network devices and choose the best path for application traffic. Because the network control plane is implemented in software, rather than the firmware of hardware devices, network traffic can be managed more dynamically and at a much more granular level.

III NETWORK MANAGEMENT PROBLEMS

In numerous respects the large enterprise networks of today are reminiscent of the islands of automation that were common in manufacturing during the 1980s and 1990s the experiment in front of manufacturers was in linking together the islands of microprocessor-based controllers, PCs, minicomputers, and other components to allow end-to-end actions such as aggregated order entries leading to automated production runs. Unique way from error-prone, tedious, physically intensive operations to software assisted automated end to end operations. The network operators needed to execute automated end to end management operations taking place their networks. An sample of this is Virtual local area ne management in which an NMS GUI provides a visual picture such as a cloud of MAC addresses, VLAN members ports, VLAN IDs. The network management system can also provide the ability to easily delete, modify and add virtual local area network members as well as indicate any faults as and when they (Tom Anderson, 2008) . An additional example is enterprise WAN management in which ATM or FR virtual circuits are used to carry the traffic from branch offices into central sites. In this situation the enterprise network manager wants to be able to easily create, delete, modify, and view any faults on the virtual circuit to the remote places. Supplementary examples include storage including SANs management and video/audio conferencing equipment management. The problems presenting menu options appropriate to a given selected NE provides abstraction for example if the user wants to add a given NE interface to an IEEE 802.1Q VLAN then that device must support this frame tagging technology. The Network management system should be able to figure this out and present the option only if the underlying hardware supports it. Through awarding only appropriate options the NMS reduces the amount of data the user must sift through to actually execute network management actions.

Overview of network problems

At hand some serious problems affecting network management. A taking beside managed data and code together is one of the central foundations of computing and network management. Accomplishing this union of data and code in a scalable fashion is a problem that gets more difficult as networks grow. Management information base tables expand as more network resident managed objects such as virtual circuits are added in managing additions to large Management information base tables (Saurav Das et.al, 2013). The grander than before size networks is matched by ever more dense devices. The originators of management systems need a rarified skill set that matches the range of technologies embedded in NEs and network supplementary emphasis is needed on solutions than on technology particularly as the components of the technology are combined in new and complex ways for instance in layer 2 and layer 3 VPNs. Resolutions should try to hide as much of the underlying network complexity as possible. Network management system technology can help in hiding unnecessary complexity. The liberal use of standards documents and linked overviews are some important tools for tackling the complexity of system development managed object derivation and explanation. Ethics documents can be used in conjunction with UML to inform and open up the development process to stakeholders.

IV LIST OF OPEN FLOW SOFTWARE'S

Ns3 is the best simulation tool for research on implementation of SDN & too for making the performs analysis on SDN.

Switch Software and Stand-Alone Open Flow Stacks

- **Open vSwitch**: (C/Python) Open v switch is a an Open Flow stack that is used both as a vs witch in virtualized environments and has been ported to multiple hardware platforms. It is now part of the Linux kernel (as of 3.3).
- **OpenFlow Reference**: (C) The OpenFlow reference implementation is a minimal OpenFlow stack that tracks the spec.
- **Pica8**: (C) An open switch software platform for hardware switching chips that includes an L2/L3 stack and support for OpenFlow.
- **Indigo**: (C) Indigo is a for-hardware-switching OpenFlow implementation based on the Stanford reference implementation.
- **POX**: (Python) Pox as a general SDN controller that supports OpenFlow. It has a high-level SDN API including a queriable topology graph and support for virtualization.
- **MUL**: (C) MūL, is an openflow (SDN) controller. It has a C based muli-threaded infrastructure at its core. It supports a multi-level north bound interface for hooking up applications. It is designed for performance and reliability which is the need of the hour for deployment in mission-**NOX**: (C++/Python) NOX was the first OpenFlow controller.
- **Jaxon**: (Java) Jaxon is a NOX-dependent Java-based OpenFlow Controller.
- **Trema**: (C/Ruby) Trema is a full-stack framework for developing OpenFlow controllers in Ruby and C.
- **Beacon**: (Java) Beacon is a Java-based controller that supports both event-based and threaded operation.
- **Floodlight**: (Java) The Floodlight controller is Java-based OpenFlow Controller. It was forked from the Beacon controller, originally developed by David Erickson at Stanford.
- **Maestro**: (Java) Maestro is an OpenFlow "operating system" for orchestrating network control applications.
- **NDDI - OESS**: OESS is an application to configure and control OpenFlow Enabled switches through a very simple and user friendly User Interface.
- **Ryu**: (Python) Ryu is an open-sourced Network Operating System (NOS) that supports OpenFlow.
- **NodeFlow** (JavaScript) NodeFlow is an OpenFlow controller written in pure JavaScript for Node.JS.
- **FlowScale** FlowScale is a project to divide and distribute traffic over multiple physical switch ports. FlowScale replicates the functionality in load balancing appliances but using a Top of Rack (ToR) switch to distribute traffic.
- **NICE-OF** NICE is a tool to test OpenFlow controller application for the NOX controller platform.

- **OFTest** OFTest is a Python based OpenFlow switch test framework and collection of test cases. It is based on unittest which is included in the standard Python distribution.
- **Mirage** Mirage is an exokernel for constructing secure, high-performance network applications across a variety of cloud computing and mobile platforms. Apparently, it supports OpenFlow.
- **Wakame VDC** (Ruby) IaaS platform that uses OpenFlow for the networking portion.
- **ENVI** ENVI is a GUI framework that was designed as an extensible platform which can provide the foundation of many interesting OpenFlow-related networking visualizations.
- **NS3** (C++/Python) NS3 is a network simulator. It has openflow support built in to emulate an openflow environment and also it can be used for real-time simulations.

V SDN STRUCTURE

The logical structure of a Software defined networking is explained as below. The main controller gets done all difficult functions, together with security checks, naming, course of action assertion, and routing. This plane set up the SDN Control Planeand be thru of one or more SDN servers. A SDN Controller describes the data flows that follow on SDN Data Plane. The bit flow through the network must first get permission from the controller verifies that the communication is permissible by the network guiding principle.

Figure 1. SDN Structure

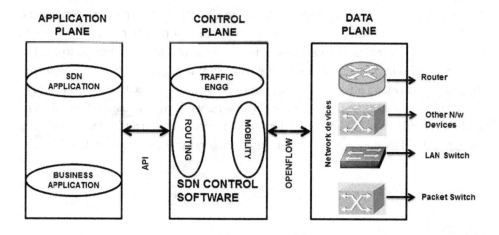

A controller allows a flow the situation computes a route for the flow to proceeds and adds an entry for that flow in each of the switches along the path. By way of means all complex functions subsumed by the controller switches simply manage flow tables whose entries can be populated individual by the control and communiqué stuck between the controller and the switches uses a standardized protocol and API. Utmost commonly this interface is the OpenFlow specification conversed consequently. SDN architecture is extraordinarily flexible. [7] It can operate with different types of switches and at different protocol layers. Software defined networking controllers and switches can be implemented for Internet routers, Ethernet switches, transport switching, or application layer switching and direction-finding. SDN be dependent on the common functions found on networking devices which essentially involve forwarding packets based on some form of flow definition. SDN architecture, a switch performs the following functions:

- Switch reduces and straight on the first packet of a flow to an SDN controller, facilitating the controller to decide whether the flow should be added to the switch flow table.
- Switch forwards received packets out the suitable port based going on flow table the flow table might include importance information dictated by the controller.
- Switch can drop packets on a specific flow in the short term or permanently, as dictated by the controller. Packet dipping can be used for security purposes limit Denial-of-Service attacks or traffic management necessities.
- In inconspicuous terms the SDN controller manages the forwarding state of the switches in the Software Defined networking.

Software defined networking domains include the following:

Scalability: The numeral of devices an SDN controller can feasibly manage is limited. Thus, a reasonably large network may need to deploy multiple SDN controllers.

Privacy: A transferor may choose to implement different privacy policies in different SDN domains. For example, an SDN domain may be dedicated to a set of customers who implement their own highly customized privacy policies, requiring that some networking information in this domain not be disclosed to an external entity.

Incremental deployment: A transporter's network may consist of portions of traditional and newer infrastructure. Dividing the network into multiple, individually manageable SDN domains allows for flexible incremental deployment.

VI IMPLEMENTATION AND PROCESS

Fig. 2 depicts the SDN architecture. There is a centralized SDN block which is a software governed protocol that manages the entire network. It is connected to the OpenFlow Controller which in turn are connected to other OpenFlow switches, routers and other networking components. It uses a unique protocol known as "Open Flow Protocol" that controls and manipulates the entire network. The open flow controller controls the open flow switches. These switches monitor the networks performance and sends periodical reports to the Network Administrator as per his preference. In case of any network failures or node failures in the network this Open Flow Controller will trigger an alert message to the network administrator.

Figure 2. SDN Design Architecture

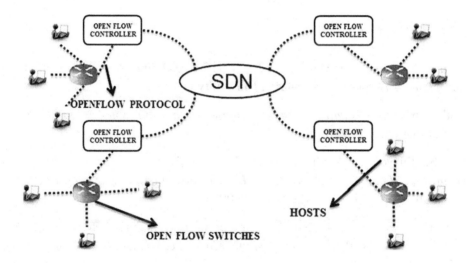

It also has the self healing process but if the issue goes beyond the scope of the Open Flow Control protocol then an alert message is triggered to the administrator. This whole process can be manipulated by the Network administrator according to the environment of the network or according to the outcome required. This structure is applicable from a small home network to a complex commercial network.

VII. CONCLUSION

This paper deals with the identification of current network problems and in addition to that the theoretical implementation of SDN technology to enhance the performance of network management. SDN benefits the Networks in many ways. It mainly simplifies the job of a Network Engineer and it also renders ubiquitous support of the network related issues. We can also set preferences like controlling the bandwidth, restricting access according to time, Probing for security issues, reporting a bug, healing the problems automatically and many more complex functions thereby reducing the work of a human labour. In this paper we have given the in depth idea about SDN and we have also shown the design architecture of SDN. As a further work the SDN will be implemented in the network which we designed and the performance analysis is done through network simulator (Ns3).

REFERENCES

Algarni, M., Nair, V., & Martin, D. (2013). *Software Defined Networking Overview And Implementation*. George Mason University IFS.

Anderson. (2008). *OpenFlow enabling innovation in campus networks*. University of Washington.

Das, S., Talayco, D., & Sherwood, R. (2013). *Software-Defined Networking and OpenFlow, Handbook of Fiber Optic Data Communication* (4th ed.). Academic Press. doi:10.1016/B978-0-12-401673-6.00017-9

Kim, H., & Feamster, N. (2013). Improving network management with software defined networking. *IEEE Communications Magazine, 51*(2), 114–119. doi:10.1109/MCOM.2013.6461195

Yazici, V., Sunay, M. O., & Ercan, A. O. (2014). Controlling a software-defined network via distributed controllers. CoRR, vol. abs/1401.7651

Chapter 11
Security Management in Mobile Ad Hoc Networks

Jhum Swain
Institute of Technical Education and Research, Bhubaneswar, India

ABSTRACT

A mobile ad hoc network (MANETs) is an assortment of a variety of portable nodes that are linked collectively in a greater number in a wireless medium that has no permanent infrastructure. Here, all the nodes in the node partake in acting as both router and host and is in charge for accelerating packets to other nodes. This chapter discusses the various attacks on different layers and on various security protocols. So, designing a secure routing protocol is a main challenge in MANET. As we all know, this is a mobile ad hoc network so nodes in the network dynamically establish paths among each other so it is vulnerable to different kinds of threats. So, in this case, we need secured communication among the nodes present in the network.

INTRODUCTION

MANETs is autonomous system of nodes which are portable in nature and are linked by wireless links. Each node not only acts as an end system but also has a router to forward packets. As nodes freely move around the system and they organize themselves into a network, so for this reason its structure changes frequently. So, a special kind of algorithm is needed to accommodate the changing topology. As nodes dynamically establish paths among each other its attractive to various types of attackers, so we need secured communication among each other. Dynamically the nodes in MANETs join and leave the network(M. S. Athulya and V. S. Sheeba, (2012) .

DOI: 10.4018/978-1-5225-9493-2.ch011

In MANET nodes openly connect and disappear at any point of time to maintain the connection. Some of the typical applications include Military applications, emergency and rescue operations, aircrafts, wireless sensor network, medical service, commercial use and personal area network. As we know MANETs is an infra-structure less type of network which consists of mobile nodes with wireless network interfaces, so nodes dynamically establish paths among one another [1]. So due to this reason it is attractive to various types of attacks. So we need a protected communication and as a result security is an important feature. So MANETS has no clear line of defence so it is accessible to both legitimate network users and malicious attackers. So one of the main challenges is to design a robust security solution that can protect MANETs from various routing attacks. Therefore, due to such disadvantages it is a challenging field to develop secured routing security in MANETs.

RELATED WORK

Many research have been done about how to securely transmit audio data, in many research, dual safety measures by encrypting and decrypting the audio at each node in the route using stream ciphering method has been presented. Secure Multiparty Computation is discussed and how modification of data is done using optimization technique. Various security problems have been explained which are adopted in the network layer. Security issues regarding data query processing and location monitoring have been illustrated in many research articles (Rashid Sheik, Mahakal Singh Chandel and Durgesh Kumar Mishra, (2012). Security solution for the routing protocol OLSR (Optimized Link State Routing) is being proposed in some articles as shown in Table 1.

Table 1. Different kinds of Secured OLSR Protocol

S.No.	Name of the protocol	Advantages	Disadvantages
1.	OLSR-SDK (Secure Data Key)	It provides sophisticated safety measures and enhanced traffic performance.	The Packet Delivery Ratio does not get better.
2.	RA-OLSR (Radio Aware)	It provides sophisticated safety measures.	It has major impact on traffic performance.
3.	COD-OLSR()	It does not generate a major traffic load.	It cannot cancel out strict attacks.
4.	Secure traffic routing OLSR	It battles against link – spoofing attack.	It has main effect on routing performance and it does not discard aggressive attacks.
5.	Trust to secure OLSR	It has fine routing presentation in a extremely damaged environment.	It does not provide sophisticated security against compound attacks.

1. Vulnerabilities Faced by MANET

The various vulnerabilities faced by MANET are listed as follows:

1.1 Lack of Centralized Management

It does not have centralized monitor server so for this reason it is not easy to keep track of the traffic in a highly vibrant and large-scale ad-hoc network (Praveen Joshi,(2011).

1.2 Scalability

As nodes move freely and openly so scaling of ad-hoc network changes frequently. So, security should be capable of handling a bulky as well as tiny network.

1.3 Resource Availability

Providing a safe and protected communication in such a changing environment and against specific threats and attacks leads to development of security schemes and architecture.

1.4 Dynamic Topology

Unstable nodes may disturb the expected relationship among nodes and trust may also be disturbed if some nodes are detected as agreement.

1.5 Limited Power Supply

Nodes in MANET behave in selfish manner if there is limited power supply.

2. Security Goals of MANET

To develop a secure MANET, we have to follow the goals that has already been standardized and we will discuss them here. They are as follows:

2.1 Availability

A node always provides services it is designed for. It concentrates crucially on Denial-of-Service attacks. Some selfish nodes make some network services unavailable.

2.2 Integrity

It refers to the process of guaranteeing the identity of the messenger. There are challenges such as malicious attacks and accidental altering. The difference between the two is intent i.e., in malicious attack, the attacker intentionally changes information whereas in accidental altering, alteration is accidentally done by a benign node.

2.3 Confidentiality

Sometimes, some information ought to be accessible only to few, who have been authorized to access it. Others, who are unauthorized, should not be get a hold of this confidential information.

2.4 Authenticity

Authenticity checks if a node is an impersonator or not [11]. It is imperative that the identities of the participants are secured by encrypting their respective codes. The opponent could mimic a gentle node and can gain a way to private resources or even distribute some harmful messages.

2.5 Non- Repudiation

Non-repudiation guarantees that the sender and the receiver of a message cannot refuse that he/she has not sent/received such a message. The instance of being compromised is established without ambiguity. For e.g. if a node identifies that the message it has received is invalid or not genuine. The node can then use the incorrect message as a proof to alert other nodes that the nodes have been compromised.

2.6 Authentication

A bonafide credentials to be issued by the appropriate authority which will be mandatory to assign access rights to users at different levels. It usually uses an authorization process.

3. Survey of Security Issues

SRAC (Secure Routing Against Collusion) i,e. an optimal routing algorithm with routing metric that combines both requirements on a node's trust worthiness and performance is measured and this protocol makes a routing decision based on its neighbouring nodes. S-AODV protocol that analyzes the adaptive strategy and delays the verification of digital signatures has been proposed by the author. A virtual subnet model i.e. used to construct secure group communication and with the help of this model the composition of groups is established by forming group of keys. SSR (Secure Source Routing) protocol i.e. used to detect transmission failures which continuously configure the operation i.e. used to avoid and tolerate data loss to ensure the availability of communication is being proposed. In RIPsec(Routing Information) protocol, that secures the MANET in operating in a hostile environment is being proposed. The author analyzed the routing information of their vicinity and the border nodes to detect sink holes more properly. The CLPKM (Certificate Less On Demand Public Key Managemet) protocol, that aims at providing the strongest verification routes for authentication purposes and it restricts the probability for a verification route by restricting its length and the strength is evaluated using end-to-end value (D. Tardioli, (2014).

An IDS system i.e. based on enhanced windowing method to carry out the collection and analysis of selected cross-layer features and this uses a model to recognize malicious packet dropping behaviour has been proposed. A game theoretic framework for stochastic multipath routing in MANET and here the source node selects a path for data communication and it switches the strategy among the multiple established paths. CMEA (Cluster Based Mobility and Energy Aware) protocol is proposed by the author and here XML technique is used for encrypting the secure data that improves the content delivery over CMEA network and it finds efficient and reliable route between source and destination node based on mobility and energy aware metric.DSR (Dynamic Source Routing) is proposed by the author which provides the solution to prevent it from entering of malicious nodes between source and destination. A game theoretic scheme that uses LCTF (Least Total Cost Factor) is used to transfer data packets from source to destination have been proposed by the author. The HMAC-SHA12 is used that provides data integrity along with authentication i.e. based on Trust-Based system. The author has presented a framework of novel routing technique in order to jointly address the problem pertaining to routing overhead and energy drainage among the mobile nodes. Here CA (Certificate Authority) distribution and trust-based threshold revocation method i.e. computed from direct and indirect trust values and CA distributes the secret key to all nodes. The author proposed QaASs (QoS Aware Adaptive Scheme) i.e. an adaptive mechanism that encounters the effect of delay overhead by adapting cryptography and multimedia properties.

4. Evaluation of Secure Routing Protocols in MANET

The author has discussed about the different secure routing protocols in MANET and this is being summarized in the table 2 below as follows:

Table 2. Secure Routing Protocols in MANET

S.No.	Name of the Protocol	Use of the Protocol	Advantages	Disadvantages
1.	Secure- Dynamic Source Routing (S-DSR)	To develop node's genuineness and to find out safety-way of sharing files without involving any malicious nodes in its way.	It provides enhanced packet delivery ratio and less transparency.	It does not sort out all the security issues related to P2P.
2.	Secure – On Demand Routing Protocol (ARIADNE)	It sends the message directly to one of the three ideas such as shared secrets between couple of nodes, communicating nodes, or by digital signature.	Any variation in the node list can be sensed.	There are some assured attacks like wormhole and cache poisoning attack that cannot be prohibited. The key replacement is complex .
3.	S-AODV (Secure- Ad Hoc On Demand Distance Vector Routing Protocol)	It is used to secure AODV and it uses digital signature to validate non-alterable fields such as RREQ and RREP.	Replay as well as delay attacks can be avoided by using sequence number.	It utilizes the public key cryptography that requires a high processing transparency. Here an intermediary node damages the route discovery. The IP portion of the S-AOADV traffic can be compromised.
4.	SAR (Secure- Aware Routing Protocol)	It takes a step to route that includes security level of nodes into traditional routing metrics.	Here security is used as a flexible set that improves the importance of the routes i.e. exposed by ad-hoc routing protocols.	It does not disclose as on how to use or employ the security level as a metric. As it does not have a proper route discovery process it fails.
5.	SEAD (Secure Efficient Distance Vector Routing Protocol)	It is a practical robust routing protocol i.e. dependent on DSDV-SQ protocol. It relies on one-way hash chain for security.	It uses proficient, cheap cryptographic techniques that defends against multiple un co-ordinated attackers, active attackers or compromised nodes which plays a significant role in computation and bandwidth- constrained nodes.	It does not offer a way to avoid an attacker from tampering with "next hop" or "destination columns".
6.	SDSDV (Securing the Destination Sequenced Distance Vector Routing Protocol)	It requires cryptographic mechanism for entity and message validation .	Data integrity is confined, data origin is validated, a route with a falsified destination can be detected, an advertised route with a falsified distance can be detected, an advertise route with a falsified next hop can be identified and route update with half-truths can be detected.	It produces high network transparency.
7.	SLSP (Secure- Link State Routing Protocol)	It is in charge for securing the discovery and distribution of a link state information.	It is less susceptible to DOS attacks. Nodes can choose to authenticate the public key or not.	It is exposed to colluding attacks.

continues on following page

Table 2. Continued

S.No.	Name of the Protocol	Use of the Protocol	Advantages	Disadvantages
8.	ARAN (Authenticated Routing for Ad- Hoc Network)	It guarantees that each node knows the right next hop on a route to the destination by public key cryptography.	It is secure as long as CA is not compromised because of public key encryption, network structure is not exposed and it is resistant to most of the attacks.	It requires extra memory. It has high processing overhead for encryption. It does not use hop count, so the discovered path may not be optimal .
9.	SPAAR (Secure Position Aided Ad hoc Routing)	It provides sophisticated security at the cost of performance.	-	The practice of certificate server and tremendous need to keep the server uncompromised. Issues still exist with compromised nodes already having valid certificates .
10.	Byzantine Failure Resilient Protocol	It recommends to flood the route requests and route replies in order to defend against Byzantine failures .	As long as there is fault free path, even in a highly adversarial controlled network, it will be discovered after bounded numbers of faults have been occurred.	It is difficult to design scheme i.e. flexible to large number of adversaries .
11.	SRP (Secure Routing Protocol)	It is used to setup security association (SA) between the source and the destination node.	It assures the correct connectivity information over an unidentified network in the existence of malicious node, confidentiality is protected, it has fewer processing transparency, route signalling cannot be spoofed, fabricated routing messages cannot be formed through malicious actions and routes cannot be redirected from shortest path through malicious nodes.	It exposes network structured with un encrypted routing path i.e. it is susceptible to "invisible node attack" .

5. Attacks on Different Layers

Attacks can be classified into two broad categories:

5.1 Passive Attacks

The attacker just spies around the network without distracting the network operation. This attack compromises the privacy of the data and says which nodes are working in immoral way.

5.2 Active Attacks

It is a type of attack in which the attacker disturbs the normal operation of the network by fabricating messages, dropping or changing packets, by repeating or channelling them to other part of the network. Basically, the content of the message is changed Z. Mi, Y. Yang and J. Y. Yang, (2015).It is of two types:

5.2.1 External Attacks

Here the attacker causes network jamming and this is done by the propagation of fake routing information. The attack disrupts the nodes to gain services.

5.2.2 Internal Attacks

Here the attacker wants to gain access to network and wants to get involved in network activities. Attacker does this by some malicious imitation to get access to the network as a new node or by directly through a current node and using it as a basis to conduct the attack.

Different attacks on different layers are summarized as follows and is given in the table 3 and 4 for reference:

6. Short Description of the Important Attacks on Network Layer

6.1 Black –Hole Attack

A pernicious hub sends fake steering data asserting as shown in figure 1 that it has an ideal course and makes other great hubs course information bundles through the malevolent hubs (V. Srovnal Jr., Z. Machacek and V. Srovnal, (2009). For e.g. In AODV, the aggressor can send a fake RREP that incorporates a fake goal arrangement number i.e. manufactured to be equivalent or higher than the one joined in the RREQ to the source hub, asserting that it has an adequately crisp course to the goal hub.

This makes the source hub select the course that goes through the assailant. In this way, all activity will be directed through the assailant and along these lines the aggressor can abuse or dispose of the movement

6.2 Cache Poisoning Attack

The assailant sends a fashioned DNS reaction, the degenerate information is given by the aggressor and it gets reserved by the genuine DNS name server. This is shown in Figure 2. It's now that the DNS reserve is considered as "harmed". Thus, future clients that endeavor to visit the degenerate space will rather be directed to the new IP address chose by the aggressor. Clients will keep on receiving inauthentic IP addresses from the DNS until the harmed store has been cleared.

Table 3. Attacks on Network layer

Attacks	Description
Worm-Hole Attack	A noxious hub gets bundles at one area in the system and passages them to another area in the system where these parcels are resent into the system. This tunnel between two colliding attackers is referred to as wormhole.
Black-Hole Attack	In this assault a malignant hub publicizes legitimate and most limited course to a casualty hub and from there on furtively drops information and control bundles as they go through it.
Byzantine Attack	It includes different aggressors that work in conspiracy to debase the system execution, for example, making circles, specifically dropping bundles, picking non-ideal ways for parcel sending.
Information Disclosure	A bargained hub may disregard the secrecy rule of security and uncover vital data like private and open keys, status of hub, passwords, ideal course to approved hubs, geographic area of hubs and other control information in parcel headers to unapproved hubs introduce in the system.
Routing Table Overflow	This assault forestalls making of new genuine courses by flooding the steering table with courses to non-existent hubs.
Routing Table Poisoning	In this assault, vindictive hub sends manufactured directing refresh and blunder messages or adjusted authentic updates to approved hubs in the system.
Routing Cache Poisoning	A malevolent hub can dispatch DOS assault on any hub by just communicating ridiculed parcels with source courses.
Replay Attack	An aggressor as opposed to changing parcels content simply replay stale bundles with a specific end goal to endeavour battery power, transfer speed and computational limitation of portable hubs.
Rushing Attack	This assault includes whole system movement to go through an assailant. The source hub can't locate any protected way without the assailant.
Jellyfish Attack	It is a specific dark gap assault in which noxious hub assaults the system by reordering the parcels or expanding jitter of the bundles that go through it keeping in mind the end goal to keep it from being identified and it appears to the system that misfortune or deferral is because of ecological reasons.

Table 4. Attacks on Different layers

Name of the layer	Attacks	Description
Application Layer	Repudiation	It is a demonstration of refusal in taking an interest in all or some portion of correspondence.
	Malicious Attacks	In this assault, a pernicious hub aggravates the ordinary operation of alternate hubs in the system by assaulting the procedure framework.
Transport Layer	Session Hijacking	In this aggressor gets a path into the session condition of a specific client by taking session ID which is utilized to get into a framework and watches the information.
	SYN Flooding	A malicious node sends a huge number of SYN packets to a victim node. The victim node sends back SYN+ACK packets and keeps the entry for the unfinished connection request.
Data Link Layer	Denial of Service	There is a solitary remote channel shared by every one of the hubs so a malignant hub keeps this channel occupied by sending false parcels to deplete hub's battery control.
	MAC Targeted Attack	The MAC procedure is interrupted.
Physical Layer	Device Tampering	Hubs in specially appointed remote systems are little, smaller and hand-held like not at all like wired gadgets so they can be effortlessly stolen or harmed.
	Jamming	The assailant screens the remote medium with a specific end goal to discover recurrence at which goal hub is getting from sender hub. An aggressor must have capable transmitter to send the signs to the goal at that recurrence, in this way meddling its operations. The most common types of jamming is random noise and pulse.

Figure 1. Black-Hole Attack

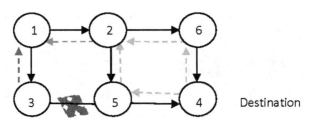

Figure 2. Cache Poisoning Attack

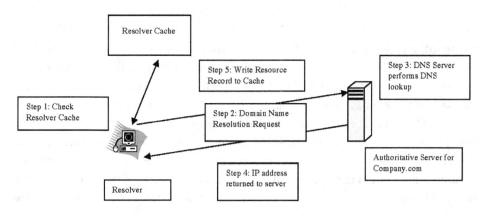

WORM-HOLE ATTACK

Here a noxious hub gets parcels at one area in the system and passages them to another area in the system where these bundles are hate into the system. It has at least one malignant hubs and it burrows between them. This passage between two impacting aggressors is alluded to as wormhole. A worm-gap assault can undoubtedly be propelled by the aggressor without knowing about the system or trading off any genuine hubs or cryptographic components. This is shown in Figure 3.

Figure 3. Worm-Hole Attack

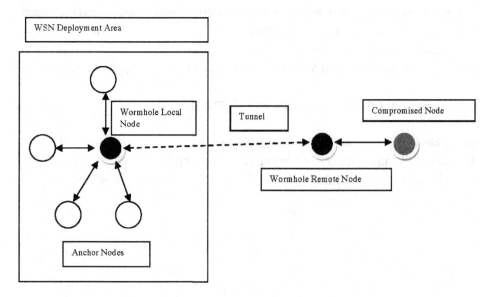

7. Intrusion Detection System

Intrusion Detection System (IDS) have turned out to be most indispensable part in the guard. It is the strategy for observing PC framework or systems for unapproved affirmation, development or document adjustment. It can likewise be utilized check arrange movement, with the goal that it can detect the framework whether it is been focused by a system assault, for example, dark opening or dissent of administration assault. IDS can accomplish location by ceaselessly observing and dissecting the system for strange movement, some exceptional assaults and action which are not same as day by day action. IDS can be ordered into two sorts relying upon information accumulation system and identification instrument. Sorts of IDS relying upon the information accumulation system incorporates Network based IDS (NIDS) and Host based IDS (HIDS). There are principally 3 sorts of IDS that goes under the class of recognition methods [69].

8 Anomaly Detection System

In this type of detection system the typical conduct or every day exercises of the client are kept inside the framework. At whatever point, a movement is performed by the client or aggressor, the framework contrasts this action and the kept information and afterward treats that action in view of the assessment whether it is a nosy action or not and as needs be reacts to the framework.

9 Misuse Detection System

In this kind of framework, it holds some outstanding assaults design and their mark. At whatever point any movement is performed, it contrasts this action and the put away example or signature and if there is any match then it is dealt with as a nosy action. E.g. Virus Detection System. It cannot identify new types of attacks.

9.1 Specification-Based Detection System

In this kind of framework, it characterizes an arrangement of principles that portray the strategy of a program or convention. At whatever point any action is performed, it checks the execution of that action with characterized set of guidelines.

10 IDS in MANET

Intrusions Interruptions are the arrangement of activities that endeavor to alter the honesty, privacy or accessibility and Intrusion Detection System (IDS) is a framework or programming application that screens organize movement, and if any suspicious action is discovered then it alarms the framework or system director. There are fundamentally 3 sorts of IDS systems that can be connected in MANET:

10.1 Stand Alone Intrusion Detection System

In this framework, an intrusion detection system runs freely on individual hub to decide interruptions. All choice taken about a specific action is reliant on data accumulate at its own hub in light of the fact that there is no coordinated effort among hubs in the system. In this manner, no data is exchanged (Zhang Hua,Gao Fei,Wen Qiaoya(2011).

10.2 Distributed and Co-Operative Intrusion Detection System

In this design, each hub has an IDS operator which recognizes interruptions locally and works together with neighboring hubs for worldwide discovery at whatever point accessible confirmation is uncertain and a more extensive pursuit is required. Every hub takes an interest in interruption location strategy and reaction as having and IDS specialist running on them (Zhijing Qin et.al,2014). The obligation of an IDS specialist is to recognize and gather nearby data and information to distinguish any assault if there is any assault in the system, and furthermore takes a reaction autonomously. It is appropriate for level system framework and not multi-layer framework.

HIERARCHICAL INTRUSION DETECTION SYSTEM

It augments the elements of dispersed and co-agent IDS framework that has been executed for multi-layer arrange foundations where the system is separated into various little systems known as bunches. The idea of multi-layering is connected to interruption discovery framework where various leveled IDS is proposed. Every IDS specialist keeps running on specific part hub and is in charge of its hub i.e. checking and choosing privately distinguished interruptions. A bunch head is capable locally for its hub and in addition universally for its group, for example, checking system activity and reporting a worldwide reaction when arrange interruption is identified.

ATTACKS ON DIFFERENT ROUTING PROTOCOLS

Here we discuss about the attacks that each routing protocol is prominent. They are summarized below:

Table 5. Attacks on routing protocol

Name of the routing protocol	Attacks
SEAD	DOS
	Tunnelling
	Spoofing
	Black-hole
	Wormhole
	Routing Table Overflows
Ariadne	DOS
	Tunnelling
	Routing Table Overflows
SRP	DOS
	Tunnelling
	Routing Table Overflows
ARAN	DOS
	Tunnelling
	Routing Table Overflows
SAR	DOS
	Tunnelling
	Routing Table Overflows
SOADV	DOS
	Tunnelling
	Routing Table Overflows

CONCLUSION

We discussed about the works done by different authors and then we discussed about the attacks on different layers and also about the attacks on different routing protocols. Further our research would be on how to develop a secured routing protocol.

REFERENCES

Athulya, M. S., & Sheeba, V. S. (2012). Security in Mobile Ad-Hoc Networks. *Computing Communication & Networking Technologies (ICCCNT), 2012 Third International Conference on*, 1-5. 10.1109/ICCCNT.2012.6396047

Hua, Z., Fei, G., & Wen, Q. (2011). A Password-Based Secure Communication Scheme in Battlefields for Internet of Things. *China Communications, 8*(1), 72–78.

Joshi, P. (2011). Security Issues in Routing protocols in MANETs at network layer. *Procedia Computer Science, 3*, 954–960. doi:10.1016/j.procs.2010.12.156

Mi, Z., Yang, Y., & Yang, J. Y. (2015). Restoring Connectivity of Mobile Robotic Sensor Networks While Avoiding Obstacles. *IEEE Sensors Journal, 15*(8), 4640-4650.

Qin, Z., Denker, G., Giannelli, C., & Bellavista, P. (2014). A Software Defined Networking architecture for the Internet-of-Things. *Network Operations and Management Symposium*, 1-9. 10.1109/NOMS.2014.6838365

Sheik, Chandel, & Mishra. (2012). *Security Issues in MANET: A Review*. IEEE.

Srovnal, V., Jr., Machacek, Z., & Srovnal, V. (2009). Wireless Communication for Mobile Robotics and Industrial Embedded Devices. *Eighth International Conference on Networks*, 253-258.

Tardioli, D. (2014). A wireless communication protocol for distributed robotics applications. *IEEE International Conference on Autonomous Robot Systems and Competitions (ICARSC)*, 253-260. 10.1109/ICARSC.2014.6849795

Chapter 12
Enhancement of Network Performance in VANET Using Dynamic Routing Strategies

Mamata Rath

ⓘ https://orcid.org/0000-0002-2277-1012
Birla School of Management, Birla Global University, India

Sushruta Mishra
Kalinga Institute of Industrial Technology, India

ABSTRACT

Vehicular ad hoc networks (VANETs) have evolved as an invigorating network system and application domain in current communication technology. In smart city applications context, there are smart vehicles embedded with sensors and dynamically programmed IoT devices, which are to be managed and controlled energetically. Progressively, vehicles are being furnished with surrounded actuators, handling signals, and wireless communication abilities. This chapter focuses on the fact that this special network has opened various possible outcomes for intense and potential extraordinary applications on security, effectiveness, comfort, confidentiality effort, and interest while they are significantly vibrant. Irrespective of many challenges such as high frequency of topology change and link failure possibility, routing management in VANET has been successful in traffic scenario during vehicle-to-vehicle communication.

DOI: 10.4018/978-1-5225-9493-2.ch012

1.INTRODUCTION

In the current technology, Vehicular Adhoc Networks (VANETs) are becoming more popular applications in road traffic management and control systems. The problem faced by smart cities in terms of traffic congestion issues can be solved better by the use of VANETs as there is a network connectivity between the vehicle and the network infrastructure (Rath et. al, 2018). Therefore, predictable information regarding road condition ahead and route information can be directed to the smart vehicles in transit and intelligent decisions can be taken sufficient time before any problem occurs. In other ways VANET in smart cities helps to reduce the problems of congestion, accidents, crime, parking problems and population overhead. Due to the overall development of the wireless technology, their applications are immense on vehicles and vehicles have been converted to smart vehicles to be accessed under smart traffic applications. The traditional driving systems and drivers have also converted to smart drivers with more technical knowledge of receiving smart signals from traffic controllers, understanding them and act accordingly(Rath et.al, 2019). VANETs support flexible communication between vehicles and traffic controlling systems in both infrastructure based and in wireless medium without fixed infrastructure. The proposed traffic congestion solution in smart city uses an improved technical explanation to the problem with powerful data analytics made by mobile agent dynamically under VANET scenario in a smart city(Rath et.al, 2018).

Vehicular Adhoc Networks (VANETS) empower the vehicles to communicate with each other vehicle and in addition with road side units. However, building up a dynamic routing policy for these networks conceivably because of the substantial portability and normal changes in these networks is a tough work. In VANETS transmission joins are at the danger of separation. Along these lines the advancement of the proficient directing component is required in VANETS. VANETS may really make to end up valuable for interstate way protection and additionally a few business purposes (Kaur K. et.al, 2016). For example; vehicular network might be utilized to educate drivers with the goal that the likelihood of guests jams; offering higher solace and also adequacy (Felipe C. et.al, 2016) . Remote association frameworks have permitted the majority of the advantages inside our lives and moreover enhanced our every day productivity additionally (Singh S et.al, 2014) . Specially appointed networks work without a characterized preset looked after foundation. VANETS working with 802.11-organized WLAN development currently got critical intrigue (M. Rath et.al, 2018). For the reason that vehicles worked with Wi-Fi gear speak to the cell nodes (S. Singh et.al, 2018). Another locale in which there is a ton prospect of remote advancements to make a gigantic impact might be the district of between vehicular interchanges (IVC)(O Sami et.al, 2017).

Figure 1. A Scenario with V2R Communication using VANET protocol

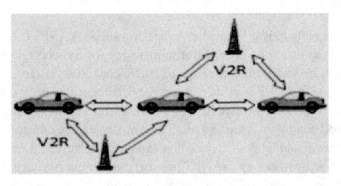

Fig.1. presents a scenario with V2R Communication using VANET protocol. Intervehicle organization (IVC) is ever-increasing huge passion from the exploration province and the engine vehicle industry; in which it will valuable in giving shrewd transportation framework (ITS) alongside drivers and explorers relate administrations (C. Trung et.al, 2014). The idea of IVC is regularly called vehicle-to-vehicle correspondences (V2V) and notwithstanding framework it alluded vehicle-to-foundation inside their transmission extend and vehicular adhoc networks (VANET) (A. Rahim et.al, 2018).

2. LITERATURE STUDY

With a sharp increment of vehicles out and about, new technology is imagined to give offices to the travelers including wellbeing application, help to the drivers, crisis cautioning and so forth. Specially appointed network is a sort of wireless correspondence network which built up by utilizing a bunch of portable nodes with wireless deciphering supplies. The network which is totally versatile which require practically no framework is begat with the term MANET (Mobile impromptu network). These MANETs have a few properties (K. Kaur et.al, 2016) like unique topologies, constrained data transfer capacity, restricted vitality and some more. Vehicular specially appointed network (VANET) is a sub class of MANET with some one of a kind properties. VANETs have rising out nowadays because of the requirement for supporting the expanded number of wireless supplies that can be utilized as a part of vehicles (S. Singh et.al, 2014) . A portion of these items are worldwide situating framework, cell phones and workstations. As versatile wireless types of gear and networks turn out to be progressively essential, the interest for Vehicle-to-Vehicle (V2V) and Vehicles-to-Roadside (V2R) Communication or Vehicle-to-Infrastructure

(V2I) Communication will keep on growing . VANETs have some unique properties then MANETs like street design limitations, no confinement on network measure, dynamic topology, versatility models, and unbounded vitality supply, restriction usefulness et cetera. Every one of these qualities made VANET condition a trying for creating productive directing conventions.

In recent years, rapid development in the quantity of vehicles out and about has expanded requests for correspondence moving. Another sort of Adhoc network with a massive change in mechanical developments is rising nowadays known as VANET (Vehicular impromptu network) (Mayada A et.al, 2017). It is a grouping of vehicular nodes that go about as portable hosts set up a transient network without the help of any brought together organization or any settled framework. In this way, it is called self-governing and self arranged network. In VANET, two sorts of correspondence should be possible to give a rundown of uses like crisis vehicle cautioning, wellbeing and so on. These are between different vehicles known as vehicle to vehicle and amongst vehicles and roadside units known as vehicle to roadside correspondence. Execution of such sort of correspondence between vehicles relies upon different directing conventions. We tend to review some of the ongoing investigation prompts steering space.

VANETs have been developed as an exciting network system and application domain. Progressively vehicles are being furnished with implanted sensors, handling and wireless correspondence abilities. F. Chunha et.al, (2016) refer that this special network has opened a various possible outcomes for intense and potential extraordinary applications on security, effectiveness, comfort, open joint effort and interest, while they are out and about. Albeit, considered as an exceptional instance of a Mobile Adhoc Network (Rath et.al, 2014), the high yet compelled portability of vehicles convey new difficulties to information correspondence and application outline in VANETs. This is because of their very dynamic and discontinuous associated topology and distinctive application's QoS necessities. A review on VANETs concentrating on their correspondence and application challenges has been accounted for.

The location-based routing protocol has been given attention as one of the effective routing approaches in vehicular adhoc network on the subject of low overhead and high adaptability(C. Trung et.al, 2014) . Its basic influence lies in playing out a pathless routing with the end objective that a node having a group communication message to its neighbour node that gives the most limited physical separation to goal and this procedure proceeds until the point when the parcel achieves goal. The issue lies in that the link connectivity of the neighbours varies to a great extent relying upon the mobility of vehicles and the natural factors that bring about shadowing and fading effects. Another connection quality expectation metric related with location based protocols to enhance the determination of next jump, that consider both the connection quality evaluation in light of the transmission achievement rate and the connection quality appraisal in view of the forecast without bounds locations of vehicles has been presented (C. Trung et.al, 2014). Recreation comes about are introduced to show the proficiency of the proposed metric.

Route Establishment (X. Wang et.al, 2018) describes that in a route with a network gap, a vehicle usually drops messages if it cannot find the next hop. In this case, communications may be disrupted. Therefore, it is very important to avoid network gaps when routing paths are established. A reliable routing scheme (Rath et.al, 2015) indicates and aims to establish routing paths without network gaps to improve the packet delivery rate. In the said scheme, a routing establishment algorithm based on road segments and virtual nodes is proposed. This algorithm can establish the routing paths without network gaps, so the packet loss caused by network gaps is avoided. Moreover, a message can be transmitted along the shortest routing path without a network gap, so the packet delivery rate is improved. Finally, this scheme is evaluated, and the data results show that the packet delivery rate is effectively improved.

Vehicular Adhoc Networks (VANET) is a standout amongst the most concrete and testing research zones in car organizations and ITS planners. The nearness of such networks opens the route for an extensive variety of uses, for example, wellbeing applications, portability and availability for both driver and travelers to abuse the vehicle frameworks in an easily, proficiently and more secure way. For wellbeing applications the best directing convention must be chosen. The three most normal directing conventions that are utilized as a part of VANET are: DSR, AODV and DSDV (Seyed M. et.al, 2018) . In reality, it is vital and fundamental to test and assess distinctive directing conventions identified with VANET framework before applying them in the genuine condition which should be possible by means of VANET recreation apparatuses. Execution of three diverse directing conventions for VANET framework has been evaluated.The execution is assessed and thought about as far as PDR, normal throughput, deferral and aggregate vitality. The purpose is to evaluate the execution of steering model for city situation. The principle objective is to locate the appropriate steering convention in a high thickness activity zone. Three directing conventions DSR, AODV and DSDV has been considered and the outcomes demonstrate the low quality of DSDV convention which is a sort from proactive steering conventions. The AODV convention (Rath et.al, 2016) accomplishes most extreme normal throughput which is equivalent to 330.07 Kbps. The base estimation of postponement acquired from utilizing DSR convention was 15.81 ms.

A Vehicular Social Network (VSN) is a rising field of correspondence where significant ideas are being obtained from two unique controls, i.e., vehicular specially appointed networks (VANETs) and versatile social networks (MSNs). This developing worldview displays new research fields for content sharing, information dispersal, and conveyance services. In light of social network examination (SNA) applications and strategies, interdependencies of network substances can be abused in VSNs for imminent applications. VSNs include social associations of suburbanites having comparative destinations, interests, or portability designs in the virtual network of vehicles, travelers, and drivers on the streets. Considering social networking in a vehicular situation (A. Rahim et.al, 2018) researches the planned uses of VSNs and correspondence engineering. VSNs advantage from the social practices and versatility of nodes to create novel suggestion frameworks and course arranging. A best in class writing audit on socially-mindful uses of VSNs, information scattering, and portability demonstrating has been accounted for.

3. PROTOCOLS AND ROUTING CHALLENGES IN VANET

The way towards routing in Vehicular Adhoc Networks (VANET) is a difficult task in metro city conditions. Finding the shortest end-to-end associated way fulfilling defer limitation and insignificant overhead is gone up against with numerous imperatives and troubles. Such troubles are because of the high portability of vehicles, the successive way disappointments, and the different obstacles, which may influence the dependability of the information transmission and directing. Business Unmanned Aerial Vehicles (UAVs) or what are usually alluded to as drones (R. Oliveira et.al, 2017) can prove to be useful in managing these imperatives. Table 1. Shows a description of different routing protocols, their use and challenges which are designed and contributed by various researchers.

3.1 Study and Analysis of Routing Protocols in VANET

Routing in a VANET is really challenging due to many factors as discussed including dynamic topology change, satisfying quality of service requirement, considering load balancing during communication and high rate of device mobility. The incessant research is in advancement to progress routing, considering various aspects and challenging features of VANETs. This part of our work projects on design approaches and contributions adopted in different projects along with their precincts. This will facilitate out new researchers of the field to understand existing state of the art of proposed schemes. Regardless of classification some randomly selected proposed routing schemes are described in the subsections below. Table in the end of each subsection shows simulation, routing and scenario parameters of the studied proposed routing scheme.Table 1. Shows a description of different routing protocols, their use and challenges which are designed and contributed by various researchers.

Table 1. Description of different routing protocols, their use and challenges with domains

Routing Protocols Study and Design in VANET				
Sl. No	Literature	Year	Scheme / Challenges/Issues	Related Domain
1	S.Singh et al.	2014	Study and Analysis of VANET Routing Protocols	VANET Protocols
2	F.Cunha et al.	2016	Applications and Issues of VANET Routing Protocols	VANET Protocols and applications
3	C.Ngo et al.	2014	Link quality prediction metric for shadowing and fading effects in VANET	Routing Metric in VANET
4	X.Wang et al.	2018	Reliable routing in IP-Based VANET with network gaps analysis	Reliable Routing in VANET
5	A.Saleh et al.	2017	Reliable Routing Protocols in VANET	Reliable Routing in VANET
6	M.Hossain et al.	2017	Novel MAC Protocol Design in VANET	Protocol Design in VANET
7	J.Filho et al.	2016	Technical Survey on DTN and VDTN Routing Protocols in VANET	Routing Protocol in VANET
8	B.Brik et al.	2016	Distributed Data Gathering Protocol Design for VANET	Routing Protocol in VANET

Reliable Protocol for VANET named as R2P has been outlined (Ahmed I., 2017), which separates the network into covering zones. For each zone, an extraordinary node is elevated to be the Master Node (MN), which keeps up an exceptional routing sheets for bury/intra-zone communication. R2P relies upon two kinds of sheets, in particular; Internal Routing Board (IRB) and External Routing Board (ERB). Two kinds of IRB are utilized, to be specific; Zone Routing Board (ZRB) that is kept up by MNs, and Private Routing Board (PRB) that is kept up by each network node. Both ZRB and PRB enroll routes among zone nodes, while ERB, which is kept up by MN, registers accessible portals to neighboring zones. R2P utilizes an extraordinary route revelation component to find accessible routes to the goal, and afterward chooses the most reliable route. It has been looked at against the ongoing VANET's routing protocols. Exploratory outcomes have demonstrated that R2P beats the others.

A developed MAC Protocol for VANET (M. Kowsar et.al, 2017) with Proficient medium access control has been proposed which acts as a key part of any remote network in communication engineering. MAC protocols are required for nodes to access the mutual remote medium effectively. Vehicular Adhoc networks (VANETs) are a rising network innovation very nearly vast scale transport. The dynamic network topologies in VANETs caused by high versatility rates of vehicles exhibits an extraordinary test in reliable information exchange. A MAC protocol that empowers fast reservation of group transmission carried out by smart vehicles that desire to send data packets is pivotal in tending to this experiment. A distributed MAC computation ResVMAC for VANETs has been introduced that outperforms existing MAC layer protocols (Rath et.al, 2018) in VANET.

After a detailed study and review it was discovered that the technical routing in VANET are best routed down by utilizing Delay Tolerant Network (DTN) (J. Filho et.al, 2016) navigation protocols that can tolerate immense delays, route instability and encourage applications to utilize a base number of roundtrip reaction affirmations. DTN steering protocols are thought to be the most appropriate other option to traditional directing protocols in VANET conditions. They are intended for putting away and sending messages through a progression of forwarders to keep up network availability. A methodical specialized review and a similar investigation of a scientific categorization of DTN directing protocols have been occurred which were stretched out and adapted to another arrangement of VDTN (VANET/DTN) navigation protocol classifications with their execution assessment.

A novel Distributed Data Gathering Protocol (DDGP) has been projected (B. Brik et.al, 2016) for dealing the delay tolerant factor in VANET routing and in addition constant information in both urban and interstate situations has been planned . The fundamental commitment of DDGP is another medium access procedure that empowers vehicles to get to the direct distributedly in light of their area data. In addition, DDGP actualizes another conglomeration scheme, which erases redundant, terminated, and undesired information. It offers an analytical confirmation of accuracy of DDGP, in addition to the execution assessment through a broad arrangement of simulation tests. The outcomes show that DDGP improves the effectiveness and the reliability of the information accumulation process by beating existing schemes as far as a few criteria, for example, delay and message overhead, collection ratio, and information re transmission rate.

3.2. Geographical Routing in VANET

Table 2. Shows a description of different geographical routing protocols, their use and challenges which are designed and contributed by various researchers.

Currently numerous steering protocols have been proposed for Vehicular Adhoc NET works (VANETs) by considering their particular qualities. The protocols in light of the vehicles' positions, named geographic steering protocols (GR) or position-based directing protocols (PBR), were appeared to be the most sufficient to the VANETs because of their strength in managing the dynamic condition changes and the high portability of the vehicles. Rather than utilizing the IP addresses, as

Table 2. Description of different Geographical routing protocols with challenges

Geographical Routing in VANET				
Sl. No	**Literature**	**Year**	**Scheme/Challenges/Issues**	**Related Domain**
1	S.Lahlah et. al	2018	Geographical Routing Protocols in VANET	Geographical Routing in VANET
2	I Halim et al.	2018	Prediction Based Protocols in VANET	Intelligent Routing in VANET
3	T.Darwish et al.	2017	Green Geographical Routing in VANET	Green Routing in VANET
4	H.Cheng et al.	2015	Road side Unit Deployment Protocol for Urban VANET	RSU Deployment Algorithm
5	D. Venkat raman et al.	2017	SDN Enabled Connectivity aware Geographical Routing Protocol for VANET	Software Defined Network

on account of Mobile Ah hoc NET works (MANETs) protocols (Rath et.al, 2018), position-construct directing protocols are situated in light of the land position of the vehicles while choosing the best way to forward the information (S. Lahlah et.al, 2018). Further, they don't trade interface state data and don't keep up set up courses as in MANET directing protocols. This makes the protocols more powerful to the regular topology changes and the high portability of the vehicles.

The high mobility of vehicles as a noteworthy normal for Vehicular Adhoc Networks (VANETs) influences distinctively the dynamic idea of the networks and results in extra overhead as far as additional messages and time delay. The future developments of the vehicles are typically unsurprising. The consistency of the vehicles future developments is an after effect of the movement conditions, the urban format, and the driving prerequisites to watch the activity obliges. Subsequently, foreseeing these future developments could assume a significant part for both building dependable vehicular correspondence protocols and comprehending a few issues of insightful transportation frameworks. There are various expectation based protocols are displayed for VANETs. Rules of methodical writing are additionally checked on to give a head and unprejudiced study of the current forecast based protocols and create novel scientific categorizations of those protocols in view of their primary expectation applications and targets. A talk on every class of the two scientific classifications is exhibited by numerous specialists, with an emphasis on the necessities, compels, and challenges (Islam T et.al, .2018).Additionally, utilization examination and execution correlations are researched so as to determine the reasonableness of every expectation goal to the different applications. Likewise, the pertinent difficulties and open research territories are distinguished to control the potential new headings of expectation based research in VANETs.

Green Geographical Routing in VANET-Information and Communication Technologies get through around 3% of the overall worldwide energy.. Subsequently, creating green communication frameworks is important for all divisions of advances. Likewise, as different remote gadgets may between speak with Vehicular Adhoc Networks (VANETs), control limitations for such gadgets must be considered. Besides, with the presentation of full and mixture electrical vehicles, proficient vitality utilization in vehicles communication is getting to be fundamental. Albeit geographical routing is a prevailing routing approach in VANETs, bundles routing through multi-jump communications devours the vast majority of the communication vitality of vehicles' remote gadgets. An examination on the advances and difficulties in green geographical routing protocols for VANETs has been done (Tasneem D et.al, 2017), while existing protocols are characterized to guide based and beaconless. Distinctive techniques utilized by the green geographical routing protocols were investigated to diminished their vitality utilizations. Notwithstanding, an impressive research work is required to enhance VANETs green geographical routing.

Vehicular Adhoc networks have risen as a promising research zone. Planning a practical scope protocol for RSU organization in vehicular networks introduces a test because of various service region, arranged versatility examples, and asset imperatives. With a specific end goal to determine these issues, a geometry-based scanty scope protocol by GeoCover (H. Cheng et.al, 2015) has been predicted, which means to consider the geometrical qualities of road networks, development examples of vehicles and asset confinements. By considering the measurements of road sections, GeoCover gives a buffering task to suit diverse kinds of road topology. By finding hotspots from follow documents, GeoCover can delineate the versatility designs and to find the most significant road zone to be secured. To take care of the asset obliged scope issue, two variations of meagre scope are given which considers the financial aspect individually. The scope issue is settled by both genetic algorithm and greedy algorithm.

3.3 Cluster Based Routing Solution in VANET

Table 3 Shows a description of different cluster based routing protocols, their use and challenges which are designed and contributed by various researchers.

Grouping of vehicles in Cluster computing is an significant strategy to decrease the high portability impact of vehicles. An analytical model utilized (R. Pal et.al, 2018) has been assessed for the execution of a clustered vehicular Adhoc network (VANET). The analytical model is created to assess three important parameters, to be specific packet delivery ratio, throughput, and delay. The outcomes acquired from the analytical model are additionally joined by simulation comes about. This model can be additionally reached out by researchers taking a shot at clustered

Table 3. Description of different Cluster based routing protocols and their details

Cluster Based Routing Solution in VANET				
Sl. No	**Literature**	**Year**	**Related Topic / Scheme**	**Related Domain**
1	K.Kaur et al.	2016	Cluster based Ant Colony Optimized Protocol in VANET	Soft Computing
2	C. Lai et al.	2018	Group Management Framework in VANET Cellular Network	Cloud Computing
3	O.Wahab et al.	2013	QoS Based Clustering Protocol in VANET	QoS & Cluster Computing
4	R.Pal et al.	2018	Analytical Model for Cluster based VANETs	Cluster & VANET
5	F.Abbas et al.	2018	Reliable Low Latency Routing Scheme using ACO Method in VANET	Soft Computing & VANET

VANET situations and will be useful in displaying their protocols or algorithms. Besides, this model can confirm the simulation comes about acquired from any network test system.

Clustering in Vehicular Adhoc Networks (VANETs) utilizing Quality of Service Optimized Link State Routing (QoS-OLSR) protocol has been reported (O. Wahab et.al, 2013). A few clustering algorithms have been proposed for VANET and MANET. In any case, the portability based algorithms disregard the Quality of Service prerequisites that are important for VANET wellbeing, emergency, and mixed media services while the QoS-based algorithms overlook the fast versatility limitations. The solution is another QoS-based (Rath et.al, 2017) clustering algorithm that considers a tradeoff between QoS necessities and fast versatility imperatives. The objective is to frame stable bunches and keep up the stability amid correspondences and connection disappointments while fulfilling the Quality of Service prerequisites.

In vehicular Adhoc networks (VANETs), message carrier links break with high frequency because of the high speed of mobile vehicles. Based on the current Adhoc on-request multipath remove vector routing scheme, another clustering-based solid low-idleness multipath routing (CRLLR) scheme (Fakhar A et.al, 2018) has been proposed by utilizing Ant Colony Optimization (ACO) system. In this the link reliability is utilized as criteria for Cluster Head (CH) determination. In a given cluster, a vehicle will be chosen as CH in the event that it has greatest link reliability. Besides, the ACO strategy is utilized to effectively figure the optimal courses among the conveying vehicles for VANETs as far as four QoS measurements, reliability, end-to-end dormancy, throughput and vitality utilization. Simulation comes about exhibit that the proposed scheme outflanks the AQRV and T-AOMDV in term of general idleness and reliability at the costs of somewhat higher vitality utilization.It is seen in writing study that the AODV-R beats all other related conventions. Yet at the same time issues identified with blockage control should be tended to. An Ant Colony Optimization based AODV-R convention for better information collection and course determination has been suggested that utilized a strategy which has indicated very huge change over accessible ones.

Secured Group Access Management (SEGM) VANET - Currently, the integrated communication between vehicles in vehicular Adhoc network (VANET) and cellular network can suit rich applications since coordinated VANET-Cellular network empowers portable administrators to furnish vehicle clients with consistent information access to the administrators' administrations. In built-in VANET-Cellular networks, the clustering algorithm partner vehicles into gatherings can give elite information transmission benefit and have been broadly examined (Rath et.al, 2019). Be that as it may, in the majority of the current writing, the security issues in such a situation can't be taken into full thought, which may diminish the dependability of the framework, principally including: 1) How to effectively and safely set up and keep up a gathering; and 2) How to play out the protected gathering access administration. For this end, a protected gathering administration structure in coordinated VANET-Cellular networks has been proposed (C. Lai et.al, 2018) which comprises of two plans, i.e., SEGM-I and SEGM-II. SEGM-I is proposed for the settled put stock in part situation, and SEGM-II is intended for a general case, i.e., dynamic untrusted part situation. The SEGM can be partitioned into two stages: gather setup and upkeep stage, and gathering access validation stage. Amid amass setup and upkeep stage, SEGM-II can dynamically keep up the gathering by utilizing contributory gathering key assention, supporting gathering joining, bunch leaving, aggregate blending, and gathering allotment. Amid assemble get to verification stage, we propose a lightweight gathering access confirmation convention in light of message validation code (MAC) accumulation strategy with DoS assaults obstruction. The proposed SEGM is correspondence productive and secure against antagonistic meddlers and additionally different assaults particular to gather settings. At long last, execution assessments exhibit its productivity regarding bunch administration and access verification overhead.

3.4 Data Communication in Smart city and Smart Application in VANET Scenario

Developing methods of vehicle-to-vehicle correspondence (V2V) in VANET takes place keeping in mind the end goal to give data and help required to maintain a strategic distance from crashes, congested road, and so forth is right now a test among scientists everywhere throughout the world. . Vehicular Adhoc Networks (VANETs) have incredible potential for giving safety and solace applications to Intelligent

Transport Systems (ITS) and to enhance movement safety on streets(Mayada A et.al, 2017). A Case investigation of movement control at Casablanca, the biggest city in Morocco which has a highway going through it has been formulated (A. Fitah et.al, 2018). Two situations have been taken,, first the highway and trunk streets and second alternate streets (essential, optional, tertiary and private streets). To exhibit these situations, different instruments were utilized, for example, OpenStreetMap and SUMO to create a sensible versatility show that gives parts of genuine vehicular activity. Created situations are added to NS-2.35 (Network Simulator) keeping in mind the end goal to break down the execution of 802.11a (Wi-Fi) and 802.11p (DSRC/WAVE).

One of the critical components from Internet of Things (IoT) (Mamata R. et.al, 2017) such as transportation vehicles are using sudden advancements of connection improvement. In Presentldays, to make the Social IoT (SIoT) a reality there are numerous on-going overall research activities and institutionalization endeavors. Internet of Vehicles (IoV) is an expansion to the idea of SIoT that proposes Social Internet of Vehicles (SIoV). SIoV is a case of SIoT, where Smart vehicles are the objects of SIoT that manufactures social relationship and trades data to improve the driving learning and gives different services to the drivers. The proposed Vehicular Social Networks in light of the VIoT (B. Jain et.al, 2018) comprise of countless that transmit information remotely. Notwithstanding, the simple high heterogeneity in equipment abilities of things and QoS necessities for various applications confines the execution of traditional layered protocol solutions and the current cross-layer solutions for remote sensor networks. The outcomes demonstrate that the proposed VSNP (Vehicular Social Network Protocol) for VIoT in light of WSN solution can accomplish a worldwide correspondence ideal and beats existing layered solutions. The novel cross-layer module is an essential advance towards giving effective activity administration and solid vehicle to vehicle correspondence in the SIoT. Table 4 shows VANET routing and successful projects in Smart City Applications

VANETs will be powerful in the event that they can guarantee security, privacy, and trust for road clients. Because of the extensive number of vehicles, a few difficulties exist for open key administration. To start with, the authentication disavowal rundown will be substantial. Likewise, there will be a great deal of correspondence overhead. An answer for this issue is to plan a plan that does not require open key testaments. Symmetric keys can prove to be useful for VANETs, in spite of the fact that they can't ensure non-renouncement. A superior option is to join both symmetric key encryption and open key cryptography. Another arrangement is to total messages to maintain a strategic distance from excess. Secure and proficient accumulation calculations are essential for associated vehicles innovation. Whenever associated vehicles innovation is to increase across the board reception, suspicious clients should be persuaded that their privacy won't be abused. The inquiry that at that

Table 4. VANET routing and successful projects in Smart City Applications

ITS and Smart City Based Smart Applications in VANET				
Sl. No	Literature	Year	Related Topic	Related Domain
1	A.Fitah et al.	2018	Intelligent Transport System in VANET	ITS VANET
2	D.Singh et al.	2016	Visual Big Data Analytics for Traffic Monitoring in Smart City	Big Data Smart City VANET
3	Z.Ding et al.	2016	Smart Transport System using parallel spatio temporal database approach	STS Database System
4	H.Qiu et al.	2015	802.11p VANET Broadcasting Performance with vehicle distribution in VANET	802.11p VANET
5	A.Fazziki et al.	2017	Agent Based Traffic Regulation System for Road side Air Quality Control	Quality Assurance VANET
6	H. Wu et al.	2017	ITS with Network Security in IoV (Internet of Vehicles)	IoV

point emerges is this: in what manner can VANETs guarantee security and trust without encroaching on privacy of road clients? Absolutely, fleeting unknown private and open keys appear to be an incredible arrangement. More research is expected to check the legitimacy and adequacy of this plan in VANETs. All the more particularly, how the authentications will get refreshed and re-filled. Existing arrangements accept Road Side Units (Rath et.al, 2018) will be all around accessible in roads. Nonetheless, there is no certification that this will be the situation. In this manner, security arrangements ought to be completely conveyed with the end goal that vehicles validate messages through between vehicular correspondence. More than whatever else, interruption identification approaches to recognize pernicious vehicles is an absolute necessity have for VANETs. Trouble making recognition systems ought to incorporate position check, flag quality detecting, and data approval. The plan must expel every single vindictive vehicle and prohibit them from adding to message spread (Rath et.al, 2018). There are as of now arrangements that can be based on and enhanced towards this end . Secure directing is additionally a critical piece of VANETs. Future advances must embrace steering conventions that are secure, as well as proficient.

Vehicular Adhoc Networks (VANETs) supports safe travel in roads with smart vehicles and intelligent driving components, transportation security, reliability and administration. A game theory based trust model for VANETs has been distinguished (M. Mohsin et.al, 2017) in late research. The proposed show depends on an attacker and protector security game to distinguish and counter the attacker/vindictive nodes.

The parameters considered for attackers and protector's strategy are lion's share feeling, betweenness centrality, and node thickness. The result of the particular game is dictated by the game lattice which contains the cost (result) values for conceivable activity response blend. Nash harmony (when connected to figure the best strategy for attacker and protector vehicles. The model is recreated in Network Simulator (ns2), and comes about demonstrate that the proposed show performs superior to anything the schemes with arbitrary vindictive nodes and existing game theory based approach regarding throughput, retransmission endeavors and information drop rate for various attacker and safeguard situations.

The way towards routing in Vehicular Adhoc Networks (VANET) is a difficult task in metro city conditions. Finding the shortest end-to-end associated way fulfilling defer limitation and insignificant overhead is gone up against with numerous imperatives and troubles. Such troubles are because of the high portability of vehicles, the successive way disappointments, and the different obstacles, which may influence the dependability of the information transmission and directing. Business Unmanned Aerial Vehicles (UAVs) or what are usually alluded to as drones(R. Oliveira et.al, 2017) can prove to be useful in managing these imperatives. A Detail ponder has been done on how UAVs working in impromptu mode can collaborate with VANET on the ground to aid the directing procedure and enhance the dependability of the information conveyance by spanning the correspondence hole at whatever point it is conceivable. An UVAR – a UAV-Assisted VANETs directing convention was proposed, which enhances information steering and availability of the vehicles on the ground using UAVs. Be that as it may, UVAR does not completely misuse UAVs in the sky for information sending since it utilizes UAVs just when the network is ineffectively thick. In this way, an augmentation of the convention has been modified by supporting two distinctive methods for steering information: (I) conveying information parcels only on the ground utilizing UVAR-G; and (ii) transmitting information bundles in the sky utilizing a responsive directing in light of UVAR-S. Recreation comes about exhibit that the half and half correspondence amongst vehicles and UAVs is in a perfect world suited for VANETs contrasted with customary vehicle-to-vehicle (V2V) interchanges.

3.5 Security and Privacy preserving scheme in VANET -

Vehicular Adhoc Networks (VANETs) are developing in ongoing decades giving constant correspondence between vehicles for a more secure and more open to driving. The fundamental thought of VANET is the way that vehicles can communicate impromptu messages, for example, activity episodes and crisis occasions. The security of such networks is very basic. Initially survey and investigation are being done by prominent specialists and after that the fundamental confirmation conspires in

VANET are contrasted and their advantages and disadvantages. Another confirmation conspire which gives secure interchanges in VANET has been introduced (Seyed M. et.al, 2018) .The method used is a mix of Road Side Unit Based (RSUB) and Tamper Proof Device Based (TPDB) plans.An original thought in NECPPA (Novel and Efficient Conditional Privacy-Preserving Authentication Scheme)is to let the keys and the principle parameters of the framework be put away in the Tamper Proof Device (TPD) of Road Side Units (RSUs). Since, there is dependably a safe and quick communicational connection amongst TA and RSU, embeddings TPD in RSUs is substantially more productive than embeddings them in OBUs. It additionally ought to be noticed that because of the way that in NECPPA scheme, the fundamental key of TA (ace mystery key) isn't put away in all OBUs, the trade off or hacking a solitary OBU does not undermine the entire network notwithstanding what occurs in TPDB conspire which influences the entire vehicles re-to enroll and change their mystery keys. In addition, the proposed plot is substantially more cost effective contrast with other on-line RSUB plans, as it needn't bother with the foundation of on-line RSUs in the entire streets. The security of the scheme has been assessed with formal evidence and ProVerif programmed examination device.

Security purpose in vehicular Adhoc networks (VANETs) depends vigorously on communication plans for the trading of status messages. Solid and on-time conveyance of safety messages is basic in VANETs that requires to a great degree effective communicate models. Model of the unwavering quality of flooding utilized as the basic information dispersal protocol for time-basic safety messages has been delineated (Saira et.al, 2018) with a multi-bounce VANET and end-to-end dependability gave by the network layer. The investigative outcomes release imperative bits of knowledge about the flooding system. For example, the dependability drops exponentially past a specific threshold estimation of parcel misfortune likelihood. Also, the effect of bounce length on dependability can be disastrous if not taken care of fittingly. Keeping in perspective of the way that vitality productive protocols is a key necessity in the forthcoming Internet of Vehicles (IoV), this plan likewise demonstrates that prohibitive flooding is more vitality proficient than plain flooding under a similar unwavering quality requirement. Hypothetical discoveries have been approved through reenactments and propose alterations to plain flooding to make it more solid.

VANET is an emanate innovation with promising future and also incredible difficulties particularly in its security. VANET security systems are introduced in three noteworthy practical parts (Hamssa H. Et.al, 2017) . The initial segment shows a broad outline of VANET security qualities and difficulties and in addition necessities. These prerequisites ought to be mulled over to empower the execution of secure VANET framework with productive correspondence between parties. We give the subtle elements of the ongoing security models and the notable security measures protocols. The second spotlights on a novel order of the diverse assaults known in the VANET writing and their related arrangements. The third area is a correlation between a portion of these arrangements in light of surely understood security criteria in VANET. At that point consideration has been attracted to various open issues and specialized difficulties identified with VANET security, which can help numerous different scientists for sometime later. Table 5. Shows details of safety and security based protocols in VANET.

Table 5. Details of safety and security based protocols in VANET

Safety and Security Based Protocols in VANET				
Sl. No	**Literature**	**Year**	**Related Topic / Scheme**	**Related Domain**
1	S.Morteza et al.	2018	Privacy Preserving Authentication scheme for VANET	Security & Privacy in VANET
2	S.Sattar et al.	2018	Energy efficiency Analysis and reliability of safety messages broadcast in VANET	Reliability Safety VANET
3	H.Hasrouny et al.	2017	VANET Security Challenges & Solutions	Security VANET
4	R. Baiad et al.	2016	Cross layer detection scheme to detect black hole attack in VANET	Network Security Safety VANET
5	S.Sharma et al.	2018	VANET Based Intrusion detection System using Cluster Head	Security Cluster Computing VANET
6	R.oliveira et al.	2017	Reliable Protocol in VANET with Safety Applications	Safety Security VANET

Existing Vehicle-to-Vehicle (V2V) and Vehicle-to-Infrastructure (V2I) based communication experiences different security and execution issues, subsequently Cluster based Communication is favored nowadays. Be that as it may, Cluster based Communication adds additional overhead and weight on the Cluster Head (CH) in deep network situations which in the long run presents delay and slows doen network execution. To decrease the overburdening of single CH, a multi cluster head scheme is proposed (in which numerous nodes in a cluster can go about as CH to share the load of single CH. For a determination of stable CH, Hybrid Fuzzy Multi-criteria Decision making approach (HF-MCDM) is proposed in which Fuzzy Analytic Hierarchy Process (AHP) and TOPSIS strategies are clubbed together for optimal basic leadership. Assist as a result of relationship of Vehicular Adhoc Network (VANET) with life-basic applications, there is a critical requirement for a security structure to identify different malicious attacks. Machine Learning based Intrusion Detection System (IDS) like Support Vector Machine (SVM) is one of the methodologies for checking such attacks. These intrusion detection based system can be joined with different existing optimization strategies to enhance their execution, and Dolphin Swarm Algorithm is one such approach. Dolphins have numerous significant organic highlights like echolocation, exchange of data, coordination, and division of work. These organic highlights joined with swarm insight can be used for streamlining the detection and precision of SVM based IDS. A Multi-Cluster Head abnormality based IDS upgraded by Dolphin Swarm Algorithm has been proposed (S. Sharma et.al, 2018) and its outcomes are contrasted and different existing Security systems as far as parameters can imagine false positive, detection rate, detection time, and so forth and it is watched that the proposed approach performs better. Miniature drones (R. Oliveira et.al, 2017) are being used in checking, transport, security and disaster administration, and different spaces. Imagining that drones frame autonomous networks consolidated into the air activity, An abnormal state engineering for the outline of a collective flying system comprising of drones with on-load up sensors and inserted preparing, coordination, and networking capacities. A multi-drone system has been proposed comprising of quad copters and exhibit its potential in disaster help, pursuit and protect, and airborne checking. Moreover, illustration of configuration difficulties and present potential solutions have been actually looked into.

4. FUTURE RESEARCH DIRECTION IN VANET

VANETs will be very powerful in next generation of wireless technology specifically due to the reason that they can guarantee security, privacy, and trust for road clients. Because of the increase in extensive number of vehicles, a few difficulties exist for open challenge and proper administration. To start with, the authentication disavowal rundown will be substantial. Likewise, there will be a great deal of correspondence overhead. An answer for this issue is to plan a plan that does not require open key testaments. Symmetric keys can prove to be useful for VANETs, in spite of the fact that they can't ensure non-renouncement. A superior option is to join both symmetric key encryption and open key cryptography. Another arrangement is to total messages to maintain a strategic distance from excess. Secure and proficient accumulation calculations are essential for associated vehicles innovation. Whenever associated vehicles innovation is to increase across the board reception, suspicious clients should be persuaded that their privacy won't be abused. The inquiry that at that point emerges is this: in what manner can VANETs guarantee security and trust without encroaching on privacy of road clients? Absolutely, fleeting unknown private and open keys appear to be an incredible arrangement. More research is expected to check the legitimacy and adequacy of this plan in VANETs. All the more particularly, how the authentications will get refreshed and re-filled. Existing arrangements accept Road Side Units will be all around accessible in roads. Nonetheless, there is no certification that this will be the situation. In this manner, security arrangements ought to be completely conveyed with the end goal that vehicles validate messages through between vehicular correspondence. More than whatever else, interruption identification approaches to recognize pernicious vehicles is an absolute necessity have for VANETs. Trouble making recognition systems ought to incorporate position check, flag quality detecting, and data approval. The plan must expel every single vindictive vehicle and prohibit them from adding to message spread. There are as of now arrangements that can be based on and enhanced towards this end. Secure directing is additionally a critical piece of VANETs. Future advances must embrace steering conventions that are secure, as well as proficient.

CONCLUSION

In current communication technology, VANETs have many important advantages compare to other cellular and committed networks. These advantages have motivated researchers and network designers to focus more on their efficiency and to develop more robust application oriented protocols.. The most significant part of VANET is the sensor devices to be implemented in control parts of the vehicle to account the situation and circumstance of the vehicle and immediately react to the external stimuli as programmed and signal the action very rapidly in an intelligent way to the driver. It should also receive and act on commands from the drivers and execute them for safety and security. VANETS are basically deigned to become useful for highway path protection as well as several business purposes basically for faster communication overcoming the network obscurity. For instance; vehicular network may be used to inform drivers so that the likelihood of visitors jams, road congestion, accident possibilities and emergency signals can be given offering higher comfort as well as effectiveness . Wireless connection systems have allowed most of the benefits inside our lives and in addition improved our daily efficiency also . Adhoc networks work without a defined preset maintained infrastructure. VANETS working with 802.11structured WLAN innovation now obtained significant interest. The above article presents a complete analysis on current state of the art of the spectacular network of VANET and the vibrant routing schemes used by different research work and provides a study platform to the new researchers to extend their study further in various challenges in routing.

REFERENCES

Abbas, F., & Fan, P. (2018). Clustering-based reliable low-latency routing scheme using ACO method for vehicular networks. *Vehicular Communications, 12*, 66-74. doi:10.1016/j.vehcom.2018.02.004

Abdel-Halim, I. T., & Hossam, M. A. F. (2018). Prediction-based protocols for vehicular Adhoc Networks: Survey and taxonomy. *Computer Networks, 130*, 34-50. doi:10.1016/j.comnet.2017.10.009

Abdel Wahab, O., Otrok, H., & Mourad, A. (2013). VANET QoS-OLSR: QoS-based clustering protocol for Vehicular Adhoc Networks. *Computer Communications, 36*(13), 1422-1435. doi:10.1016/j.comcom.2013.07.003

Abdelgadir, M., Saeed, R. A., & Babiker, A. (2017). Mobility Routing Model for Vehicular Adhoc Networks (VANETs), Smart City Scenarios. *Vehicular Communications, 9*, 154-161.

Baiad, R., Alhussein, O., Otrok, H., & Muhaidat, S. (2016). Novel cross layer detection schemes to detect blackhole attack against QoS-OLSR protocol in VANET. *Vehicular Communications, 5*, 9-17. doi:10.1016/j.vehcom.2016.09.001

Boussoufa-Lahlah, S., Semchedine, F., & Bouallouche-Medjkoune, L. (2018). Geographic routing protocols for Vehicular Adhoc NETworks (VANETs): A survey. *Vehicular Communications, 11*, 20-31. doi:10.1016/j.vehcom.2018.01.006

Brik, B., Lagraa, N., Lakas, A., & Cheddad, A. (2016). DDGP: Distributed Data Gathering Protocol for vehicular networks. *Vehicular Communications, 4*, 15-29. doi:10.1016/j.vehcom.2016.01.001

Cheng, H., & Fei, X. (2018). An efficient sparse coverage protocol for RSU deployment over urban VANETs. *Adhoc Networks, 24*(B), 85-102. doi:10.1016/j.adhoc.2014.07.022

Cunha, Villas, Boukerche, Maia, Viana, Mini, & Loureiro. (2016). Data communication in VANETs: Protocols, applications and challenges. *Adhoc Networks, 44*, 90-103.

Darwish, T., Abu Bakar, K., & Hashim, A. (2016). Green geographical routing in vehicular Adhoc networks: Advances and challenges. *Computers & Electrical Engineering, 64*, 436-449. .compeleceng.2016.09.030 doi:10.1016/j

Feukeu, E. A., & Zuva, T. (2017). DBSMA Approach for Congestion Mitigation in VANETs. *Procedia Computer Science, 109*, 42-49. .05.293 doi:10.1016/j.procs.2017

Filho, J. G., Patel, A., Bruno, L. A. B., & Celestino, J. (2016). A systematic technical survey of DTN and VDTN routing protocols. *Computer Standards & Interfaces, 48*, 139-159. doi:10.1016/j.csi.2016.06.004

Fitah, A., Badri, A., Moughit, M., & Sahel, A. (2018). Performance of DSRC and WIFI for Intelligent Transport Systems in VANET. *Procedia Computer Science, 127*, 360-368. doi:1016/j.procs.2018.01.133

Hasrouny, H., Samhat, A. E., Bassil, C., & Laouiti, A. (2017). VANet security challenges and solutions: A survey. *Vehicular Communications, 7*, 7-20. doi:10.1016/j.vehcom.2017.01.002

Hossain, Datta, Hossain, & Edmonds. (2017). ResVMAC: A Novel Medium Access Control Protocol for Vehicular Adhoc Networks. *Procedia Computer Science, 109*, 432-439. doi:10.1016/j.procs.2017.05.413

Jain, B., Brar, G., Malhotra, J., Rani, S., & Ahmed, S. H. (2018). A cross layer protocol for traffic management in Social Internet of Vehicles. *Future Generation Computer Systems, 82*, 707-714. doi:10.1016/j.future.2017.11.019

Kaur, K., & Kad, S. (2016). Enhanced clustering based AODV-R protocol using Ant Colony Optimization in VANETS. *IEEE 1st International Conference on Power Electronics, Intelligent Control and Energy Systems (ICPEICES)*, 1-5. 10.1109/ICPEICES.2016.7853381

Lai, C., Zheng, D., Zhao, Q., & Jiang, X. (2018). SEGM: A secure group management framework in integrated VANET-cellular networks. *Vehicular Communications, 11*, 33-45.

Mamata, R. (2018). An Analytical Study of Security and Challenging Issues in Social Networking as an Emerging Connected Technology. In *Proceedings of 3rd International Conference on Internet of Things and Connected Technologies*. Malaviya National Institute of Technology. https://ssrn.com/abstract=3166509

Mamata, R. B. P. (2018). Communication Improvement and Traffic Control Based on V2I in Smart City Framework. *International Journal of Vehicular Telematics and Infotainment Systems, 2*(1).

Mehdi, M. M., Raza, I., & Hussain, S. A. (2017). A game theory based trust model for Vehicular Adhoc Networks (VANETs). *Computer Networks, 121*, 152-172. doi:10.1016/j.comnet.2017.04.024

Ngo & Oh. (2014). A Link Quality Prediction Metric for Location based Routing Protocols under Shadowing and Fading Effects in Vehicular Adhoc Networks. *Procedia Computer Science, 34*, 565-570. doi:10.1016/j.procs.2014.07.071

Oliveira, R., Montez, C., Boukerche, A., & Wangham, M. S. (2017). Reliable data dissemination protocol for VANET traffic safety applications. *Adhoc Networks, 63*, 30-44. doi:10.1016/j.adhoc.2017.05.002

Oubbati, O. S., Lakas, A., Zhou, F., Güneş, M., Lagraa, N., & Yagoubi, M. B. (2017). Intelligent UAV-assisted routing protocol for urban VANETs. *Computer Communications, 107*, 93-111. doi:1016/j.comcom.2017.04.001

Pal, R., Prakash, A., Tripathi, R., & Singh, D. (2018). Analytical model for clustered vehicular Adhoc network analysis. *ICT Express*. doi:10.1016/ j.icte.2018.01.001

Pattanayak, B., & Rath, M. (2014). A Mobile Agent Based Intrusion Detection System Architecture For Mobile Adhoc Networks. *Journal of Computational Science, 10*(6), 970–975. doi:10.3844/jcssp.2014.970.975

Pournaghi, S. M., Zahednejad, B., Bayat, M., & Farjami, Y. (2018). NECPPA: A novel and efficient conditional privacy-preserving authentication scheme for VANET. *Computer Networks, 134*, 78-92. .01.015 doi:10.1016/ j.comnet.2018

Rahim, A., Kong, X., Xia, F., Ning, Z., Ullah, N., Wang, J., & Das, S. K. (2018). Vehicular Social Networks: A survey. *Pervasive and Mobile Computing, 43*, 96-113. doi:10.1016/j.pmcj.2017.12.004

Rath & Oreku. (2018). Security Issues in Mobile Devices and Mobile Adhoc Networks. In Mobile Technologies and Socio-Economic Development in Emerging Nations. IGI Global. DOI: doi:10.4018/978-1-5225-4029-8.ch009

Rath. (2018). Effective Routing in Mobile Adhoc Networks With Power and End-to-End Delay Optimization: Well Matched With Modern Digital IoT Technology Attacks and Control in MANET. In *Advances in Data Communications and Networking for Digital Business Transformation*. IGI Global. DOI: doi:10.4018/978-1-5225-5323-6.ch007

Rath, M. (2017). Resource provision and QoS support with added security for client side applications in cloud computing. *International Journal of Information Technology, 9*(3), 1–8.

Rath, M. (2018). An Exhaustive Study and Analysis of Assorted Application and Challenges in Fog Computing and Emerging Ubiquitous Computing Technology. *International Journal of Applied Evolutionary Computation, 9*(2), 17-32. www. igi-global.com/ijaec

Rath, M. (2018). A Methodical Analysis of Application of Emerging Ubiquitous Computing Technology With Fog Computing and IoT in Diversified Fields and Challenges of Cloud Computing. *International Journal of Information Communication Technologies and Human Development, 10*(2). DOI: doi:10.4018/978-1-5225-4100-4.ch002

Rath, M. (2018). Smart Traffic Management System for Traffic Control using Automated Mechanical and Electronic Devices. *Mater. Sci. Eng., 377.* /377/1/01220110.1088/1757-899X/377/1/012201

Rath, M., & Panda, M. R. (2017). MAQ system development in mobile Adhoc networks using mobile agents. *IEEE 2nd International Conference on Contemporary Computing and Informatics (IC3I),* 794-798.

Rath, M., & Panigrahi, C. (2016). Prioritization of Security Measures at the Junction of MANET and IoT. In *Second International Conference on Information and Communication Technology for Competitive Strategies.* ACM Publication. 10.1145/2905055.2905187

Rath, M., & Pati, B. (2017). *Load balanced routing scheme for MANETs with power and delay optimisation, International Journal of Communication Network and Distributed Systems* , 19.

Rath, M., Pati, B., Panigrahi, C. R., & Sarkar, J. L. (2019). QTM: A QoS Task Monitoring System for Mobile Adhoc Networks. In P. Sa, S. Bakshi, I. Hatzilygeroudis, & M. Sahoo (Eds.), *Recent Findings in Intelligent Computing Techniques. Advances in Intelligent Systems and Computing* (Vol. 707). Springer. doi:10.1007/978-981-10-8639-7_57

Rath, M., Pati, B., & Pattanayak, B. (2015). Energy Competent Routing Protocol Design in MANET with Real time Application Provision. *International Journal of Business Data Communications and Networking, IGI Global, 11*(1), 50–60. doi:10.4018/IJBDCN.2015010105

Rath, M., Pati, B., & Pattanayak, B. (2015). Delay and power based network assessment of network layer protocols in MANET. *2015 International Conference on Control, Instrumentation, Communication and Computational Technologies (IEEE ICCICCT),* 682-686. 10.1109/ICCICCT.2015.7475365

Rath, M., Pati, B., & Pattanayak, B. (2016). QoS Satisfaction in MANET Based Real Time Applications. *International Journal of Control Theory and Applications, 9*(7), 3069-3083.

Rath, M., Pati, B., & Pattanayak, B. (2016). Energy Efficient MANET Protocol Using Cross Layer Design for Military Applications. *Defence Science Journal, 66*(2), 146. doi:10.14429/dsj.66.9705

Rath, M., Pati, B., & Pattanayak, B. (2016). Comparative analysis of AODV routing protocols based on network performance parameters in Mobile Adhoc Networks. In Foundations and Frontiers in Computer, Communication and Electrical Engineering, (pp. 461-466). CRC Press, Taylor & Francis.

Rath, M., Pati, B., & Pattanayak, B. (2016). Resource Reservation and Improved QoS for Real Time Applications in MANET. *Indian Journal of Science and Technology, 9*(36). doi:10.17485/ijst/2016/v9i36/100910

Rath, M., Pati, B., & Pattanayak, B. K. (2016). Inter-Layer Communication Based QoS Platform for Real Time Multimedia Applications in MANET. Wireless Communications, Signal Processing and Networking (IEEE WiSPNET), 613-617. doi:10.1109/WiSPNET.2016.7566203

Rath, M., Pati, B., & Pattanayak, B. K. (2017). Cross layer based QoS platform for multimedia transmission in MANET. *11th International Conference on Intelligent Systems and Control (ISCO)*, 402-407. 10.1109/ISCO.2017.7856026

Rath, M., Pati, B., & Pattanayak, B. K. (2018). Relevance of Soft Computing Techniques in the Significant Management of Wireless Sensor Networks. In Soft Computing in Wireless Sensor Networks (pp. 86-106). Chapman and Hall/CRC, Taylor & Francis Group. doi:10.1201/9780429438639-4

Rath, M., Pati, B., & Pattanayak, B. K. (2019). Mobile Agent-Based Improved Traffic Control System in VANET. In A. Krishna, K. Srikantaiah, & C. Naveena (Eds.), *Integrated Intelligent Computing, Communication and Security. Studies in Computational Intelligence* (Vol. 771). Springer. doi:10.1007/978-981-10-8797-4_28

Rath, M., & Pattanayak, B. (2016). A Contemporary Survey and Analysis of Delay and Power Based Routing Protocols in MANET. *Journal of Engineering and Applied Sciences (Asian Research Publishing Network), 11*(1), 536–540.

Rath, M., & Pattanayak, B. (2017). MAQ:A Mobile Agent Based QoS Platform for MANETs. *International Journal of Business Data Communications and Networking, IGI Global, 13*(1), 1–8. doi:10.4018/IJBDCN.2017010101

Rath, M., & Pattanayak, B. (2018). Technological improvement in modern health care applications using Internet of Things (IoT) and proposal of novel health care approach. *International Journal of Human Rights in Healthcare*. doi:10.1108/IJHRH-01-2018-0007

Rath, M., & Pattanayak, B. (2018). Technological improvement in modern health care applications using Internet of Things (IoT) and proposal of novel health care approach. *International Journal of Human Rights in Healthcare*. doi:10.1108/IJHRH-01-2018-0007

Rath, M., & Pattanayak, B. K. (2014). A methodical survey on real time applications in MANETS: Focussing On Key Issues. *International Conference on, High Performance Computing and Applications (IEEE ICHPCA)*, 1-5, 22-24. 10.1109/ICHPCA.2014.7045301

Rath, M., & Pattanayak, B. K. (2018). SCICS: A Soft Computing Based Intelligent Communication System in VANET. Smart Secure Systems – IoT and Analytics Perspective. *Communications in Computer and Information Science*, *808*, 255–261. doi:10.1007/978-981-10-7635-0_19

Rath, M., & Pattanayak, B. K. (2018). Monitoring of QoS in MANET Based Real Time Applications. In Information and Communication Technology for Intelligent Systems (ICTIS 2017) - Volume 2. ICTIS 2017. Smart Innovation, Systems and Technologies (vol. 84, pp. 579-586). Springer. doi:10.1007/978-3-319-63645-0_64

Rath, M., & Pattanayak, B. K. (2018). Monitoring of QoS in MANET Based Real Time Applications. In Information and Communication Technology for Intelligent Systems Volume 2. ICTIS. Smart Innovation, Systems and Technologies (vol. 84, pp. 579-586). Springer. doi:10.1007/978-3-319-63645-0_64

Rath, M., Pattanayak, B. K., & Pati, B. (2017). *Energetic Routing Protocol Design for Real-time Transmission in Mobile Adhoc Network. In Computing and Network Sustainability, Lecture Notes in Networks and Systems* (Vol. 12). Springer.

Rath, M., Rout, U. P., & Pujari, N. (2017). *Congestion Control Mechanism for Real Time Traffic in Mobile Adhoc Networks, Computer Communication, Networking and Internet Security. In Lecture Notes in Networks and Systems* (Vol. 5). Springer.

Rath, M., Swain, J., Pati, B., & Pattanayak, B. K. (2018). *Attacks and Control in MANET. In Handbook of Research on Network Forensics and Analysis Techniques*. IGI Global.

Rtah, M. (2018). Big Data and IoT-Allied Challenges Associated With Healthcare Applications in Smart and Automated Systems. *International Journal of Strategic Information Technology and Applications*, 9(2), 18–34. Advance online publication. doi:10.4018/IJSITA.2018040102

Rtah, M. (2018). Big Data and IoT-Allied Challenges Associated With Healthcare Applications in Smart and Automated Systems. *International Journal of Strategic Information Technology and Applications*, 9(2), 18–34. Advance online publication. doi:10.4018/IJSITA.2018040102

Sahoo, J., & Rath, M. (2017). Study and Analysis of Smart Applications in Smart City Context. *2017 International Conference on Information Technology (ICIT)*, 225-228. 10.1109/ICIT.2017.38

Saleh, Gamel, & Abo-Al-Ez. (2017). A Reliable Routing Protocol for Vehicular Adhoc Networks. *Computers & Electrical Engineering, 64*, 473-495. doi:10.1016/j.compeleceng.2016.11.011

Sattar, S., Qureshi, H. K., Saleem, M., Mumtaz, S., & Rodriguez, J. (2018). Reliability and energy-efficiency analysis of safety message broadcast in VANETs. *Computer Communications, 119*, 118-126. doi:10.1016/j.comcom.2018.01.006

Sharma, S. (2018). Hybrid fuzzy multi-criteria decision making based multi cluster head dolphin swarm optimized IDS for VANET. *Vehicular Communications, 12*, 23-38. doi:10.1016/ j.vehcom.2017.12.003

Singh, S., & Agrawal, S. (2014). *VANET routing protocols: Issues and challenges. In 2014 Recent Advances in Engineering and Computational Sciences*. RAECS.

Singh, S., Negi, S., Verma, S. K., & Panwar, N. (2018). Comparative Study of Existing Data Scheduling Approaches and Role of Cloud in VANET Environmen. *Procedia Computer Science, 125*, 925-934. .procs.2017.12.118 doi:10.1016/j

Wang, X., Wang, D., & Sun, Q. (2018). Reliable routing in IP-based VANET with network gaps. *Computer Standards & Interfaces, 55*, 80-94. doi:10.1016/j.csi.2017.05.002

Compilation of References

10 Gigabit Ethernet - Wikipedia. (n.d.). Retrieved July 10, 2020, from https://en.wikipedia.org/wiki/10_Gigabit_Ethernet

30 Eye-Opening Big Data Statistics for 2020: Patterns Are Everywhere. (n.d.). Retrieved July 10, 2020, from https://kommandotech.com/statistics/big-data-statistics/

39 Big Data Statistics for 2020. (n.d.). Retrieved July 10, 2020, from https://leftronic.com/big-data-statistics/

3 GPP. (2015, September). *RWS-150002: views on next generation wireless access.* 3GPP RAN Workshop on 5G, 9/2015, Lenovo, Motorola Mobility.

5G Gear Up - Huawei. (n.d.). Retrieved July 10, 2020, from https://carrier.huawei.com/en/spotlight/5g

Abbas, F., & Fan, P. (2018). Clustering-based reliable low-latency routing scheme using ACO method for vehicular networks. *Vehicular Communications, 12*, 66-74. doi:10.1016/j.vehcom.2018.02.004

Abdel Wahab, O., Otrok, H., & Mourad, A. (2013). VANET QoS-OLSR: QoS-based clustering protocol for Vehicular Adhoc Networks. *Computer Communications, 36*(13), 1422-1435. doi:10.1016/j.comcom.2013.07.003

Abdelgadir, M., Saeed, R. A., & Babiker, A. (2017). Mobility Routing Model for Vehicular Adhoc Networks (VANETs), Smart City Scenarios. *Vehicular Communications, 9*, 154-161.

Abdel-Halim, I. T., & Hossam, M. A. F. (2018). Prediction-based protocols for vehicular Adhoc Networks: Survey and taxonomy. *Computer Networks, 130*, 34-50. doi:10.1016/j.comnet.2017.10.009

Abdullahi, I., Arif, S., & Hassan, S. (2015). Survey on caching approaches in Information Centric Networking. *Journal of Network and Computer Applications, 56*(2015), 48–59. doi:10.1016/j.jnca.2015.06.011

Abusch-Magder, D., Bosch, P., Klein, T. E., Polakos, P. A., Samuel, L. G., & Viswanathan, H. (2007). 911-NOW: A network on wheels for emergency response and disaster recovery operations. *Bell Labs Technical Journal, 11*(4), 113–133. doi:10.1002/bltj.20199

Afanasyev, A., Halderman, J. A., Ruoti, S., Seamons, K., Yu, Y., Zappala, D., & Zhang, L. (2016). Content-based security for the web. In *New Security Paradigms Workshop* (pp. 49–60). Academic Press.

Afanasyev, A., Jiang, X., Yu, Y., Tan, J., Xia, Y., Mankin, A., & Zhang, L. (2017). NDNS: A DNS-like name service for NDN. *International Conference on Computer Communications and Networks (ICCCN)*. 10.1109/ICCCN.2017.8038461

Afifi, A., Elsayed, K. M., & Khattab, A. (2013). Interference-aware radio resource management framework for the 3GPP LTE uplink with QoS constraints. In *2013 IEEE Symposium on Computers and Communications (ISCC)* (pp. 693–698). IEEE. 10.1109/ISCC.2013.6755029

Ahlgren, B., Dannewitz, C., Imbrenda, C., Kutscher, D., & Ohlman, B. (2012). A Survey of Information-Centric Networking. *IEEE Communications Magazine, 50*(July), 1–26. doi:10.1109/MCOM.2012.6231276

Ahmad, Ayaz, & Assaad, M. (2011a). *Canonical dual method for resource allocation and adaptive modulation in uplink SC-FDMA systems.* ArXiv Preprint ArXiv:1103.4547

Ahmad, Ayaz, & Assaad, M. (2011b). Polynomial-complexity optimal resource allocation framework for uplink SC-FDMA systems. In *2011 IEEE Global Telecommunications Conference-GLOBECOM 2011* (pp. 1–5). IEEE.

Ahmad, Ayaz, & Assaad, M. (2011c). Power efficient resource allocation in uplink SC-FDMA systems. In *2011 IEEE 22nd International Symposium on Personal, Indoor and Mobile Radio Communications* (pp. 1351–1355). IEEE.

Ahmad, A. (2014). Resource allocation and adaptive modulation in uplink SC-FDMA systems. *Wireless Personal Communications, 75*(4), 2217–2242. doi:10.100711277-013-1464-6

Ahmad, A. (2015). Power allocation for uplink SC-FDMA systems with arbitrary input distribution. *Electronics Letters, 52*(2), 111–113. doi:10.1049/el.2015.1779

Ahmad, A., & Anwar, M. (2016). Resource Allocation for OFDMA Based Cognitive Radio Networks with Arbitrarily Distributed Finite Power Inputs. *Wireless Personal Communications, 88*(4), 839–854. doi:10.100711277-016-3208-x

Ahmad, A., & Khan, R. (2017). Resource allocation for SC-FDMA based cognitive radio systems. *International Journal of Communication Systems, 30*(5), e3046. doi:10.1002/dac.3046

Ahmad, A., & Shah, N. (2015). A joint resource optimization and adaptive modulation framework for uplink single-carrier frequency-division multiple access systems. *International Journal of Communication Systems, 28*(3), 437–456. doi:10.1002/dac.2677

Ahmed, I., & Mohamed, A. (2011). Fairness Aware Group Proportional Frequency Domain Resource Allocation in L-SC-FDMA Based Uplink. *International Journal of Communications. Network and System Sciences, 4*(08), 487–494. doi:10.4236/ijcns.2011.48060

Ahmed, I., & Mohamed, A. (2015). Power control and group proportional fairness for frequency domain resource allocation in L-SC-FDMA based LTE uplink. *Wireless Networks*, *21*(6), 1819–1834. doi:10.100711276-014-0845-4

Ahmed, I., Mohamed, A., & Shakeel, I. (2010). *On the group proportional fairness of frequency domain resource allocation in L-SC-FDMA based LTE uplink. In 2010 IEEE Globecom Workshops*. IEEE.

Ahmed, S. H., Bouk, S. H., & Kim, D. (2015). RUFS: RobUst forwarder selection in vehicular content-centric networks. *IEEE Communications Letters*, *19*(9), 1616–1619. doi:10.1109/LCOMM.2015.2451647

Ahmed, S. H., Bouk, S. H., Yaqub, M. A., Kim, D., & Song, H. (2017). DIFS: Distributed Interest Forwarder Selection in Vehicular Named Data Networks. *IEEE Transactions on Intelligent Transportation Systems*.

Aijaz, A., Nakhai, M. R., & Aghvami, A. H. (2014). Power efficient uplink resource allocation in LTE networks under delay QoS constraints. In *2014 IEEE Global Communications Conference* (pp. 1239–1244). IEEE. 10.1109/GLOCOM.2014.7036978

Akande, K., Iqbal, N., Zerguine, A., Al-Dhahir, N., & Zidouri, A. (2016). Frequency domain soft-constraint multimodulus blind equalization for uplink SC-FDMA. *EURASIP Journal on Advances in Signal Processing*, *2016*(1), 23. doi:10.118613634-016-0317-3

Al-Ashban, A., & Burney, M. A. (2001). Customer adoption of tele-banking technology: The case of Saudi Arabia. *International Journal of Bank Marketing*, *19*(5), 191–204. doi:10.1108/02652320110399683

Algarni, M., Nair, V., & Martin, D. (2013). *Software Defined Networking Overview And Implementation*. George Mason University IFS.

Al-Imari, M., Xiao, P., Imran, M. A., & Tafazolli, R. (2014). Radio resource allocation for uplink OFDMA systems with finite symbol alphabet inputs. *IEEE Transactions on Vehicular Technology*, *63*(4), 1917–1921. doi:10.1109/TVT.2013.2287809

Alizadeh, M., Greenberg, A., Maltz, D., Padhye, J., Patel, P., Prabhakar, B., Sengupta, S., & Sridharan, M. (2010). *DCTCP: Efficient packet transport for the commoditized data center*. Academic Press.

Allman, M, Glover, D., & Sanchez, L. (1999). *RFC2488: Enhancing TCP Over Satellite Channels using Standard Mechanisms*. RFC Editor.

Allman, M., Paxson, V., & Stevens, W. (1999). *TCP congestion control*. Academic Press.

Allman, M., & Paxson, V. (2001). On estimating end-to-end network path properties. *Computer Communication Review*, *31*(2, supplement), 124–151. doi:10.1145/844193.844203

Al-Sultan, S., Al-Doori, M. M., Al-Bayatti, A. H., & Zedan, H. (2014). A comprehensive survey on vehicular ad hoc network. *Journal of Network and Computer Applications*, *37*, 380–392. doi:10.1016/j.jnca.2013.02.036

Amditis, A., Bertolazzi, E., Bimpas, M., Biral, F., Bosetti, P., Da Lio, M., Danielsson, L., Gallione, A., Lind, H., Saroldi, A., & Sjogren, A. (2010). A holistic approach to the integration of safety applications: The INSAFES subproject within the European framework programme 6 integrating project PReVENT. *IEEE Transactions on Intelligent Transportation Systems*, *11*(3), 554–566. doi:10.1109/TITS.2009.2036736

Anckar, B., & D'Incau, D. (2002). Value creation in mobile commerce: Findings from a consumer survey. *Journal of Information Technology Theory and Application*, *4*(1), 41–62.

Anderson. (2008). *OpenFlow enabling innovation in campus networks*. University of Washington.

Andre, T., Hummel, K. A., Schoellig, A. P., Yanmaz, E., Asadpour, M., Bettstetter, C., Grippa, P., Hellwagner, H., Sand, S., & Zhang, S. (2014). Application-driven design of aerial communication networks. *IEEE Communications Magazine*, *52*(5), 129–137. doi:10.1109/MCOM.2014.6815903

Antonino, P. O., & Morgenstern, A. (2018, April 30). Straightforward specification of adaptation-architecture-significant requirements of IoT-enabled cyber-physical systems. *2018 IEEE International Conference on Software Architecture Companion (ICSA-C)*. 10.1109/ICSA-C.2018.00012

Athulya, M. S., & Sheeba, V. S. (2012). Security in Mobile Ad-Hoc Networks. *Computing Communication & Networking Technologies (ICCCNT), 2012 Third International Conference on*, 1-5. 10.1109/ICCCNT.2012.6396047

Baheti, R., & Gill, H. (2011). *Cyber-physical Systems*. Chapter. In T. Samad & A. M. Annaswamy (Eds.), *The impact of control technology*. IEEE Control Systems Society. Available at www.ieeecss.org

Baiad, R., Alhussein, O., Otrok, H., & Muhaidat, S. (2016). Novel cross layer detection schemes to detect blackhole attack against QoS-OLSR protocol in VANET. *Vehicular Communications*, *5*, 9-17. doi:10.1016/j.vehcom.2016.09.001

Baracca, P., Tomasin, S., & Benvenuto, N. (2013). Resource allocation with multicell processing, interference cancelation and backhaul rate constraint in single carrier FDMA systems. *Physical Communication*, *8*, 69–80. doi:10.1016/j.phycom.2012.09.003

Barbará, D. (1999). Mobile Computing and Databases - A Survey. *IEEE Transactions on Knowledge and Data Engineering*, *11*(1), 108–117. doi:10.1109/69.755619

Basudeo, S., & Jasmine, K. (2012). Comparative Study on Various Methods and types of mobile payment system. *Proc. International Conference on Advances in Mobile Network, Communication and Its Applications*, 10.

Berardinelli, G., de Temino, L. A. M. R., Frattasi, S., Rahman, M. I., & Mogensen, P. (2008). OFDMA vs. SC-FDMA: Performance comparison in local area IMT-A scenarios. *IEEE Wireless Communications*, *15*(5), 64–72. doi:10.1109/MWC.2008.4653134

Bernardini, C., Asghar, M. R., & Crispo, B. (2017). *Security and privacy in vehicular communications: Challenges and opportunities*. Vehicular Communications.

Bian, C., Zhao, T., Li, X., & Yan, W. (2015a). Boosting named data networking for data dissemination in urban VANET scenarios. *Vehicular Communications*, 2(4), 195–207. doi:10.1016/j.vehcom.2015.08.001

Bian, C., Zhao, T., Li, X., & Yan, W. (2015b). Boosting named data networking for efficient packet forwarding in urban VANET scenarios. In *IEEE International Workshop on Local and Metropolitan Area Networks (LANMAN)* (pp. 1–6). 10.1109/LANMAN.2015.7114718

Biaz, S., & Vaidya, N. H. (2003). Differentiated services: A new direction for distinguishing congestion losses from wireless losses. Technical report, University of Auburn.

Bicket, J., Aguayo, D., Biswas, S., & Morris, R. (2005). Architecture and evaluation of an unplanned 802.11 b mesh network. *Proceedings of the 11th Annual International Conference on Mobile Computing and Networking*, 31–42.

Black, N. J., Lockett, A., Winklhofer, H., & Ennew, C. (2001). The adoption of Internet financial services: A qualitative study. *International Journal of Retail & Distribution Management*, 29(8), 390–398. doi:10.1108/09590550110397033

Blasch, E., & Kadar, I. (2017). Panel summary of Cyber-Physical Systems (CPS) and Internet of Things (IoT) opportunities with information fusion. *Proceedings Volume 10200, Signal Processing, Sensor/Information Fusion, and Target Recognition XXVI*. doi:10.1117/12.2264683

Booysen, M. J., Zeadally, S., & Van Rooyen, G.-J. (2011). Survey of media access control protocols for vehicular ad hoc networks. *IET Communications*, 5(11), 1619–1631. doi:10.1049/iet-com.2011.0085

Bouk, S. H., Ahmed, S. H., Hussain, R., & Eun, Y. (2018). Named Data Networking's Intrinsic Cyber-Resilience for Vehicular CPS. *IEEE Access*. Retrieved from https://ieeexplore.ieee.org/abstract/document/8492520/

Bouk, S. H., Ahmed, S. H., & Kim, D. (2014). Hierarchical and hash-based naming scheme for vehicular information centric networks. In *International Conference on Connected Vehicles and Expo (ICCVE)* (pp. 765–766). 10.1109/ICCVE.2014.7297653

Bouk, S. H., Ahmed, S. H., & Kim, D. (2015). Hierarchical and hash based naming with Compact Trie name management scheme for Vehicular Content Centric Networks. *Computer Communications*, 71, 73–83. doi:10.1016/j.comcom.2015.09.014

Boussoufa-Lahlah, S., Semchedine, F., & Bouallouche-Medjkoune, L. (2018). Geographic routing protocols for Vehicular Adhoc NETworks (VANETs): A survey. *Vehicular Communications, 11*, 20-31. doi:10.1016/j.vehcom.2018.01.006

Bouwman, H., Carlsson, C., Molina-Castillo, F. J., & Walden, P. (2007). Barriers and drivers in the adoption of current and future mobile services in Finland. *Telematics and Informatics*, 24(2), 145–160. doi:10.1016/j.tele.2006.08.001

Brakmo, L. S., & Peterson, L. L. (1995). TCP Vegas: End to end congestion avoidance on a global Internet. *IEEE Journal on Selected Areas in Communications, 13*(8), 1465–1480. doi:10.1109/49.464716

Brik, B., Lagraa, N., Lakas, A., & Cheddad, A. (2016). DDGP: Distributed Data Gathering Protocol for vehicular networks. *Vehicular Communications, 4*, 15-29. doi:10.1016/j.vehcom.2016.01.001

Buchegger, S., & Le Boudec, J. Y. (2004). A robust reputation system for peer-to-peer and mobile ad-hoc networks. In *P2PEcon 2004* (pp. 2004–2009). No. LCA-CONF.

Burg, A., Chattopadhyay, A., & Lam, K.-Y. (2018). Wireless communication and security issues for cyber–physical systems and the internet-of-things. *Proceedings of the IEEE, 106*(1), 38–60. doi:10.1109/JPROC.2017.2780172

Bu, S., Yu, F. R., Liu, X. P., Mason, P., & Tang, H. (2011a). Distributed combined authentication and intrusion detection with data fusion in high-security mobile ad hoc networks. *IEEE Transactions on Vehicular Technology, 60*(3), 1025–1036. doi:10.1109/TVT.2010.2103098

Bu, S., Yu, F. R., Liu, X. P., & Tang, H. (2011b). Structural results for combined continuous user authentication and intrusion detection in high security mobile ad-hoc networks. *IEEE Transactions on Wireless Communications, 10*(9), 3064–3073. doi:10.1109/TWC.2011.071411.102123

Caini, C., & Firrincieli, R. (2004). TCP Hybla: A TCP enhancement for heterogeneous networks. *International Journal of Satellite Communications and Networking, 22*(5), 547–566. doi:10.1002at.799

Campista, M. E. M., Rubinstein, M. G., Moraes, I. M., Costa, L. H. M. K., & Duarte, O. C. M. B. (2014). Challenges and research directions for the future internetworking. *IEEE Communications Surveys and Tutorials, 16*(2), 1050–1079. doi:10.1109/SURV.2013.100213.00143

Cao, Y., Yu, D., Zeng, L., Liu, Q., Wu, F., Gui, X., & Huang, M. (2019). Towards Efficient Parallel Multipathing: A Receiver-Centric Cross-Layer Solution to Aid Multipath TCP. *2019 IEEE 25th International Conference on Parallel and Distributed Systems (ICPADS)*, 790–797.

Car to Car Communication Consortium. (n.d.). *CAR 2 CAR Communication Consortium Manifesto*. Author.

Cavalleri, M., Prudentino, R., Pozzoli, U., & Veni, G. (2000). A set of tools for building PostgreSQL distributed database in biomedical environment. *Proc.* 22nd Annual *International conference on Engineering in Medicine & Biology*, 540-544. 10.1109/IEMBS.2000.900796

Cerf, V. G., & Icahn, R. E. (2005). A protocol for packet network intercommunication. *Computer Communication Review, 35*(2), 71–82. doi:10.1145/1064413.1064423

Chandran, K., Raghunathan, S., Venkatesan, S., & Prakash, R. (2001). A feedback-based scheme for improving TCP performance in ad hoc wireless networks. *IEEE Personal Communications, 8*(1), 34–39. doi:10.1109/98.904897

Chang, C.-H., Chao, H.-L., & Liu, C.-L. (2011). Sum throughput-improved resource allocation for LTE uplink transmission. In 2011 IEEE Vehicular Technology Conference (VTC Fall) (pp. 1–5). IEEE. doi:10.1109/VETECF.2011.6093138

Chapin, J. M., & Chan, V. W. (2011, November). The next 10 years of DoD wireless networking research. In *Military Communications Conference, 2011-MILCOM 2011* (pp. 2238–2245). IEEE. doi:10.1109/MILCOM.2011.6127653

Chen, B., Jamieson, K., Balakrishnan, H., & Morris, R. (2002). Span: An energy-efficient coordination algorithm for topology maintenance in ad hoc wireless networks. *Wireless Networks, 8*(5), 481–494. doi:10.1023/A:1016542229220

Cheng, H., & Fei, X. (2018). An efficient sparse coverage protocol for RSU deployment over urban VANETs. *Adhoc Networks, 24*(B), 85-102. doi:10.1016/j.adhoc.2014.07.022

Chen, G., Song, S. H., & Letaief, K. B. (2011). *A low-complexity precoding scheme for PAPR reduction in SC-FDMA systems. In 2011 IEEE Wireless Communications and Networking Conference.* IEEE.

Chen, J., & Chang, Z. (2016). Hierarchical Data Gathering Scheme for Energy Efficient Wireless Sensor Network. *International Journal of Future Generation Communication and Networking, 9*(3), 189–200. doi:10.14257/ijfgcn.2016.9.3.18

Chen, P. Y., & Ht, L. M. (2002). Measuring switching costs and the determinants of customer retention in Internet-enabled businesses: A study of the online brokerage industry. *Information Systems Research, 13*(3), 255–274. doi:10.1287/isre.13.3.255.78

Chen, Q., Xie, R., Yu, F. R., Liu, J., Huang, T., & Liu, Y. (2016). Transport control strategies in named data networking: A survey. *IEEE Communications Surveys and Tutorials, 18*(3), 2052–2083. doi:10.1109/COMST.2016.2528164

Chilamkurti, N. K., & Soh, B. (2002). A simulation study on multicast congestion control algorithms with multimedia traffic. *Global Telecommunications Conference, 2002. GLOBECOM'02. IEEE, 2*, 1779–1783. 10.1109/GLOCOM.2002.1188504

Chodorek, A., & Chodorek, R. R. (2020). Light-Weight Congestion Control for the DCCP: Implementation in the Linux Kernel. In Data-Centric Business and Applications (pp. 245–267). Springer.

Cho, J. H., Swami, A., & Chen, R. (2011). A survey on trust management for mobile ad hoc networks. *IEEE Communications Surveys and Tutorials, 13*(4), 562–583. doi:10.1109/SURV.2011.092110.00088

Chowdhury, M., Gawande, A., & Wang, L. (2017). Secure Information Sharing among Autonomous Vehicles in NDN. In *International Conference on Internet-of-Things Design and Implementation (IoTDI)* (pp. 15–26). 10.1145/3054977.3054994

Ciochina, C., Castelain, D., Mottier, D., & Sari, H. (2009). New PAPR-preserving mapping methods for single-carrier FDMA with space-frequency block codes. *IEEE Transactions on Wireless Communications*, 8(10), 5176–5186. doi:10.1109/TWC.2009.081231

Cisco Visual Networking Index Predicts Annual Internet Traffic to Grow More Than 20 Percent (reaching 1.6 Zettabytes) by 2018 I The Network. (n.d.). Retrieved July 10, 2020, from https://newsroom.cisco.com/press-release-content?articleId=1426270

Cisco. (2017, September 15). *Cisco visual networking index: forecast and methodology, 2016–2021*. Cisco.

Cole, D. T., Thompson, P., Göktoğan, A. H., & Sukkarieh, S. (2010). System development and demonstration of a cooperative UAV team for mapping and tracking. *The International Journal of Robotics Research*, 29(11), 1371–1399. doi:10.1177/0278364910364685

Comer, D. E., & Yavatkar, R. S. (1990). A rate-based congestion avoidance and control scheme for packet switched networks. *Proceedings 10th International Conference on Distributed Computing Systems*, 390–397. 10.1109/ICDCS.1990.89307

Congestion Control video lecture by Prof Ajit Pal of IIT Kharagpur. (n.d.). Retrieved July 10, 2020, from https://freevideolectures.com/course/2278/data-communication/23

Constantinides, E. (2002). The 4S Web-marketing mix model, Electronic Commerce Research and Applications. *Elsevier Science, 1*(1), 57–76.

Cunha, Villas, Boukerche, Maia, Viana, Mini, & Loureiro. (2016). Data communication in VANETs: Protocols, applications and challenges. *Adhoc Networks, 44*, 90-103.

Cunha, F., Villas, L., Boukerche, A., Maia, G., Viana, A., Mini, R. A. F., & Loureiro, A. A. F. (2016). Data communication in VANETs: Protocols, applications and challenges. *Ad Hoc Networks, 44*, 90–103. doi:10.1016/j.adhoc.2016.02.017

Dannewitz, C., Kutscher, D., Ohlman, B., Farrell, S., Ahlgren, B., & Karl, H. (2013). Network of Information (NetInf) - An information-centric networking architecture. *Computer Communications, 36*(7), 721–735. doi:10.1016/j.comcom.2013.01.009

Darwish, T., Abu Bakar, K., & Hashim, A. (2016). Green geographical routing in vehicular Adhoc networks: Advances and challenges. *Computers & Electrical Engineering, 64*, 436-449. .compeleceng.2016.09.030 doi:10.1016/j

Das, A., & Islam, M. M. (2012). Secured Trust: A dynamic trust computation model for secured communication in multiagent systems. *IEEE Transactions on Dependable and Secure Computing, 9*(2), 261–274. doi:10.1109/TDSC.2011.57

Das, S., Talayco, D., & Sherwood, R. (2013). *Software-Defined Networking and OpenFlow, Handbook of Fiber Optic Data Communication* (4th ed.). Academic Press. doi:10.1016/B978-0-12-401673-6.00017-9

Davaslioglu, K., & Ayanoglu, E. (2014). Efficiency and fairness trade-offs in SC-FDMA schedulers. *IEEE Transactions on Wireless Communications, 13*(6), 2991–3002. doi:10.1109/TWC.2014.042914.131176

Davies, D. (1972). The Control of Congestion in Packet Switch Networks. *IEEE Transactions on Communications, 20*(3), 546–550. doi:10.1109/TCOM.1972.1091198

Dechene, D. J., & Shami, A. (2014). Energy-aware resource allocation strategies for LTE uplink with synchronous HARQ constraints. *IEEE Transactions on Mobile Computing, 13*(2), 422–433. doi:10.1109/TMC.2012.256

Demers, A., Keshav, S., & Shenker, S. (1989). Analysis and simulation of a fair queueing algorithm. *Computer Communication Review, 19*(4), 1–12. doi:10.1145/75247.75248

Deng, G., Wang, L., Li, F., & Li, R. (2016). Distributed Probabilistic Caching strategy in VANETs through Named Data Networking. *IEEE INFOCOM,* 314–319. doi:10.1109/INFCOMW.2016.7562093

Deng, G., Xie, X., Shi, L., & Li, R. (2015). Hybrid information forwarding in VANETs through named data networking. In *International Symposium on Personal, Indoor, and Mobile Radio Communications (PIMRC)* (pp. 1940–1944). 10.1109/PIMRC.2015.7343616

Din, I. U., Hassan, S., Khan, M. K., Guizani, M., Ghazali, O., & Habbal, A. (2018). Caching in Information-Centric Networking: Strategies, Challenges, and Future Research Directions. *IEEE Communications Surveys and Tutorials, 20*(c), 1–1. doi:10.1109/COMST.2017.2787609

Djeloual, Amira, & Bensaali. (2018). Compressive Sensing-Based IoT Applications: A Review. *Journal of Sensor and Actuator Network.*

Dong, P., Gao, K., Xie, J., Tang, W., Xiong, N., & Vasilakos, A. V. (2019). Receiver-side TCP countermeasure in cellular networks. *Sensors (Basel), 19*(12), 2791. doi:10.339019122791 PMID:31234375

Donoho David, L. (2006). Compressed sensing. *IEEE Transactions on Information Theory, 52*(4), 1289–1306. doi:10.1109/TIT.2006.871582

Drira, W., & Filali, F. (2014). NDN-Q: An NDN query mechanism for efficient V2X data collection. In *IEEE International Conference on Sensing, Communication, and Networking Workshops (SECON Workshops)* (pp. 13–18). 10.1109/SECONW.2014.6979698

Duarte, J. M., Braun, T., & Villas, L. A. (2018). Source Mobility in Vehicular Named-Data Networking: An Overview. In Ad Hoc Networks (pp. 83–93). Springer.

Duarte, M. F., Shen, G., Ortega, A., & Baraniuk, R. G. (2012). Signal compression in wireless sensor networks. *Philosophical Transactions of the Royal Society A: Mathematical, Physical and Engineering Sciences, 370*(1958), 118-135.

Dukkipati, N., Kobayashi, M., Zhang-Shen, R., & McKeown, N. (2005). Processor sharing flows in the internet. *International Workshop on Quality of Service,* 271–285.

Dukkipati, N., Refice, T., Cheng, Y., Chu, J., Herbert, T., Agarwal, A., Jain, A., & Sutin, N. (2010). An argument for increasing TCP's initial congestion window. *Computer Communication Review, 40*(3), 26–33. doi:10.1145/1823844.1823848

Faber, T., Landweber, L. H., & Mukherjee, A. (1992). Dynamic Time Windows: packet admission control with feedback. *Conference Proceedings on Communications Architectures & Protocols,* 124–135. 10.1145/144179.144255

Fahmi, A., Astuti, R. P., Meylani, L., Asvial, M., & Gunawan, D. (2016). A Combined User-order and Chunk-order Algorithm to Minimize the Average BER for Chunk Allocation in SC-FDMA Systems. *KOMNIKA Telecommunication, Computing. Electronics and Control, 14*(2), 574–587.

Falconer, D., Ariyavisitakul, S. L., Benyamin-Seeyar, A., & Eidson, B. (2002). Frequency domain equalization for single-carrier broadband wireless systems. *IEEE Communications Magazine, 40*(4), 58–66. doi:10.1109/35.995852

Fang, Y., Zhu, X., & Zhang, Y. (2009). Securing resource-constrained wireless ad hoc networks. *IEEE Wireless Communications, 16*(2).

Fan, J., Lee, D., Li, G. Y., & Li, L. (2015). Multiuser scheduling and pairing with interference mitigation for LTE uplink cellular networks. *IEEE Transactions on Vehicular Technology, 64*(2), 481–492. doi:10.1109/TVT.2014.2321679

Fantacci, R., Marabissi, D., & Papini, S. (2004a). Multiuser interference cancellation receivers for OFDMA uplink communications with carrier frequency offset. In *IEEE Global Telecommunications Conference, 2004. GLOBECOM '04.* (Vol. 5, pp. 2808–2812). IEEE. 10.1109/GLOCOM.2004.1378866

Farrell, S., & Jensen, C. (2004). A Flexible Interplanetary Internet. *Tools and Technologies for Future Planetary Exploration, 543,* 87–94.

Feng, B., Zhou, H., & Xu, Q. (2016). Mobility support in Named Data Networking: A survey. *EURASIP Journal on Wireless Communications and Networking, 2016*(1), 220. Advance online publication. doi:10.118613638-016-0715-0

Feukeu, E. A., & Zuva, T. (2017). DBSMA Approach for Congestion Mitigation in VANETs. *Procedia Computer Science, 109,* 42-49. .05.293 doi:10.1016/j.procs.2017

Filho, J. G., Patel, A., Bruno, L. A. B., & Celestino, J. (2016). A systematic technical survey of DTN and VDTN routing protocols. *Computer Standards & Interfaces, 48,* 139-159. doi:10.1016/j.csi.2016.06.004

Fitah, A., Badri, A., Moughit, M., & Sahel, A. (2018). Performance of DSRC and WIFI for Intelligent Transport Systems in VANET. *Procedia Computer Science, 127,* 360-368. doi:1016/j.procs.2018.01.133

Floyd, S, Mahdavi, J., Mathis, M., & Podolsky, M. (2000). *RFC2883: An Extension to the Selective Acknowledgement (SACK) Option for TCP.* RFC Editor.

Floyd, S. (2003). *RFC3649: HighSpeed TCP for large congestion windows*. RFC Editor.

Floyd, S., Handley, M., Padhye, J., & Widmer, J. (2008). TCP friendly rate control (TFRC): Protocol Specification. IETF RFC 5348.

Floyd, S.,, Henderson, T., & Gurtov, A. (2004). *RFC3782: The newreno modification to TCP's fast recovery algorithm*. RFC Editor.

Fotiou, N., Nikander, P., Trossen, D., Polyzos, G. C., & Associates. (2010). Developing Information Networking Further: From PSIRP to PURSUIT. In Broadnets (pp. 1–13).

Fox, J. (2000). A river of money will flow through the Wireless Web in coming years. All the big players want is a piece of the action. *Fortune, 142*(8), 140–146.

Fu, Z., Zerfos, P., Luo, H., Lu, S., Zhang, L., & Gerla, M. (2003). The impact of multihop wireless channel on TCP throughput and loss. *IEEE INFOCOM 2003. Twenty-Second Annual Joint Conference of the IEEE Computer and Communications Societies (IEEE Cat. No. 03CH37428), 3*, 1744–1753.

Fu, C. P., & Liew, S. C. (2003). TCP Veno: TCP enhancement for transmission over wireless access networks. *IEEE Journal on Selected Areas in Communications, 21*(2), 216–228. doi:10.1109/JSAC.2002.807336

Fushiki, M., Ohseki, T., & Konishi, S. (2011). Throughput gain of fractional frequency reuse with frequency selective scheduling in SC-FDMA uplink cellular system. In 2011 IEEE Vehicular Technology Conference (VTC Fall) (pp. 1–5). IEEE. doi:10.1109/VETECF.2011.6093053

Gehring, O., & Fritz, H. (1997). Practical results of a longitudinal control concept for truck platooning with vehicle to vehicle communication. In *Intelligent Transportation System, 1997. ITSC'97., IEEE Conference on* (pp. 117–122). 10.1109/ITSC.1997.660461

Golestani, S. J. (1991). Congestion-free communication in high-speed packet networks. *IEEE Transactions on Communications, 39*(12), 1802–1812. doi:10.1109/26.120166

Grassi, G., Pesavento, D., Pau, G., Zhang, L., & Fdida, S. (2015). Navigo: Interest forwarding by geolocations in vehicular Named Data Networking. In *International Symposium on World of Wireless, Mobile and Multimedia Networks (WoWMoM)* (pp. 1–10). 10.1109/WoWMoM.2015.7158165

Gray, M. (1996). *Web Growth Summary*. https://stuff.mit.edu/people/mkgray/net/web-growth-summary.html

Griffor. E. (2017, December 12). *Reference architecture for cyber-physical systems*. NIST (National Institute of Standards and Technology).

Guo, S., & Zeng, D. (Eds.). (2019). *Cyber-Physical Systems: Architecture*. Security and Application, EAI/Springer Innovations in Communication and Computing.

Haas, Z. (1991). Adaptive admission congestion control. *Computer Communication Review, 21*(5), 58–76. doi:10.1145/122431.122436

Ha, H. K., Tuan, H. D., & Nguyen, H. H. (2013). Joint optimization of source power allocation and cooperative beamforming for SC-FDMA multi-user multi-relay networks. *IEEE Transactions on Communications*, *61*(6), 2248–2259. doi:10.1109/TCOMM.2013.041113.120480

Han, Y., Chang, Y., Cui, J., & Yang, D. (2010). A novel inter-cell interference coordination scheme based on dynamic resource allocation in LTE-TDD systems. In *2010 IEEE 71st Vehicular Technology Conference* (pp. 1–5). IEEE. 10.1109/VETECS.2010.5494073

Harrison, T. (2000). *Financial Services Marketing*. Prentice Hall.

Ha, S., Rhee, I., & Xu, L. (2008). CUBIC: A new TCP-friendly high-speed TCP variant. *Operating Systems Review*, *42*(5), 64–74. doi:10.1145/1400097.1400105

Hasrouny, H., Samhat, A. E., Bassil, C., & Laouiti, A. (2017). VANet security challenges and solutions: A survey. *Vehicular Communications, 7*, 7-20. doi:10.1016/ j.vehcom.2017.01.002

Hassija, V., Chamola, V., Saxena, V., Jain, D., Goyal, P., & Sikdar, B. (2019). A Survey on IoT security: Application Areas, Security Threats, and Solution Architectures (vol. 7). IEEE Access.

Hatoum, R., Hatoum, A., Ghaith, A., & Pujolle, G. (2014). Qos-based joint resource allocation with link adaptation for SC-FDMA uplink in heterogeneous networks. In *Proceedings of the 12th ACM international symposium on Mobility management and wireless access* (pp. 59–66). ACM. 10.1145/2642668.2642673

Hayat, S., Yanmaz, E., & Muzaffar, R. (2016). Survey on Unmanned Aerial Vehicle Networks for Civil Applications: A Communications Viewpoint. *IEEE Communications Surveys and Tutorials*, *18*(4), 2624–2661. doi:10.1109/COMST.2016.2560343

Hedrick, J. K., Tomizuka, M., & Varaiya, P. (1994). Control issues in automated highway systems. *IEEE Control Systems*, *14*(6), 21–32. doi:10.1109/37.334412

He, H., & Maple, C. (2016, July 24). The security challenges in the IoT enabled cyber-physical systems and opportunities for evolutionary computing & other computational intelligence. *2016 IEEE Congress on Evolutionary Computation (CEC)*. 10.1109/CEC.2016.7743900

Heinzelman, W. B. (2000). *Application-specific protocol architectures for wireless networks* (Doctoral dissertation). Massachusetts Institute of Technology.

Holland, G., & Vaidya, N. (2002). Analysis of TCP performance over mobile ad hoc networks. *Wireless Networks*, *8*(2–3), 275–288. doi:10.1023/A:1013798127590

Hossain, Datta, Hossain, & Edmonds. (2017). ResVMAC: A Novel Medium Access Control Protocol for Vehicular Adhoc Networks. *Procedia Computer Science, 109*, 432-439. doi:10.1016/j. procs.2017.05.413

Hsieh, H.-Y., Kim, K.-H., Zhu, Y., & Sivakumar, R. (2003). A receiver-centric transport protocol for mobile hosts with heterogeneous wireless interfaces. *Proceedings of the 9th Annual International Conference on Mobile Computing and Networking*, 1–15. 10.1145/938985.938987

Hsu, L.-H., Chao, H.-L., & Liu, C.-L. (2013). Window-based frequency-domain packet scheduling with QoS support in LTE uplink. In *2013 IEEE 24th Annual International Symposium on Personal, Indoor, and Mobile Radio Communications (PIMRC)* (pp. 1805–1810). IEEE.

Hu, L., & Xie, N. (2012, April). Review of cyber-physical system architecture. *2012 IEEE 15th International Symposium on Object/Component/Service-Oriented Real-Time Distributed Computing Workshops.* doi:10.1109/ISORCW.2012.15

Hua, F., & Lua, Y. (2016, March). Robust cyber–physical systems: Concept, models, and implementation. *Future Generation Computer Systems, 56,* 449–475. doi:10.1016/j.future.2015.06.006

Huang, G., Nix, A., & Armour, S. (2007). Impact of radio resource allocation and pulse shaping on PAPR of SC-FDMA signals. In *2007 IEEE 18th International Symposium on Personal, Indoor and Mobile Radio Communications* (pp. 1–5). IEEE. 10.1109/PIMRC.2007.4394297

Hua, Z., Fei, G., & Wen, Q. (2011). A Password-Based Secure Communication Scheme in Battlefields for Internet of Things. *China Communications, 8*(1), 72–78.

Hui, R., Zhu, B., Huang, R., Allen, C., Demarest, K., & Richards, D. (2001). 10-Gb/s SCM fiber system using optical SSB modulation. *IEEE Photonics Technology Letters, 13*(8), 896–898. doi:10.1109/68.935840

Hu, Y., & Ci, S. (2015). QoE-driven Joint Resource Allocation and User-paring in Virtual MIMO SC-FDMA Systems. *Transactions on Internet and Information Systems (Seoul), 9*(10), 3831–3850.

IEEE Standards Association. (n.d.). *Wireless access in vehicular environments.* IEEE.

Infographic: How Much Data is Generated Each Day? (n.d.). Retrieved July 10, 2020, from https://www.visualcapitalist.com/how-much-data-is-generated-each-day/

Intel labs Berkeley data. (n.d.). http://db.csul.mit.edu/www.select.cs.cmu.edu/data/labapp3/

International Society of Automation (ISA). (2018). *ISA-95, Enterprise-control System Integration.* https://www.isa.org/isa95/

Internet Growth Statistics 1995 to 2019 - the Global Village Online. (n.d.). Retrieved July 10, 2020, from https://www.internetworldstats.com/emarketing.htm

Internet Top 20 Countries - Internet Users 2020. (n.d.). Retrieved July 10, 2020, from https://internetworldstats.com/top20.htm

Iqbal, Olaleye, & Bayoumi. (2016). A Review on Internet of Things (IoT): Security and Privacy Requirements and the Solution Approaches. *Global Journal of Computer Science and Technology: E Network, Web & Security, 16.*

Irshad, M. (2016). A Systematic Review of Information Security Frameworks in the Internet of things. In *18th International Conference on High Performance Computing and Communications,* (pp. 1270-1275). IEEE. 10.1109/HPCC-SmartCity-DSS.2016.0180

ISO/IEC (2013, April). *High Efficiency Video Coding (HEVC), ISO/IEC 23008-2 MPEG-H Part 2/ITU-T H.265*. ISO/IEC Moving Picture Experts Group (MPEG) and ITU-T Video Coding Experts Group (VCEG).

Iwamoto, D., Sugahara, D., Bandai, M., & Yamamoto, M. (2018). Adaptive Congestion Control for Handover in Heterogeneous Mobile Content-Centric Networking. *2018 Eleventh International Conference on Mobile Computing and Ubiquitous Network (ICMU)*, 1–6. 10.23919/ICMU.2018.8653625

Jacobson, V. (1988). Congestion avoidance and control. *Computer Communication Review, 18*(4), 314–329. doi:10.1145/52325.52356

Jain, B., Brar, G., Malhotra, J., Rani, S., & Ahmed, S. H. (2018). A cross layer protocol for traffic management in Social Internet of Vehicles. *Future Generation Computer Systems, 82*, 707-714. doi:10.1016/j.future.2017.11.019

Jain, R. (1986). A timeout-based congestion control scheme for window flow-controlled networks. *IEEE Journal on Selected Areas in Communications, 4*(7), 1162–1167. doi:10.1109/JSAC.1986.1146431

Jain, R. (n.d.). Congestion Control in Computer Networks. *Issues (Chicago, Ill.)*.

Jar, M., & Fettweis, G. (2012). Throughput maximization for LTE uplink via resource allocation. In *2012 International Symposium on Wireless Communication Systems (ISWCS)* (pp. 146–150). IEEE. 10.1109/ISWCS.2012.6328347

Jayawardhena, C., & Foley, P. (2000). Changes in the Banking sector - the case of Internet Banking in the UK'. *Internet Research: Electronic Networking Applications and Policy, 10*(1), 19–30. doi:10.1108/10662240010312048

Jiang, J.-R. (2018, June). An Improved cyber-physical systems architecture for Industry 4.0 smart factories. *Advances in Mechanical Engineering, 10*(6). Advance online publication. doi:10.1177/1687814018784192

Jin, C., Wei, D., Low, S. H., Bunn, J., Choe, H. D., Doylle, J. C., Newman, H., Ravot, S., Singh, S., & Paganini, F. (2005). FAST TCP: From theory to experiments. *IEEE Network, 19*(1), 4–11. doi:10.1109/MNET.2005.1383434

Jones. (2018). Security Review On The Internet of Things. In *Third International Conference on Fog and Mobile Edge Computing (FMEC)*. IEEE.

Joshi, P. (2011). Security Issues in Routing protocols in MANETs at network layer. *Procedia Computer Science, 3*, 954–960. doi:10.1016/j.procs.2010.12.156

Kaddour, F. Z., Pischella, M., Martins, P., Vivier, E., & Mroueh, L. (2013). Opportunistic and efficient resource block allocation algorithms for LTE uplink networks. In 2013 IEEE Wireless Communications and Networking Conference (WCNC) (pp. 487–492). IEEE. doi:10.1109/WCNC.2013.6554612

Kalil, M., Shami, A., & Al-Dweik, A. (2013a). Power-efficient QoS scheduler for LTE uplink. In *2013 IEEE International Conference on Communications (ICC)* (pp. 6200–6204). IEEE. 10.1109/ICC.2013.6655598

Kandris, D., Tsagkaropoulos, M., Politis, I., Tzes, A., & Kotsopoulos, S. (2011). Energy efficient and perceived QoS aware video routing over wireless multimedia sensor networks. *Ad Hoc Networks*, 9(4), 591–607. doi:10.1016/j.adhoc.2010.09.001

Kaneko, K., Fujikawa, T., Su, Z., & Katto, J. (2007). TCP-Fusion: A hybrid congestion control algorithm for high-speed networks. *Proc. PFLDnet*, 7, 31–36.

Kargl, F., Papadimitratos, P., Buttyan, L., Müter, M., Schoch, E., Wiedersheim, B., Thong, T.-V., Calandriello, G., Held, A., Kung, A., & Hubaux, J.-P. (2008). Secure vehicular communication systems: Implementation, performance, and research challenges. *IEEE Communications Magazine*, 46(11), 110–118. doi:10.1109/MCOM.2008.4689253

Karnouskos. (2004). Mobile Payment: A Journey through Existing Procedures & Standardization Initiatives. *IEEE Communications Surveys & Tutorials*, 44-66.

Katabi, M. H. C. R. (2002). Internet Congestion Control for Future High Bandwidth-Delay Product Environments. *ACM SIGCOMM 2002.*

Katabi, D., Handley, M., & Rohrs, C. (2002). Congestion control for high bandwidth-delay product networks. *Proceedings of the 2002 Conference on Applications, Technologies, Architectures, and Protocols for Computer Communications*, 89–102. 10.1145/633025.633035

Kaur, K., & Kad, S. (2016). Enhanced clustering based AODV-R protocol using Ant Colony Optimization in VANETS. *IEEE 1st International Conference on Power Electronics, Intelligent Control and Energy Systems (ICPEICES)*, 1-5. 10.1109/ICPEICES.2016.7853381

Kelly, T. (2003). Scalable TCP: Improving performance in highspeed wide area networks. *Computer Communication Review*, 33(2), 83–91. doi:10.1145/956981.956989

Kent, C. A., & Mogul, J. C. (1987). *Fragmentation considered harmful* (Vol. 17). Academic Press.

Keshav, S. (1991). *Congestion Control in Computer Networks PhD Thesis*. UC Berkeley TR-654.

Kessab, A., Kaddour, F. Z., Vivier, E., Mroueh, L., Pischella, M., & Martins, P. (2012). Gain of multi-resource block allocation and tuning in the uplink of LTE networks. In *2012 International Symposium on Wireless Communication Systems (ISWCS)* (pp. 321–325). IEEE. 10.1109/ISWCS.2012.6328382

Khan, M. H., & Barman, P. C. (2015). 5G - Future Generation Technologies of Wireless Communication. Revolution 2020. *American Journal of Engineering Research, 4*(5), 206-215.

Khan, A., Yanmaz, E., & Rinner, B. (2015). Information exchange and decision making in micro aerial vehicle networks for cooperative search. *IEEE Transactions on Control of Network Systems*, 2(4), 335–347. doi:10.1109/TCNS.2015.2426771

Khan, M. S., Midi, D., Khan, M. I., & Bertino, E. (2017). Fine-Grained Analysis of Packet Loss in MANETs. *IEEE Access: Practical Innovations, Open Solutions, 5,* 7798–7807. doi:10.1109/ACCESS.2017.2694467

Khelifi, H., Luo, S., Nour, B., Moungla, H., & Ahmed, S. H. (2018). Reputation-based Blockchain for Secure NDN Caching in Vehicular Networks. In *IEEE Conference on Standards for Communications and Networking (CSCN)* (pp. 1–6). Paris, France: IEEE. 10.1109/CSCN.2018.8581849

Khelifi, H., Luo, S., Nour, B., Sellami, A., Moungla, H., & Naït-Abdesselam, F. (2018). An Optimized Proactive Caching Scheme based on Mobility Prediction for Vehicular Networks. In *IEEE Global Communications Conference (IEEE GLOBECOM)* (pp. 1–6). 10.1109/GLOCOM.2018.8647898

Khelifi, H., Luo, S., Nour, B., & Shah, C. S. (2018). *Security & Privacy Issues in Vehicular Named Data Networks: An Overview.* Mobile Information Systems. doi:10.1155/2018/5672154

Kho, L. C., Défago, X., Lim, A. O., & Tan, Y. (2013). A taxonomy of congestion control techniques for tcp in wired and wireless networks. *2013 IEEE Symposium on Wireless Technology & Applications (ISWTA),* 147–152. 10.1109/ISWTA.2013.6688758

Khorov, E., Kiryanov, A., Lyakhov, A., & Bianchi, G. (2019). A tutorial on IEEE 802.11ax high efficiency WLANs. *IEEE Communications Surveys and Tutorials, 21*(1), 197–216. doi:10.1109/COMST.2018.2871099

Kiefhaber, R., Satzger, B., Schmitt, J., Roth, M., & Ungerer, T. (2010, December). Trust measurement methods in organic computing systems by direct observation. In *2010 IEEE/IFIP International Conference on Embedded and Ubiquitous Computing* (pp. 105-111). IEEE. 10.1109/EUC.2010.25

Kim, D., Kim, J., Kim, H., Kim, K., & Han, Y. (2010). An efficient scheduler for uplink single carrier FDMA system. In *21st Annual IEEE International Symposium on Personal, Indoor and Mobile Radio Communications* (pp. 1348–1353). IEEE.

Kim, D., Toh, C.-K., & Choi, Y. (2001). TCP-BuS: Improving TCP performance in wireless ad hoc networks. *Journal of Communications and Networks (Seoul), 3*(2), 1–12. doi:10.1109/JCN.2001.6596860

Kim, H., Chung, M. Y., Lee, T.-J., Kim, M., & Choo, H. (2015). Scheduling Based on Maximum PF Selection with Contiguity Constraint for SC-FDMA in LTE Uplink. *Journal of Information Science and Engineering, 31*(4), 1455–1473.

Kim, H., & Feamster, N. (2013). Improving network management with software defined networking. *IEEE Communications Magazine, 51*(2), 114–119. doi:10.1109/MCOM.2013.6461195

Kim, J., Hwang, I. S., & Kang, C. G. (2015). Scheduling for virtual MIMO in single carrier FDMA (SC-FDMA) system. *Journal of Communications and Networks (Seoul), 17*(1), 27–33. doi:10.1109/JCN.2015.000006

King, R., Baraniuk, R., & Riedi, R. (2005). TCP-Africa: An adaptive and fair rapid increase rule for scalable TCP. *Proceedings IEEE 24th Annual Joint Conference of the IEEE Computer and Communications Societies, 3*, 1838–1848. 10.1109/INFCOM.2005.1498463

Kiran, P., & Jibukumar, M. G. (2016). Dynamic Multiuser Scheduling with Interference Mitigation in SC-FDMA-Based Communication Systems. In *Proceedings of the Second International Conference on Computer and Communication Technologies* (pp. 289–297). Springer. 10.1007/978-81-322-2523-2_27

Koponen, T., Chawla, M., Chun, B.-G., Ermolinskiy, A., Kim, K. H., Shenker, S., & Stoica, I. (2007). A data-oriented (and beyond) network architecture. *Computer Communication Review, 37*(4), 181–192. doi:10.1145/1282427.1282402

Kopparty, S., Krishnamurthy, S. V., & Faloutsos, M., S. T. (2002). Split TCP for mobile ad hoc networks. *IEEE Global Telecommunications Conference GLOBECOM'02*, 138–142.

Koutsoukos, X., Karsai, G., Laszka, A., Neema, H., Potteiger, B., Volgyesi, P., Vorobeychik, Y., & Sztipanovits, J. (2018, January). SURE: A modeling and simulation integration platform for evaluation of SecUre and REsilient cyber-physical systems. *Proceedings of the IEEE, 106*(1), 93–112. doi:10.1109/JPROC.2017.2731741

Kraounakis, S., Demetropoulos, I. N., Michalas, A., Obaidat, M. S., Sarigiannidis, P. G., & Louta, M. D. (2015). A robust reputation-based computational model for trust establishment in pervasive systems. *IEEE Systems Journal, 9*(3), 878–891. doi:10.1109/JSYST.2014.2345912

Kuai, M., Hong, X., & Yu, Q. (2016). Density-Aware Delay-Tolerant Interest Forwarding in Vehicular Named Data Networking. In *IEEE Vehicular Technology Conference (VTC-Fall)* (pp. 1–5). 10.1109/VTCFall.2016.7880953

Kuzmanovic, A., & Knightly, E. W. (2006). TCP-LP: Low-priority service via end-point congestion control. *IEEE/ACM Transactions on Networking, 14*(4), 739–752. doi:10.1109/TNET.2006.879702

Kwan, R., & Leung, C. (2010). A survey of scheduling and interference mitigation in LTE. *Journal of Electrical and Computer Engineering, 2010*, 1–10. doi:10.1155/2010/273486

Lafuente-Martínez, J., Hernández-Solana, Á., & Valdovinos, A. (2011). Sector-based radio resource management for SC-FDMA cellular systems. In *2011 8th International Symposium on Wireless Communication Systems* (pp. 750–754). IEEE. 10.1109/ISWCS.2011.6125369

Lafuente-Martinez, J., Hernandez-Solana, A., Guio, I., & Valdovinos, A. (2011). Inter-cell interference management in SC-FDMA cellular systems. In *2011 IEEE 73rd Vehicular Technology Conference (VTC Spring)* (pp. 1–5). IEEE. 10.1109/VETECS.2011.5956184

Lafuente-Martínez, J., Hernández-Solana, Á., Guío, I., & Valdovinos, A. (2011). *Radio resource strategies for uplink inter-cell interference fluctuation reduction in SC-FDMA cellular systems. In 2011 IEEE Wireless Communications and Networking Conference.* IEEE.

Lai, C., Zheng, D., Zhao, Q., & Jiang, X. (2018). SEGM: A secure group management framework in integrated VANET-cellular networks. *Vehicular Communications, 11*, 33-45.

Lam, S., & Reiser, M. (1979). Congestion Control of Store-and-Forward Networks by Input Buffer Limits - An Analysis. *IEEE Transactions on Communications*, *27*(1), 127–134. doi:10.1109/TCOM.1979.1094280

Lan, K. C., & Wei, M. Z. (2017). A compressibility-based clustering algorithm for hierarchical compressive data gathering. *IEEE Sensors Journal*, *17*(8), 2550–2562. doi:10.1109/JSEN.2017.2669081

Laroui, M., Sellami, A., Nour, B., Moungla, H., Afifi, H., & Boukli-Hacéne, S. (2018). Driving Path Stability in VANETs. *IEEE Global Communications Conference (IEEE GLOBECOM)*.

Lee, D. L., Lee, W.-C., Xu, J., & Zheng, B. (2002). Data management in location-dependent information services: Challenges and issues. *IEEE Pervasive Computing*, *1*(3), 65–72. doi:10.1109/MPRV.2002.1037724

Lee, E. A. (2015, February). Present and future of cyber-physical systems: A focus on models. *Sensors (Basel)*, *15*(3), 4837–4869. doi:10.3390150304837 PMID:25730486

Lee, J., Bagheri, B., & Kao, H.-A. (2015). A Cyber-physical systems architecture for Industry 4.0-based manufacturing systems. *Manufacturing Letters*, *3*, 18–23. doi:10.1016/j.mfglet.2014.12.001

Lee, S.-B., Pefkianakis, I., Meyerson, A., Xu, S., & Lu, S. (2009). Proportional fair frequency-domain packet scheduling for 3GPP LTE uplink. In *IEEE INFOCOM 2009* (pp. 2611–2615). IEEE. doi:10.1109/INFCOM.2009.5062197

Lei, H., & Li, X. (2009). System level study of LTE uplink employing SC-FDMA and virtual MU-MIMO. In *2009 IEEE International Conference on Communications Technology and Applications* (pp. 152–156). IEEE.

Leinonen, M., Codreanu, M., & Juntti, M. (2015). Sequential compressed sensing with progressive signal reconstruction in wireless sensor networks. *IEEE Transactions on Wireless Communications*, *14*(3), 1622–1635. doi:10.1109/TWC.2014.2371017

Letichevsky, A. A., & Letychevsky, O. O. (2017, November). Cyber-physical systems. *Cybernetics and Systems Analysis*, *53*(6), 821–834. Advance online publication. doi:10.100710559-017-9984-9

Li, Z., Yin, C., & Yue, G. (2009). Delay-bounded power-efficient packet scheduling for uplink systems of lte. In *2009 5th International Conference on Wireless Communications, Networking and Mobile Computing* (pp. 1–4). IEEE. 10.1109/WICOM.2009.5303491

Lim, H.-J., Kim, T.-K., & Im, G.-H. (2010). A proportional fair scheduling algorithm for SC-FDMA with iterative multiuser detection. In *2010 International Conference on Information and Communication Technology Convergence (ICTC)* (pp. 243–244). IEEE. 10.1109/ICTC.2010.5674669

Ling, Z., Luo, J., Xu, Y., & Gao, C. (2017). *Security Vulnerabilities of Internet of Things: A case study of the Smart Plug System*. IEEE Internet of Things Journal.

Liu, Z., Xing, W., Zeng, B., Wang, Y., & Lu, D. (2013, March). Distributed spatial correlation-based clustering for approximate data collection in WSNs. In *Advanced Information Networking and Applications (AINA), 2013 IEEE 27th International Conference on* (pp. 56-63). IEEE.

Liu, C., Wu, K., & Pei, J. (2007). An energy-efficient data collection framework for wireless sensor networks by exploiting spatiotemporal correlation. *IEEE Transactions on Parallel and Distributed Systems*, *18*(7), 1010–1023. doi:10.1109/TPDS.2007.1046

Liu, J., & Singh, S. (2001). ATCP: TCP for mobile ad hoc networks. *IEEE Journal on Selected Areas in Communications*, *19*(7), 1300–1315. doi:10.1109/49.932698

Liu, L. C., Xie, D., Wang, S., & Zhang, Z. (2015). CCN-based cooperative caching in VANET. In *International Conference on Connected Vehicles and Expo (ICCVE)* (pp. 198–203). 10.1109/ICCVE.2015.24

Liu, Q., & Hwang, J.-N. (2003). End-to-end available bandwidth estimation and time measurement adjustment for multimedia QoS. *2003 International Conference on Multimedia and Expo. ICME'03. Proceedings (Cat. No. 03TH8698), 3*.

Liu, R., Chen, Y., & Xiong, X. (2015). Efficient Resources Allocation for Femtocells in Heterogeneous Cellular Networks. In *First International Conference on Information Sciences, Machinery, Materials and Energy*. Atlantis Press. 10.2991/icismme-15.2015.255

Liu, S., Başar, T., & Srikant, R. (2008). TCP-Illinois: A loss-and delay-based congestion control algorithm for high-speed networks. *Performance Evaluation*, *65*(6–7), 417–440. doi:10.1016/j.peva.2007.12.007

Liu, W., & Yu, M. (2014). AASR: Authenticated anonymous secure routing for MANETs in adversarial environments. *IEEE Transactions on Vehicular Technology*, *63*(9), 4585–4593. doi:10.1109/TVT.2014.2313180

Liu, X., Nicolau, M. J., Costa, A., Macedo, J., & Santos, A. (2016). A geographic opportunistic forwarding strategy for vehicular named data networking. *Intelligent Distributed Computing*, *9*, 509–521.

Li, Y.-T., Leith, D., & Shorten, R. N. (2007). Experimental evaluation of TCP protocols for high-speed networks. *IEEE/ACM Transactions on Networking*, *15*(5), 1109–1122. doi:10.1109/TNET.2007.896240

Loguinov, D., & Radha, H. (2002). Increase-decrease congestion control for real-time streaming: Scalability. *Proceedings. Twenty-First Annual Joint Conference of the IEEE Computer and Communications Societies, 2*, 525–534. 10.1109/INFCOM.2002.1019297

Lozano, A., Tulino, A. M., & Verdú, S. (2008). Optimum power allocation for multiuser OFDM with arbitrary signal constellations. *IEEE Transactions on Communications*, *56*(5), 828–837. doi:10.1109/TCOMM.2008.060211

Luo, J., & Hubaux, J.-P. (2004). *A survey of inter-vehicle communication*. Academic Press.

Luo, J., Xiang, L., & Rosenberg, C. (2010, May). Does compressed sensing improve the throughput of wireless sensor networks? In Communications (ICC), 2010 IEEE international conference on (pp. 1-6). IEEE. doi:10.1109/ICC.2010.5502565

Luo, C., Wu, F., Sun, J., & Chen, C. W. (2009, September). Compressive data gathering for large-scale wireless sensor networks. In *Proceedings of the 15th annual international conference on Mobile computing and networking* (pp. 145-156). ACM. 10.1145/1614320.1614337

Lu, X., Yang, K., Li, W., Qiu, S., & Zhang, H. (2016). Joint user grouping and resource allocation for uplink virtual MIMO systems. *Science China. Information Sciences*, *59*(2), 1–14. doi:10.100711432-015-5514-4

Madan, R., & Ray, S. (2011). Uplink resource allocation for frequency selective channels and fractional power control in LTE. In *2011 IEEE International Conference on Communications (ICC)* (pp. 1–5). IEEE. 10.1109/icc.2011.5963354

Mahmoud, M. M., Lin, X., & Shen, X. S. (2015). Secure and reliable routing protocols for heterogeneous multihop wireless networks. *IEEE Transactions on Parallel and Distributed Systems*, *26*(4), 1140–1153. doi:10.1109/TPDS.2013.138

Mallat, N. (2007). Exploring Consumer Adoption of Mobile Payments - A Qualitative Study. *The Journal of Strategic Information Systems*, *16*(4), 413–432. doi:10.1016/j.jsis.2007.08.001

Mamata, R. (2018). An Analytical Study of Security and Challenging Issues in Social Networking as an Emerging Connected Technology. In *Proceedings of 3rd International Conference on Internet of Things and Connected Technologies*. Malaviya National Institute of Technology. https://ssrn.com/abstract=3166509

Mamata, R. B. P. (2018). Communication Improvement and Traffic Control Based on V2I in Smart City Framework. *International Journal of Vehicular Telematics and Infotainment Systems*, *2*(1).

Manoj, V. & Bramhe. (2011). SMS Based Secure Mobile Technology. *International Journal of Engineering and Technology, 3*(6), 472-479.

Marfia, G., Palazzi, C. E., Pau, G., Gerla, M., & Roccetti, M. (2010). TCP Libra: Derivation, analysis, and comparison with other RTT-fair TCPs. *Computer Networks*, *54*(14), 2327–2344. doi:10.1016/j.comnet.2010.02.014

Mascolo, S., Casetti, C., Gerla, M., Sanadidi, M. Y., & Wang, R. (2001). TCP westwood: Bandwidth estimation for enhanced transport over wireless links. *Proceedings of the 7th Annual International Conference on Mobile Computing and Networking*, 287–297. 10.1145/381677.381704

Massey, D. (2017, November 3). Applying cybersecurity challenges to medical and vehicular cyber physical systems. *Proceeding of SafeConfig '17, 2017 Workshop on Automated Decision Making for Active Cyber Defense*, 39-39. doi:10.1145/3140368.3140379

Mathis, M., Mahdavi, J., Floyd, S., & Romanow, A. (1996). *RFC2018: TCP selective acknowledgement options*. RFC Editor.

Mauri, G., Gerla, M., Bruno, F., Cesana, M., & Verticale, G. (2017). Optimal Content Prefetching in NDN Vehicle-to-Infrastructure Scenario. *IEEE Transactions on Vehicular Technology, 66*(3), 2513–2525. doi:10.1109/TVT.2016.2580586

Mehdi, M. M., Raza, I., & Hussain, S. A. (2017). A game theory based trust model for Vehicular Adhoc Networks (VANETs). *Computer Networks, 121*, 152-172. doi:10.1016/j.comnet.2017.04.024

Meneghello, F., Calore, M., Zucchetto, D., Polese, M., & Zanella, A. (2019). *IoT: Internet of Threats. In A survey of practical security vulnerabilities in real IoT devices.* IEEE Internet of Things Journal.

Merino, L., Caballero, F., Martínez-de Dios, J. R., Ferruz, J., & Ollero, A. (2006). A cooperative perception system for multiple UAVs: Application to automatic detection of forest fires. *Journal of Field Robotics, 23*(3-4), 165–184. doi:10.1002/rob.20108

Mi, Z., Yang, Y., & Yang, J. Y. (2015). Restoring Connectivity of Mobile Robotic Sensor Networks While Avoiding Obstacles. IEEE Sensors Journal, 15(8), 4640-4650.

Miller, H. J. (2004). Tobler's first law and spatial analysis. *Annals of the Association of American Geographers, 94*(2), 284–289. doi:10.1111/j.1467-8306.2004.09402005.x

Minoli, D. (1983). A new design criterion for store-and-forward networks. *Computer Networks, 7*(1), 9–15. doi:10.1016/0376-5075(83)90003-X

Minoli, D. (2012). *Mobile video with Mobile IPv6.* Wiley., doi:10.1002/9781118647059.

Minoli, D. (2013). *Building the Internet of Things with IPv6 and MIPv6.* Wiley.

Minoli, D., Occhiogrosso, B., & ... . (2017). IoT considerations, requirements, and architectures for insurance applications. In Q. Hassan, A. R. Khan, & S. A. Madani (Eds.), *Internet of Things: challenges, advances and applications.* CRC Press.

Mishra, P. P., & Kanakia, H. (1992). A hop by hop rate-based congestion control scheme. *Conference Proceedings on Communications Architectures & Protocols,* 112–123. 10.1145/144179.144254

Modesto, F. M., & Boukerche, A. (2017a). An analysis of caching in information-centric vehicular networks. In *IEEE International Conference on Communications (ICC)* (pp. 1–6). 10.1109/ICC.2017.7997019

Modesto, F. M., & Boukerche, A. (2017b). SEVeN: A novel service-based architecture for information-centric vehicular network. *Computer Communications.*

Mogul, J. C., Kent, C. A., Partridge, C., & McCloghrie, K. (1988). *RFC1063: IP MTU discovery options.* RFC Editor.

Monostori, L. (2014). *Cyber-physical production systems: roots, expectations and R&D challenges. In Procedia CIRP* (Vol. 17). Elsevier. doi:10.1016/j.procir.2014.03.115

Muchtar, F., Al-Adhaileh, M. H., Alubady, R., Singh, P. K., Ambar, R., & Stiawan, D. (2020). Congestion Control for Named Data Networking-Based Wireless Ad Hoc Network BT. In *Proceedings of First International Conference on Computing, Communications, and Cyber-Security (IC4S 2019)*. Springer Singapore.

Muharemovic, T., & Shen, Z. (2008). *Power Settings for the Sounding Reference signal and the Scheduled Transmission in Multi-Channel Scheduled Systems*. Academic Press.

Mukherji, U. (1986). *A schedule-based approach for flow-control in data communication networks*. Massachusetts Inst of Tech Cambridge Lab for Information And Decision Systems.

Müller, H. A., & Litoiu, M. (2016). Engineering cybersecurity into cyber physical systems. In CASCON 2016. Markham, Canada: ACM.

Myung, H. G., Oh, K., Lim, J., & Goodman, D. J. (2008). *Channel-dependent scheduling of an uplink SC-FDMA system with imperfect channel information. In 2008 IEEE Wireless Communications and Networking Conference*. IEEE.

Nagle, J. (1984). *Congestion control in IP/TCP internetworks. Request For Comment 896*. Network Working Group.

Nagle, J. (1987). On packet switches with infinite storage. *IEEE Transactions on Communications, 35*(4), 435–438. doi:10.1109/TCOM.1987.1096782

Nakada, M., Obara, T., Yamamoto, T., & Adachi, F. (2012). Power allocation for direct/cooperative AF relay switched SC-FDMA. In *2012 IEEE 75th Vehicular Technology Conference (VTC Spring)* (pp. 1–5). IEEE.

Narendiran, C., Albert Rabara, S., & Rajendran, N. (2009). Public Key Infrastructure for Mobile Banking Security. *Global Mobile Congress*, 1-6. 10.1109/GMC.2009.5295898

Neshenko, N., Bou-Harb, E., Crichigno, J., Kaddoum, G., & Ghani, N. (2019). *Demystifying IoT Security: An Exhaustive Survey on IoT Vulnerabilities and a First Empirical Look on Internet-scale IoT Exploitations*. IEEE.

News Release 060929a. (n.d.). Retrieved July 10, 2020, from https://www.ntt.co.jp/news/news06e/0609/060929a.html

Ngo & Oh. (2014). A Link Quality Prediction Metric for Location based Routing Protocols under Shadowing and Fading Effects in Vehicular Adhoc Networks. *Procedia Computer Science, 34*, 565-570. doi:10.1016/j.procs.2014.07.071

Nguyen, A. G., & Hwang, J.-N. (2002). SPEM online rate control for realtime streaming video. *Proceedings. International Conference on Information Technology: Coding and Computing*, 65–70. 10.1109/ITCC.2002.1000361

Nguyen, M. T., & Rahnavard, N. (2013, November). Cluster-based energy-efficient data collection in wireless sensor networks utilizing compressive sensing. In *Military Communications Conference, MILCOM 2013-2013 IEEE* (pp. 1708-1713). IEEE. 10.1109/MILCOM.2013.289

Nguyen, M. T., Teague, K. A., & Rahnavard, N. (2016). CCS: Energy-efficient data collection in clustered wireless sensor networks utilizing block-wise compressive sensing. *Computer Networks*, *106*, 171–185. doi:10.1016/j.comnet.2016.06.029

Noh, J.-H., & Oh, S.-J. (2009). Distributed SC-FDMA resource allocation algorithm based on the Hungarian method. In *2009 IEEE 70th Vehicular Technology Conference Fall* (pp. 1–5). IEEE. 10.1109/VETECF.2009.5378857

Nour, B., Sharif, K., Li, F., Moungla, H., Kamal, A. E., & Afifi, H. (2018). NCP: A Near ICN Cache Placement Scheme for IoT-based Traffic Class. In *IEEE Global Communications Conference (GLOBECOM)* (pp. 1–6). 10.1109/GLOCOM.2018.8647629

Nour, B., Sharif, K., Li, F., Moungla, H., & Liu, Y. (2017). M2HAV: A Standardized ICN Naming Scheme for Wireless Devices in Internet of Things. In *International Conference on Wireless Algorithms, Systems, and Applications (WASA)* (pp. 289–301). Springer International Publishing. 10.1007/978-3-319-60033-8_26

Number of internet users worldwide | Statista. (n.d.). Retrieved July 10, 2020, from https://www.statista.com/statistics/273018/number-of-internet-users-worldwide/

Nwamadi, O., Zhu, X., & Nandi, A. K. (2011). Dynamic physical resource block allocation algorithms for uplink long term evolution. *IET Communications*, *5*(7), 1020–1027. doi:10.1049/iet-com.2010.0316

Oehlmann, F. (2013). Content-Centric Networking. *Network (Bristol, England)*, *43*, 11–18.

Oliveira, R., Montez, C., Boukerche, A., & Wangham, M. S. (2017). Reliable data dissemination protocol for VANET traffic safety applications. *Adhoc Networks, 63*, 30-44. doi:10.1016/j.adhoc.2017.05.002

Optical Carrier transmission rates - Wikipedia. (n.d.). Retrieved July 10, 2020, from https://en.wikipedia.org/wiki/Optical_Carrier_transmission_rates

Ou, W., Wang, X., Han, W., & Wang, Y. (2009, December). Research on Trust Evaluation Model Based on TPM. In *Frontier of Computer Science and Technology, 2009. FCST'09. Fourth International Conference on* (pp. 593-597). IEEE. 10.1109/FCST.2009.10

Oubbati, O. S., Lakas, A., Zhou, F., Güneş, M., Lagraa, N., & Yagoubi, M. B. (2017). Intelligent UAV-assisted routing protocol for urban VANETs. *Computer Communications, 107*, 93-111. doi:1016/j.comcom.2017.04.001

Pal, R., Prakash, A., Tripathi, R., & Singh, D. (2018). Analytical model for clustered vehicular Adhoc network analysis. *ICT Express*. doi:10.1016/ j.icte.2018.01.001

Pan, J., Paul, S., & Jain, R. (2011). A survey of the research on future internet architectures. *IEEE Communications Magazine*, *49*(7), 26–36. doi:10.1109/MCOM.2011.5936152

Pao, W.-C., Chen, Y.-F., Tsai, M.-G., Lou, W.-T., & Chang, Y.-J. (2012). A multiuser subcarrier and power allocation scheme in localized SC-FDMA systems. In *2012 IEEE 23rd International Symposium on Personal, Indoor and Mobile Radio Communications-(PIMRC)* (pp. 210–214). IEEE. 10.1109/PIMRC.2012.6362703

Park, K. (1993). Warp control: A dynamically stable congestion protocol and its analysis. *Journal of High Speed Networks*, *2*(4), 373–404. doi:10.3233/JHS-1993-2404

Park, N. J., & Song, Y. J. (2001). M-Commerce security platform based on WTLS and J2ME. *International Symposium on Industrial Electronics (ISIE)*.

Partridge, C., & Shepard, T. J. (1997). TCP/IP performance over satellite links. *IEEE Network*, *11*(5), 44–49. doi:10.1109/65.620521

Pattanayak, B., & Rath, M. (2014). A Mobile Agent Based Intrusion Detection System Architecture For Mobile Adhoc Networks. *Journal of Computational Science*, *10*(6), 970–975. doi:10.3844/jcssp.2014.970.975

Paxson, V. (1997). End-to-End Internet Packet Dynamics. *Proceedings of the ACM SIGCOMM '97 Conference on Applications, Technologies, Architectures, and Protocols for Computer Communication*, 139–152. 10.1145/263105.263155

Paxson, V. (1999). End-to-end internet packet dynamics. *IEEE/ACM Transactions on Networking*, *7*(3), 277–292. doi:10.1109/90.779192

Perazzo, P., Sorbelli, F. B., Conti, M., Dini, G., & Pinotti, C. M. (2017). Drone Path Planning for Secure Positioning and Secure Position Verification. *IEEE Transactions on Mobile Computing*, *16*(9), 2478–2493. doi:10.1109/TMC.2016.2627552

Pillai, B., & Singhai, R. (2020). *Congestion Control Using Fuzzy-Based Node Reliability and Rate Control BT*. In K. N. Das, J. C. Bansal, K. Deep, A. K. Nagar, P. Pathipooranam, & R. C. Naidu (Eds.), *Soft Computing for Problem Solving* (pp. 67–75). Springer Singapore.

Postel, J. (1981a). *Internet control message protocol*. RFC Editor.

Postel, J. (1981b). *Rfc0793: Transmission control protocol*. RFC Editor.

Pournaghi, S. M., Zahednejad, B., Bayat, M., & Farjami, Y. (2018). NECPPA: A novel and efficient conditional privacy-preserving authentication scheme for VANET. *Computer Networks*, *134*, 78-92. .01.015 doi:10.1016/ j.comnet.2018

Prasad, M. R., Gyani, J., & Murti, P. R. K. (2012). Mobile Cloud Computing: Implications and Challenges. *Journal of Information Engineering and Applications, 2*(7), 7 - 15.

Qazi, I. A., Andrew, L. L. H., & Znati, T. (2009). Congestion control using efficient explicit feedback. *IEEE INFOCOM, 2009*, 10–18.

Qin, Z., Denker, G., Giannelli, C., & Bellavista, P. (2014). A Software Defined Networking architecture for the Internet-of-Things. *Network Operations and Management Symposium*, 1-9. 10.1109/NOMS.2014.6838365

Quan, W., Xu, C., Guan, J., Zhang, H., & Grieco, L. A. (2014). Social cooperation for information-centric multimedia streaming in highway VANETs. In *International Symposium on a World of Wireless, Mobile and Multimedia Networks (WoWMoM)* (pp. 1–6). 10.1109/WoWMoM.2014.6918992

Quaritsch, M., Kruggl, K., Wischounig-Strucl, D., Bhattacharya, S., Shah, M., & Rinner, B. (2010). Networked UAVs as aerial sensor network for disaster management applications. *Elektrotechnik und Informationstechnik, 127*(3), 56-63.

Qu, D., Wang, X., Huang, M., Li, K., Das, S. K., & Wu, S. (2018). A cache-aware social-based QoS routing scheme in Information Centric Networks. *Journal of Network and Computer Applications, 121*, 20–32. doi:10.1016/j.jnca.2018.07.002

Quintanilla, F. G., & Cardin, O. (2016). Implementation framework for cloud-based holonic control of cyber-physical production systems. In *IEEE Proceedings of the 14th International Conference on Industrial Informatics*. Poitiers, France: IEEE.

Raghunath, K., & Chockalingam, A. (2009). SC-FDMA versus OFDMA: Sensitivity to large carrier frequency and timing offsets on the uplink. In *GLOBECOM 2009-2009 IEEE Global Telecommunications Conference* (pp. 1–6). IEEE.

Rahim, A., Kong, X., Xia, F., Ning, Z., Ullah, N., Wang, J., & Das, S. K. (2018). Vehicular Social Networks: A survey. *Pervasive and Mobile Computing, 43*, 96-113. doi:10.1016/j.pmcj.2017.12.004

Rajappa, S., Bülthoff, H., & Stegagno, P. (2017). Design and implementation of a novel architecture for physical human-UAV interaction. *The International Journal of Robotics Research, 36*(5–7), 800–819. doi:10.1177/0278364917708038

Ramakrishnan, K, & Floyd, S. (1999). *RFC2481: A Proposal to add Explicit Congestion Notification (ECN) to IP*. RFC Editor.

Ramakrishnan, K., & Floyd, S. (1999). *A proposal to add explicit congestion notification (ECN) to IP*. RFC 2481.

Ramakrishnan, K. K., & Jain, R. (1990). A binary feedback scheme for congestion avoidance in computer networks. *ACM Transactions on Computer Systems, 8*(2), 158–181. doi:10.1145/78952.78955

Ramakrishnan, K., & Jain, R. (1991). *A Selective Binary Feedback Scheme for General Topologies*.

Rath & Oreku. (2018). Security Issues in Mobile Devices and Mobile Adhoc Networks. In *Mobile Technologies and Socio-Economic Development in Emerging Nations*. IGI Global. Doi:10.4018/978-1-5225-4029-8.ch009

Rath, M. (2018). A Methodical Analysis of Application of Emerging Ubiquitous Computing Technology With Fog Computing and IoT in Diversified Fields and Challenges of Cloud Computing. *International Journal of Information Communication Technologies and Human Development, 10*(2). Doi:10.4018/978-1-5225-4100-4.ch002

Rath, M. (2018). An Exhaustive Study and Analysis of Assorted Application and Challenges in Fog Computing and Emerging Ubiquitous Computing Technology. *International Journal of Applied Evolutionary Computation, 9*(2), 17-32. www.igi-global.com/ijaec

Rath, M. (2018). Smart Traffic Management System for Traffic Control using Automated Mechanical and Electronic Devices. *Mater. Sci. Eng., 377.* /377/1/01220110.1088/1757-899X/377/1/012201

Rath, M., & Panda, M. R. (2017). MAQ system development in mobile Adhoc networks using mobile agents. *IEEE 2nd International Conference on Contemporary Computing and Informatics (IC3I),* 794-798.

Rath, M., & Panigrahi, C. (2016). Prioritization of Security Measures at the Junction of MANET and IoT. In *Second International Conference on Information and Communication Technology for Competitive Strategies.* ACM Publication. 10.1145/2905055.2905187

Rath, M., & Pattanayak, B. (2018). Technological improvement in modern health care applications using Internet of Things (IoT) and proposal of novel health care approach. *International Journal of Human Rights in Healthcare.* doi:10.1108/IJHRH-01-2018-0007

Rath, M., & Pattanayak, B. K. (2014). A methodical survey on real time applications in MANETS: Focussing On Key Issues. *International Conference on, High Performance Computing and Applications (IEEE ICHPCA),* 1-5, 22-24. 10.1109/ICHPCA.2014.7045301

Rath, M., & Pattanayak, B. K. (2018). Monitoring of QoS in MANET Based Real Time Applications. In Information and Communication Technology for Intelligent Systems (ICTIS 2017) - Volume 2. ICTIS 2017. Smart Innovation, Systems and Technologies (vol. 84, pp. 579-586). Springer. doi:10.1007/978-3-319-63645-0_64

Rath, M., Pati, B., & Pattanayak, B. (2015). Delay and power based network assessment of network layer protocols in MANET. *2015 International Conference on Control, Instrumentation, Communication and Computational Technologies (IEEE ICCICCT),* 682-686. 10.1109/ICCICCT.2015.7475365

Rath, M., Pati, B., & Pattanayak, B. (2016). Comparative analysis of AODV routing protocols based on network performance parameters in Mobile Adhoc Networks. In Foundations and Frontiers in Computer, Communication and Electrical Engineering, (pp. 461-466). CRC Press, Taylor & Francis.

Rath, M., Pati, B., & Pattanayak, B. (2016). QoS Satisfaction in MANET Based Real Time Applications. *International Journal of Control Theory and Applications, 9*(7), 3069-3083.

Rath, M., Pati, B., & Pattanayak, B. K. (2016). Inter-Layer Communication Based QoS Platform for Real Time Multimedia Applications in MANET. Wireless Communications, Signal Processing and Networking (IEEE WiSPNET), 613-617. doi:10.1109/WiSPNET.2016.7566203

Rath, M., Pati, B., & Pattanayak, B. K. (2018). Relevance of Soft Computing Techniques in the Significant Management of Wireless Sensor Networks. In Soft Computing in Wireless Sensor Networks (pp. 86-106). Chapman and Hall/CRC, Taylor & Francis Group. doi:10.1201/9780429438639-4

Rath. (2018). Effective Routing in Mobile Adhoc Networks With Power and End-to-End Delay Optimization: Well Matched With Modern Digital IoT Technology Attacks and Control in MANET. In *Advances in Data Communications and Networking for Digital Business Transformation*. IGI Global. Doi:10.4018/978-1-5225-5323-6.ch007

Rathinam, S., Zennaro, M., Mak, T., & Sengupta, R. (2004). An architecture for UAV team control. *IFAC Proceedings Volumes, 37*(8), 573-578.

Rath, M. (2017). Resource provision and QoS support with added security for client side applications in cloud computing. *International Journal of Information Technology, 9*(3), 1–8.

Rath, M., & Pati, B. (2017). *Load balanced routing scheme for MANETs with power and delay optimisation, International Journal of Communication Network and Distributed Systems* , 19.

Rath, M., Pati, B., Panigrahi, C. R., & Sarkar, J. L. (2019). QTM: A QoS Task Monitoring System for Mobile Adhoc Networks. In P. Sa, S. Bakshi, I. Hatzilygeroudis, & M. Sahoo (Eds.), *Recent Findings in Intelligent Computing Techniques. Advances in Intelligent Systems and Computing* (Vol. 707). Springer. doi:10.1007/978-981-10-8639-7_57

Rath, M., Pati, B., & Pattanayak, B. (2015). Energy Competent Routing Protocol Design in MANET with Real time Application Provision. *International Journal of Business Data Communications and Networking, IGI Global, 11*(1), 50–60. doi:10.4018/IJBDCN.2015010105

Rath, M., Pati, B., & Pattanayak, B. (2016). Energy Efficient MANET Protocol Using Cross Layer Design for Military Applications. *Defence Science Journal, 66*(2), 146. doi:10.14429/dsj.66.9705

Rath, M., Pati, B., & Pattanayak, B. (2016). Resource Reservation and Improved QoS for Real Time Applications in MANET. *Indian Journal of Science and Technology, 9*(36). doi:10.17485/ijst/2016/v9i36/100910

Rath, M., Pati, B., & Pattanayak, B. K. (2017). Cross layer based QoS platform for multimedia transmission in MANET. *11th International Conference on Intelligent Systems and Control (ISCO)*, 402-407. 10.1109/ISCO.2017.7856026

Rath, M., Pati, B., & Pattanayak, B. K. (2019). Mobile Agent-Based Improved Traffic Control System in VANET. In A. Krishna, K. Srikantaiah, & C. Naveena (Eds.), *Integrated Intelligent Computing, Communication and Security. Studies in Computational Intelligence* (Vol. 771). Springer. doi:10.1007/978-981-10-8797-4_28

Rath, M., & Pattanayak, B. (2016). A Contemporary Survey and Analysis of Delay and Power Based Routing Protocols in MANET. *Journal of Engineering and Applied Sciences (Asian Research Publishing Network)*, *11*(1), 536–540.

Rath, M., & Pattanayak, B. (2017). MAQ:A Mobile Agent Based QoS Platform for MANETs. *International Journal of Business Data Communications and Networking, IGI Global*, *13*(1), 1–8. doi:10.4018/IJBDCN.2017010101

Rath, M., & Pattanayak, B. K. (2018). SCICS: A Soft Computing Based Intelligent Communication System in VANET. Smart Secure Systems – IoT and Analytics Perspective. *Communications in Computer and Information Science*, *808*, 255–261. doi:10.1007/978-981-10-7635-0_19

Rath, M., Pattanayak, B. K., & Pati, B. (2017). *Energetic Routing Protocol Design for Real-time Transmission in Mobile Adhoc Network. In Computing and Network Sustainability, Lecture Notes in Networks and Systems* (Vol. 12). Springer.

Rath, M., Rout, U. P., & Pujari, N. (2017). *Congestion Control Mechanism for Real Time Traffic in Mobile Adhoc Networks, Computer Communication, Networking and Internet Security. In Lecture Notes in Networks and Systems* (Vol. 5). Springer.

Rath, M., Swain, J., Pati, B., & Pattanayak, B. K. (2018). *Attacks and Control in MANET. In Handbook of Research on Network Forensics and Analysis Techniques.* IGI Global.

Reichardt, D., Miglietta, M., Moretti, L., Morsink, P., & Schulz, W. (2002). CarTALK 2000: Safe and comfortable driving based upon inter-vehicle-communication. In *Intelligent Vehicle Symposium*, 2002. *IEEE* (Vol. 2, pp. 545–550). IEEE.

Renuka, B. (2012). Location Based Services on Mobile E-Commerce. *International Journal of Computer Science and Information Technologies*, *3*(1), 3147–3315.

Robinson, J., Friedman, D., & Steenstrup, M. (1989). Congestion control in BBN packet-switched networks. *Computer Communication Review*, *20*(1), 76–90. doi:10.1145/86587.86592

Romanovsky, A., & Ishikawa, F. (Eds.). (2017). *Trustworthy cyber-physical systems engineering.* CRC.

Romanow, A., & Floyd, S. (1995). Dynamics of TCP traffic over ATM networks. *IEEE Journal on Selected Areas in Communications*, *13*(4), 633–641. doi:10.1109/49.382154

Rose, O. (1992). The Q-bit scheme: Congestion avoidance using rate-adaptation. *Computer Communication Review*, *22*(2), 29–42. doi:10.1145/141800.141803

Rtah, M. (2018). Big Data and IoT-Allied Challenges Associated With Healthcare Applications in Smart and Automated Systems. *International Journal of Strategic Information Technology and Applications*, *9*(2), 18–34. Advance online publication. doi:10.4018/IJSITA.2018040102

Ruby, R., Leung, V. C., & Michelson, D. G. (2015). Uplink scheduler for SC-FDMA-based heterogeneous traffic networks with QoS assurance and guaranteed resource utilization. *IEEE Transactions on Vehicular Technology*, *64*(10), 4780–4796. doi:10.1109/TVT.2014.2367007

Rumney, M. (2008). *3GPP LTE: Introducing Single-Carrier FDMA*. de Agilent Technologies. Inc.

Safa, H., El-Hajj, W., & Tohme, K. (2013). A QoS-aware uplink scheduling paradigm for LTE networks. In *2013 IEEE 27th International Conference on Advanced Information Networking and Applications (AINA)* (pp. 1097–1104). IEEE. 10.1109/AINA.2013.38

Sahoo, J., & Rath, M. (2017). Study and Analysis of Smart Applications in Smart City Context. *2017 International Conference on Information Technology (ICIT)*, 225-228. 10.1109/ICIT.2017.38

Saino, L., Cocora, C., & Pavlou, G. (2013). Cctcp: A scalable receiver-driven congestion control protocol for content centric networking. *2013 IEEE International Conference on Communications (ICC)*, 3775–3780. 10.1109/ICC.2013.6655143

Saleh, Gamel, & Abo-Al-Ez. (2017). A Reliable Routing Protocol for Vehicular Adhoc Networks. *Computers & Electrical Engineering, 64*, 473-495. doi:10.1016/j.compeleceng.2016.11.011

Salman, A. M., & Lakshmi, R. (2008). Disconnected Modes of Operations in Mobile Environments. *Proc. INDIACom-2008, 2nd National Conference on Computing for Nation Development*, 253-256.

Salman, A.M. & Lakshmi, R. (2010). Replication Strategies in Mobile Environments. *BVICAM'S International Journal of Information Technology, 2*(1).

Sandanalaksmi, R., Manivanan, K., Manikandan, S., Barathi, R., & Devanathan, D. (2009). Fair channel aware packet scheduling algorithm for fast UL HARQ in UTRAN LTE. In *2009 International Conference on Control, Automation, Communication and Energy Conservation* (pp. 1–5). IEEE.

Sato, Y., Ryusuke, M., Obara, T., & Adachi, F. (2012). Nash bargaining solution based subcarrier allocation for uplink SC-FDMA distributed antenna network. In *2012 3rd IEEE International Conference on Network Infrastructure and Digital Content* (pp. 76–80). IEEE. 10.1109/ICNIDC.2012.6418715

Sattar, S., Qureshi, H. K., Saleem, M., Mumtaz, S., & Rodriguez, J. (2018). Reliability and energy-efficiency analysis of safety message broadcast in VANETs. *Computer Communications, 119*, 118-126. doi:10.1016/j.comcom.2018.01.006

Saxena, D., Raychoudhury, V., Suri, N., Becker, C., & Cao, J. (2016). Named Data Networking: A survey. *Computer Science Review, 19*, 15–55. doi:10.1016/j.cosrev.2016.01.001

Scherer, J., Yahyanejad, S., Hayat, S., Yanmaz, E., Andre, T., Khan, A., & Rinner, B. (2015, May). An autonomous multi-UAV system for search and rescue. In *Proceedings of the First Workshop on Micro Aerial Vehicle Networks, Systems, and Applications for Civilian Use* (pp. 33-38). ACM. 10.1145/2750675.2750683

Seeling, P., & Reisslein, M. (2014). Video Traffic characteristics of modern encoding standards: H.264/AVC with SVC and MVC extensions and H.265/HEVC. *TheScientificWorldJournal, 2014*, 189481. Advance online publication. doi:10.1155/2014/189481 PMID:24701145

Seema, A., & Reisslein, M. (2011). Towards efficient wireless video sensor networks: A survey of existing node architectures and proposal for a Flexi-WVSNP Design. *IEEE Communications Surveys and Tutorials, 13*(3), 462–486. doi:10.1109/SURV.2011.102910.00098

Shah, S. T., Gu, J., Hasan, S. F., & Chung, M. Y. (2015). SC-FDMA-based resource allocation and power control scheme for D2D communication using LTE-A uplink resource. *EURASIP Journal on Wireless Communications and Networking, 2015*(1), 137. doi:10.118613638-015-0340-3

Shakshuki, E. M., Kang, N., & Sheltami, T. R. (2013). EAACK—A secure intrusion-detection system for MANETs. *IEEE Transactions on Industrial Electronics, 60*(3), 1089–1098. doi:10.1109/TIE.2012.2196010

Shakya, R. K., Singh, Y. N., & Verma, N. K. (2013). Generic correlation model for wireless sensor network applications. *IET Wireless Sensor Systems, 3*(4), 266–276. doi:10.1049/iet-wss.2012.0094

Sharma, A. (2011). Adoption Analysis of Mobile banking in India. In *Proc. International Conference SPIN-2011 (Speech Processing and Integrated Networks)*. Amity College of Engineering.

Sharma, A. (2012). M-Commerce Technology Adoption and Trust Challenges. *International Journal of Scientific and Research Publications, 2*(2), 1-5.

Sharma, A., & Kansal, V. (2013). An Evaluation Framework of Replication Protocols in Mobile Environment. *International Journal of Database Management Systems, 5*(1), 45-51.

Sharma, S. (2018). Hybrid fuzzy multi-criteria decision making based multi cluster head dolphin swarm optimized IDS for VANET. *Vehicular Communications, 12*, 23-38. doi:10.1016/j.vehcom.2017.12.003

Sharma, A., Kansal, V., & Tomar, R.P.S. (2015). Location Based Services in M- Commerce: Customer Trust and Transaction Security Issue. *International Journal of Computer Science and Security, 9*(2), 11–21.

Sharma, V. K., & Kumar, M. (2019). Adaptive load distribution approach based on congestion control scheme in ad-hoc networks. *International Journal of Electronics, 106*(1), 48–68. doi:10.1080/00207217.2018.1501613

Sheik, Chandel, & Mishra. (2012). *Security Issues in MANET: A Review*. IEEE.

Shemsi, I., & Kadam, P. (2017). Named Data Networking in VANET: A Survey. *International Journal of Scientific Engineering and Science*, 1–5.

Shu, Y., Ge, W., Jiang, N., Kang, Y., & Luo, J. (2008). Mobile-Host-Centric Transport Protocol forEAST Experiment. *IEEE Transactions on Nuclear Science, 55*(1), 209–216. doi:10.1109/TNS.2007.914319

Singh, J., & Lambay, M. A. (n.d.). *Network Border Patrol: Preventing Congestion Collapse and Promoting Fairness in the Network*. Academic Press.

Singh, S., & Agrawal, S. (2014). *VANET routing protocols: Issues and challenges. In 2014 Recent Advances in Engineering and Computational Sciences*. RAECS.

Sing, J., & Soh, B. (2005). TCP New Vegas: improving the performance of TCP Vegas over high latency links. *Fourth IEEE International Symposium on Network Computing and Applications*, 73–82. 10.1109/NCA.2005.52

Sivakumar, M., & Sadagopan, C. (2016, August). Wireless sensor network to cyber physical systems: addressing mobility challenges for energy efficient data aggregation using dynamic nodes. *Sensor Letters*, *14*(8), 852-857. doi:10.11661.2016.3624

Sofer, E., & Segal, Y. (2005). *Tutorial on multi-access OFDM (OFDMA) technology*. DOC: IEEE 802.22-05-0005r0.

Sokmen, F. I., & Girici, T. (2010). Uplink resource allocation algorithms for single-carrier FDMA systems. In *2010 European Wireless Conference (EW)* (pp. 339–345). IEEE. 10.1109/EW.2010.5483441

Song, H., & Rawat, D. B. (2017). *Cyber-physical systems: foundations, principles and applications*. Academic Press/Elsevier.

Sreekumari, P., & Chung, S.-H. (2011). TCP NCE: A unified solution for non-congestion events to improve the performance of TCP over wireless networks. *EURASIP Journal on Wireless Communications and Networking*, *2011*(1), 23. doi:10.1186/1687-1499-2011-23

Sridharan, G., & Lim, T. J. (2012). Performance analysis of SC-FDMA in the presence of receiver phase noise. *IEEE Transactions on Communications*, *60*(12), 3876–3885. doi:10.1109/TCOMM.2012.082812.110879

Srijith, K. N., Jacob, L., & Ananda, A. L. (2005). TCP Vegas-A: Improving the performance of TCP Vegas. *Computer Communications*, *28*(4), 429–440. doi:10.1016/j.comcom.2004.08.016

Srovnal, V., Jr., Machacek, Z., & Srovnal, V. (2009). Wireless Communication for Mobile Robotics and Industrial Embedded Devices. *Eighth International Conference on Networks*, 253-258.

Stankovic, J. A. (2017, February). Research directions for cyber physical systems in wireless and mobile healthcare. *ACM Transactions on Cyber-Physical Systems*, *1*(1). doi:10.1145/2899006

Stevens, W. (1997). *RFC2001: TCP slow start, congestion avoidance, fast retransmit, and fast recovery algorithms*. RFC Editor.

Sullivan, G. J., Ohm, J.-R., Han, W.-J., & Wiegand, T. (2012). Overview of the High Efficiency Video Coding (HEVC) Standard. *IEEE Transactions on Circuits and Systems for Video Technology*, *22*(12), 1649–1668. doi:10.1109/TCSVT.2012.2221191

Sun, D., & Man, H. (2001). ENIC-an improved reliable transport scheme for mobile ad hoc networks. *GLOBECOM'01. IEEE Global Telecommunications Conference (Cat. No. 01CH37270)*, *5*, 2852–2856.

Szafir, D., Mutlu, B., & Fong, T. (2017). Designing planning and control interfaces to support user collaboration with flying robots. *The International Journal of Robotics Research, 36*(5–7), 514–542. doi:10.1177/0278364916688256

Taga, K., Karlsson, J., & Arthur, D. (2004). Little Global M-Payment Report. Academic Press.

Takahashi, A., & Asanuma, N. (2000). Introduction of Honda ASV-2 (advanced safety vehicle-phase 2). In *Intelligent Vehicles Symposium, 2000. IV 2000. Proceedings of the IEEE* (pp. 694–701). 10.1109/IVS.2000.898430

Tanenbaum, A. S. (1981). *Computer networks*. Prentice-Hall.

Tang, H., Hong, P., Xue, K., & Peng, J. (2012). Cluster-based resource allocation for interference mitigation in LTE heterogeneous networks. In 2012 IEEE Vehicular Technology Conference (VTC Fall) (pp. 1–5). IEEE. doi:10.1109/VTCFall.2012.6398901

Tan, K., Song, J., Zhang, Q., & Sridharan, M. (2006). A compound TCP approach for high-speed and long distance networks. *Proceedings - IEEE INFOCOM*, 1–12. doi:10.1109/INFOCOM.2006.188

Tan, S., Li, X., & Dong, Q. (2016). A Trust Management System for Securing Data Plane of Ad-Hoc Networks. *IEEE Transactions on Vehicular Technology, 65*(9), 7579–7592. doi:10.1109/TVT.2015.2495325

Tardioli, D. (2014). A wireless communication protocol for distributed robotics applications. *IEEE International Conference on Autonomous Robot Systems and Competitions (ICARSC)*, 253-260. 10.1109/ICARSC.2014.6849795

Theodorakopoulos, G., & Baras, J. S. (2006). On trust models and trust evaluation metrics for ad-hoc networks. *IEEE Journal on Selected Areas in Communications, 24*, 318-328.

Thornton, J., & White, L. (2001). Customer Orientations and Usage of Financial Distribution Channels. *Journal of Services Marketing, 15*(3), 168–185. doi:10.1108/08876040110392461

Thramboulidis, K., & Christoulakis, F. (2016, October). UML4IoT - A UML-based approach to exploit IoT In cyber-physical manufacturing systems. *Computers in Industry, Elsevier, 82*, 259–272. doi:10.1016/j.compind.2016.05.010

Toh, C.-K. (1997). Associativity-based routing for ad hoc mobile networks. *Wireless Personal Communications, 4*(2), 103–139. doi:10.1023/A:1008812928561

Tomar, R.P.S., & Sharma, A. (2014). Mathematical Model to Study the Cost Effects and Mobile-Users Trust on Location based Data Access in Mobile-Commerce Transactions. *International Journal of Computer Applications, 105*(12), 22-26.

Toor, Y., Muhlethaler, P., Laouiti, A., & La Fortelle, A. (2008). Vehicle ad hoc networks: Applications and related technical issues. *IEEE Communications Surveys and Tutorials, 10*(3), 74–88. doi:10.1109/COMST.2008.4625806

Total number of Websites - Internet Live Stats. (n.d.). Retrieved July 10, 2020, from https://www.internetlivestats.com/total-number-of-websites/

Tsaoussidis, V., & Zhang, C. (2002). TCP-Real: Receiver-oriented congestion control. *Computer Networks*, *40*(4), 477–497. doi:10.1016/S1389-1286(02)00291-8

Tsiropoulou, E. E., & Papavassiliou, S. (2011). Utility-based uplink joint power and subcarrier allocation in SC-FDMA wireless networks. *International Journal of Electronics*, *98*(11), 1581–1587. doi:10.1080/00207217.2011.589741

Tsiropoulou, E. E., Ziras, I., & Papavassiliou, S. (2015). Service differentiation and resource allocation in SC-FDMA wireless networks through user-centric Distributed non-cooperative Multilateral Bargaining. In *International Conference on Ad Hoc Networks* (pp. 42–54). Springer. 10.1007/978-3-319-25067-0_4

Twyman-Saint Victor, C., Rech, A. J., Maity, A., Rengan, R., Pauken, K. E., Stelekati, E., ... Odorizzi, P. M. (2015). Radiation and dual checkpoint blockade activate non-redundant immune mechanisms in cancer. *Nature*, *520*(7547), 373–377. doi:10.1038/nature14292 PMID:25754329

Varakulsiripunth, R., Shiratori, N., & Noguchi, S. (1986). A congestion-control policy on the internetwork gateway. *Computer Networks and ISDN Systems*, *11*(1), 43–58. doi:10.1016/0169-7552(86)90028-0

Varghese, G. (1996). On avoiding congestion collapse. *Viewgraphs, Washington University Workshop on the Integration of IP and ATM.*

Vasudevan, V., Phanishayee, A., Shah, H., Krevat, E., Andersen, D. G., Ganger, G. R., Gibson, G. A., & Mueller, B. (2009). Safe and effective fine-grained TCP retransmissions for datacenter communication. *Computer Communication Review*, *39*(4), 303–314. doi:10.1145/1594977.1592604

Vehicle Infrastructure Integration (VII) program of US Federal and State departments of transportation and automobile manufacturers. (n.d.).

Velloso, P. B., Laufer, R. P., Cunha, D. D. O., Duarte, O. C. M., & Pujolle, G. (2010). Trust management in mobile ad hoc networks using a scalable maturity-based model. *IEEE eTransactions on Network and Service Management*, *7*(3), 172–185. doi:10.1109/TNSM.2010.1009.I9P0339

Venkataramani, A., Kokku, R., & Dahlin, M. (2002). TCP Nice: A mechanism for background transfers. *ACM SIGOPS Operating Systems Review, 36*(SI), 329–343.

Vorakulpipat, R. Thaenkaew, & Hai. (2018). Recent Challenges, Trends and Concerns Related to IoT Security: An Evolutionary Study. IEEE.

Wang, C., & Zhu, Y. (2018, September). A dependable time series analytic framework for cyber-physical systems of IoT-based smart grid. *ACM Transactions on Cyber-Physical Systems*, *3*(1). doi:. doi:10.1145/3145623

Wang, J., Wakikawa, R., & Zhang, L. (2010). DMND: Collecting data from mobiles using named data. In IEEE Vehicular networking conference (VNC) (pp. 49–56). doi:10.1109/VNC.2010.5698270

Wang, X., & Konishi, S. (2010). Optimization formulation of packet scheduling problem in LTE uplink. In *2010 IEEE 71st Vehicular Technology Conference* (pp. 1–5). IEEE. 10.1109/VETECS.2010.5493797

Wang, X., Wang, D., & Sun, Q. (2018). Reliable routing in IP-based VANET with network gaps. *Computer Standards & Interfaces, 55*, 80-94. doi:10.1016/j.csi.2017.05.002

Wang, D., Lin, L., & Xu, L. (2011). A study of subdividing hexagon-clustered WSN for power saving: Analysis and simulation. *Ad Hoc Networks, 9*(7), 1302–1311. doi:10.1016/j.adhoc.2011.03.001

Wang, F., & Zhang, Y. (2002). Improving TCP performance over mobile ad-hoc networks with out-of-order detection and response. *Proceedings of the 3rd ACM International Symposium on Mobile Ad Hoc Networking & Computing*, 217–225. 10.1145/513800.513827

Wang, L., Wakikawa, R., Kuntz, R., Vuyyuru, R., & Zhang, L. (2012). Data naming in vehicle-to-vehicle communications. In *IEEE Conference on Computer Communications Workshops (INFOCOM WKSHPS)* (pp. 328–333). 10.1109/INFCOMW.2012.6193515

Wang, L., Waltari, O., & Kangasharju, J. (2013). Mobiccn: Mobility support with greedy routing in content-centric networks. In *IEEE Global Communications Conference (GLOBECOM)* (pp. 2069–2075). 10.1109/GLOCOM.2013.6831380

Wang, R., Valla, M., Sanadidi, M. Y., Ng, B. K. F., & Gerla, M. (2002). Efficiency/friendliness tradeoffs in TCP Westwood. *Proceedings ISCC 2002 Seventh International Symposium on Computers and Communications*, 304–311. 10.1109/ISCC.2002.1021694

Wang, S., Lei, T., Zhang, L., Hsu, C.-H., & Yang, F. (2016, August). Offloading mobile data traffic for QoS-aware service provision in vehicular cyber-physical systems. *Future Generation Computer Systems, Elsevier, 61*, 118–127. doi:10.1016/j.future.2015.10.004

Wang, W., & Minoli, D. (2017, June 6). Multimedia IoT systems and applications. In *Global IoT Summit, GIoTS-2017*. IEEE.

Wang, W., & Wang, Q. (2017, April). Multimedia Sensing As A Service (MSaaS): Cloud-edge IoTs and fogs. *IEEE IoT Journal, 4*(2), 487–495. doi:10.1109/JIOT.2016.2578722

Wang, Y., Yu, F. R., Tang, H., & Huang, M. (2014). A mean field game theoretic approach for security enhancements in mobile ad hoc networks. *IEEE Transactions on Wireless Communications, 13*(3), 1616–1627. doi:10.1109/TWC.2013.122313.131118

Wang, Z., & Crowcroft, J. (1991). A new congestion control scheme: Slow start and search (Tri-S). *Computer Communication Review, 21*(1), 32–43. doi:10.1145/116030.116033

Wang, Z., & Crowcroft, J. (1992). Eliminating periodic packet losses in the 4.3-Tahoe BSD TCP congestion control algorithm. *Computer Communication Review*, 22(2), 9–16. doi:10.1145/141800.141801

Wei, Z., Tang, H., Yu, F. R., Wang, M., & Mason, P. (2014a, June). Trust establishment with data fusion for secure routing in MANETs. In *Communications (ICC), 2014 IEEE International Conference on* (pp. 671-676). IEEE 10.1109/ICC.2014.6883396

Wei, Z., Tang, H., Yu, F. R., Wang, M., & Mason, P. C. (2014b). Security Enhancements for Mobile Ad Hoc Networks with Trust Management Using Uncertain Reasoning. *IEEE Transactions on Vehicular Technology*, 63(9), 4647–4658. doi:10.1109/TVT.2014.2313865

Weyrich, M., & Ebert, C. (2016, February). Reference architectures for the Internet of Things. *IEEE Software*.

Williamson, C. L. (1993). Optimizing file transfer response time using the loss-load curve congestion control mechanism. *Computer Communication Review*, 23(4), 117–126. doi:10.1145/167954.166249

Wong, I. C., Oteri, O., & McCoy, W. (2009). Optimal resource allocation in uplink SC-FDMA systems. *IEEE Transactions on Wireless Communications*, 8(5), 2161–2165. doi:10.1109/TWC.2009.061038

Wu, D., Hou, Y. T., & Zhang, Y.-Q. (2000). Transporting real-time video over the Internet: Challenges and approaches. *Proceedings of the IEEE*, 88(12), 1855–1877. doi:10.1109/5.899055

Wu, P., Schober, R., & Bhargava, V. K. (2013). Optimal power allocation for wideband cognitive radio networks employing SC-FDMA. *IEEE Communications Letters*, 17(4), 669–672. doi:10.1109/LCOMM.2013.021913.122708

Wu, X., Xiong, Y., Yang, P., Wan, S., & Huang, W. (2014). Sparsest random scheduling for compressive data gathering in wireless sensor networks. *IEEE Transactions on Wireless Communications*, 13(10), 5867–5877. doi:10.1109/TWC.2014.2332344

Wu, Z., Li, J., Zuo, J., & Li, S. (2018). Path Planning of UAVs Based on Collision Probability and Kalman Filter. *IEEE Access: Practical Innovations, Open Solutions*, 6, 34237–34245. doi:10.1109/ACCESS.2018.2817648

Xiang, L., Luo, J., & Vasilakos, A. (2011, June). Compressed data aggregation for energy efficient wireless sensor networks. In *Sensor, mesh and ad hoc communications and networks (SECON), 2011 8th annual IEEE communications society conference on* (pp. 46-54). IEEE. 10.1109/SAHCN.2011.5984932

Xiaoronga, C., Mingxuan, L., & Suc, L. (2012). Study on clustering of wireless sensor network in distribution network monitoring system. *Physics Procedia*, 25, 1689–1695. doi:10.1016/j.phpro.2012.03.296

Xie, Y., Yi, F., & Jamieson, K. (2020). *PBE-CC: Congestion Control via Endpoint-Centric, Physical-Layer Bandwidth Measurements.* ArXiv Preprint ArXiv:2002.03475

Xie, R., & Jia, X. (2014). Transmission-efficient clustering method for wireless sensor networks using compressive sensing. *IEEE Transactions on Parallel and Distributed Systems*, *25*(3), 806–815. doi:10.1109/TPDS.2013.90

Xu, K., Gerla, M., Qi, L., & Shu, Y. (2005). TCP unfairness in ad hoc wireless networks and a neighborhood RED solution. *Wireless Networks*, *11*(4), 383–399. doi:10.100711276-005-1764-1

Xu, L., Harfoush, K., & Rhee, I. (2004). Binary increase congestion control (BIC) for fast long-distance networks. *IEEE INFOCOM*, *2004*(4), 2514–2524.

Xylomenos, G., Ververidis, C. N., Siris, V. A., Fotiou, N., Tsilopoulos, C., Vasilakos, X., Katsaros, K. V., & Polyzos, G. C. (2014). A Survey of Information-Centric Networking Research. *IEEE Communications Surveys and Tutorials*, *16*(2), 1024–1049. doi:10.1109/SURV.2013.070813.00063

Yahyanejad, S., & Rinner, B. (2015). A fast and mobile system for registration of low-altitude visual and thermal aerial images using multiple small-scale UAVs. *ISPRS Journal of Photogrammetry and Remote Sensing*, *104*, 189–202. doi:10.1016/j.isprsjprs.2014.07.015

Yang, B., Niu, K., He, Z., Xu, W., & Huang, Y. (2013). Improved proportional fair scheduling algorithm in LTE uplink with single-user MIMO transmission. In *2013 IEEE 24th Annual International Symposium on Personal, Indoor, and Mobile Radio Communications (PIMRC)* (pp. 1789–1793). IEEE.

Yang, C.-Q., & Reddy, A. V. S. (1995). A taxonomy for congestion control algorithms in packet switching networks. *IEEE Network*, *9*(4), 34–45. doi:10.1109/65.397042

Yang, G., Wang, R., Sanadidi, M. Y., & Gerla, M. (2003). TCPW with bulk repeat in next generation wireless networks. *IEEE International Conference on Communications, 2003. ICC'03.*, *1*, 674–678. 10.1109/ICC.2003.1204260

Yang, K., Martin, S., & Yahiya, T. A. (2015). LTE uplink interference aware resource allocation. *Computer Communications*, *66*, 45–53. doi:10.1016/j.comcom.2015.04.002

Yang, L., Seah, W. K. G., & Yin, Q. (2003). Improving fairness among TCP flows crossing wireless ad hoc and wired networks. *Proceedings of the 4th ACM International Symposium on Mobile Ad Hoc Networking & Computing*, 57–63. 10.1145/778415.778423

Yang, Q., & Yoo, S. (2018). Optimal UAV Path Planning: Sensing Data Acquisition Over IoT Sensor Networks Using Multi-Objective Bio-Inspired Algorithms. *IEEE Access: Practical Innovations, Open Solutions*, *6*, 13671–13684. doi:10.1109/ACCESS.2018.2812896

Yang, X., Ge, L., & Wang, Z. (2009). F-ECN: A Loss Discrimination Based on Fuzzy Logic Control. *2009 Sixth International Conference on Fuzzy Systems and Knowledge Discovery*, *7*, 546–549. 10.1109/FSKD.2009.140

Yanmaz, E., Kuschnig, R., Quaritsch, M., Bettstetter, C., & Rinner, B. (2011, April). On path planning strategies for networked unmanned aerial vehicles. In *Computer Communications Workshops (INFOCOM WKSHPS), IEEE Conference on* (pp. 212-216). IEEE. 10.1109/INFCOMW.2011.5928811

Yazici, V., Sunay, M. O., & Ercan, A. O. (2014). Controlling a software-defined network via distributed controllers. CoRR, vol. abs/1401.7651

Yin & Li. (1990). Stern Congestion control for packet voice by selective packet discarding. *IEEE Trans. Comm, 38*(5), 674–683.

Yuan, J., & Chen, H. (2009, September). The optimized clustering technique based on spatial-correlation in wireless sensor networks. In *Information, Computing and Telecommunication, 2009. YC-ICT'09. IEEE Youth Conference on* (pp. 411-414). IEEE.

Yuen, C. H. G., & Farhang-Boroujeny, B. (2012). Analysis of the optimum precoder in SC-FDMA. *IEEE Transactions on Wireless Communications, 11*(11), 4096–4107. doi:10.1109/TWC.2012.090412.120105

Yu, F. R., Tang, H., Bu, S., & Zheng, D. (2013). Security and quality of service (QoS) co-design in cooperative mobile ad hoc networks. *EURASIP Journal on Wireless Communications and Networking, 2013*(1), 188. doi:10.1186/1687-1499-2013-188

Yu, F. R., Tang, H., Mason, P. C., & Wang, F. (2010). A hierarchical identity based key management scheme in tactical mobile ad hoc networks. *IEEE eTransactions on Network and Service Management, 7*(4), 258–267. doi:10.1109/TNSM.2010.1012.0362

Yu, X., & Xue, Y. (2016, May). Smart grids: A cyber-physical systems perspective. *Proceedings of the IEEE, 104*(5), 1058–1070. doi:10.1109/JPROC.2015.2503119

Yu, Y.-T., Gerla, M., & Sanadidi, M. Y. (2015). Scalable VANET content routing using hierarchical bloom filters. *Wireless Communications and Mobile Computing, 15*(6), 1001–1014. doi:10.1002/wcm.2495

Zalak, P., Aemi, K., Raval, G., Ukani, V., & Valiveti, S. (2017). Open Issues in Named Data Networking—A Survey. In *International Conference on Information and Communication Technology for Intelligent Systems* (pp. 285–292). Academic Press.

Zaman, M., Quadri, S. M. K., & Butt, M. A. (2012). Information Translation: A Practitioners Approach. *Proceedings of the World Congress on Engineering and Computer Science, 1.*

Zanero, S. (2017, April). Cyber-physical systems. *Computer, 50*(4), 14–16. doi:10.1109/MC.2017.105

Zhang, J., Yang, L.-L., & Hanzo, L. (2009). Multi-user performance of the amplify-and-forward single-relay assisted SC-FDMA uplink. In *2009 IEEE 70th Vehicular Technology Conference Fall* (pp. 1–5). IEEE. 10.1109/VETECF.2009.5378760

Zhang, J., Yuan, Y. F., & Archer, N. (2002). Driving Forces for M-Commerce success. Michael G. De Groote School of Business, McMaster University. doi:10.1007/0-306-47548-0_4

Zhang, L., Estrin, D., Burke, J., Jacobson, V., Thornton, J. D., Smetters, D. K., … others. (2010). *Named Data Networking (NDN) Project.* Relatório Técnico NDN-0001, Xerox Palo Alto Research Center-PARC.

Zhang, C., Zhang, X., Li, O., Yang, Y., & Liu, G. (2017). Dynamic clustering and compressive data gathering algorithm for Energy-efficient wireless sensor networks. *International Journal of Distributed Sensor Networks*, *13*(10), 1550147717738905. doi:10.1177/1550147717738905

Zhang, J., Yang, L.-L., & Hanzo, L. (2011). Energy-efficient channel-dependent cooperative relaying for the multiuser SC-FDMA uplink. *IEEE Transactions on Vehicular Technology*, *60*(3), 992–1004. doi:10.1109/TVT.2011.2104985

Zhang, J., Yang, L.-L., & Hanzo, L. (2013). Energy-efficient dynamic resource allocation for opportunistic-relaying-assisted SC-FDMA using turbo-equalizer-aided soft decode-and-forward. *IEEE Transactions on Vehicular Technology*, *62*(1), 235–246. doi:10.1109/TVT.2012.2220162

Zhang, L. (1990). Virtual clock: A new traffic control algorithm for packet switching networks. *Proceedings of the ACM Symposium on Communications Architectures & Protocols*, 19–29. 10.1145/99508.99525

Zhang, M., & Zhu, Y. (2013). An enhanced greedy resource allocation algorithm for localized SC-FDMA systems. *IEEE Communications Letters*, *17*(7), 1479–1482. doi:10.1109/LCOMM.2013.052013.130716

Zhao & Ge. (2013). A Survey on the Internet of Things Security. *9th International Conference on Computational Intelligence and Security*.

Zhao, J., Gao, F., Kuang, L., Wu, Q., & Jia, W. (2018). Channel Tracking with Flight Control System for UAV mmWave MIMO Communications. *IEEE Communications Letters*, *22*(6), 1224–1227. doi:10.1109/LCOMM.2018.2824800

Zhao, W., Qin, Y., Gao, D., Foh, C. H., & Chao, H.-C. (2017). An Efficient Cache Strategy in Information Centric Networking Vehicle-to-Vehicle Scenario. *IEEE Access: Practical Innovations, Open Solutions*, *5*, 12657–12667. doi:10.1109/ACCESS.2017.2714191

Zhen, C., Liu, W., Liu, Y., & Yan, A. (2014). Energy-efficient sleep/wake scheduling for acoustic localization wireless sensor network node. *International Journal of Distributed Sensor Networks*, *10*(2), 970524. doi:10.1155/2014/970524

Zheng, D., Tang, H., & Yu, F. R. (2012, October). A game theoretic approach for security and quality of service (QoS) co-design in MANETs with cooperative communications. In *Military Communications Conference, 2012-MILCOM 2012* (pp. 1–6). IEEE. doi:10.1109/MILCOM.2012.6415562

Zheng, K., Wei, M., Long, H., Liu, Y., & Wang, W. (2009). Impacts of amplifier nonlinearities on uplink performance in 3G LTE systems. In *2009 Fourth International Conference on Communications and Networking in China* (pp. 1–5). IEEE. 10.1109/CHINACOM.2009.5339757

Zhou, J., Shi, B., & Zou, L. (2003). Improve TCP performance in Ad hoc network by TCP-RC. *14th IEEE Proceedings on Personal, Indoor and Mobile Radio Communications, 2003. PIMRC 2003*, *1*, 216–220.

About the Contributors

Ayaz Ahmad received his B.Sc degree in Electrical Engineering from N-W.F.P University of Engineering and Technology, Peshawar Pakistan, in 2006. After getting his B.Sc degree, he was associated with the department of Electrical Engineering, National University of Computer and Emerging Sciences where he was involved in research and teaching activities. In 2008, he got his Master degree in Wireless Communication Systems from the Department of Telecommunications, Ecole Superieure d'Electricite (Supelec), Gif-sur-Yvette, France, and in 2011 he earned his Ph.D. degree in Telecommunications from the same institution. Since 2012, he is serving as Assistant Professor in the department of Electrical Engineering COMSATS Institute of Information Technology, Wah Cantt., Pakistan. Ayaz has several years of research experience and has authored/co-authored several scientific publications in various refereed international journals and conferences. He has also published several book chapters and is the leading co-editor of the book entitled, "Smart Grid as a Solution for Renewable and Efficient Energy" published by IGI Global, USA in 2016. He is currently an Associate Editor with IEEE Access and Springer Human Centric Computing and Information Sciences. He has also twice served as a guest editor of special issue for IEEE Access. He is regularly serving as TPC member for several international conferences including IEEE GLOBECOM, IEEE ICC, and IEEE PIMRC, and as reviewer for several renowned international journals. He is member of IEEE Communication Society and Senior Member of IEEE, USA. His research interests include resource allocation in wireless communication systems, energy management in smart grid, and application of optimization methods to engineering problems.

Raheel Ahmad completed his Bachelor of Engineering degree form COMSATS University Islamabad, Wah Campus. Currently he is a Master student at COMSATS University Islamabad.

Syed Hassan Ahmed (S'13, M'17, SM'18) is currently an Assistant Professor in the Computer Science Department of Georgia Southern University (GSU) at Statesboro, USA. Before starting at GSU, Dr. Hassan was a Post-Doctoral Fellow in the Department of Electrical and Computer Engineering, University of Central Florida, Orlando. Previously, he completed his B.S in Computer Science from Kohat University of Science and Technology (KUST), Pakistan and Master combined Ph.D. Degree from School of Computer Science and Engineering (SCSE), Kyungpook National University (KNU), Republic of Korea. In summer 2015, he was a visiting Ph.D. student at the Georgia Tech, Atlanta, USA. Collectively, Dr. Hassan authored/co-authored over 100 international publications including Journal articles, Conference Proceedings, Book Chapters, and 03 books. From the year 2014 to 2016, he consequently won the Research Contribution awards by SCSE at KNU, Korea. In 2016, his work on robust content retrieval in future vehicular networks lead him to win the Qualcomm Innovation Award at KNU, Korea. Dr. Hassan is an active IEEE Senior Member and ACM Professional Member and his research interests include Sensor and Ad hoc Networks, Cyber-Physical Systems, Vehicular Communications and Future Internet.

Rajiv Kumar Gupta received his PhD from the department of Electrical Engineering, IIT Bombay, 2010, MS (By Research) Communication Engineering from IIT Delhi in 2011 and BTech in Electronics and Communication Engineering from NIT Kurukshetra in 1988. He is currently working as Professor in the department of Electronics and Telecommunication at Terna College of Engineering. His research interest includes Antenna, RF communication circuits and wireless sensors and network. He has published around 90 papers in conference and journals. His work has been cited in more than 500 research papers. He has 5 patents to his credit and also a author of a book Multilayer directional and planar omni-directional antennas.

Mirza Waseem Hussain received his MCA degrees from the University of Kashmir in 2013. Completed his Ph.D. degree in 2017 from the Department of Computer Sciences at Baba Ghulam Shah Badshah University, Rajouri, J&K, India. Currently working as lecturer In the PG department of Computer Sciences. He has published more than 14 papers in various International and peer reviewed journals. He is also member of The Society of Digital Information and Wireless Communications (SDIWC), International Association of Engineers (IAENG), IEEE. His current research interests include wireless networks and congestion control.

Muhammad Irfan obtained his Master's degree from COMSATS University Islamabad, Wah Campus. Currently, he is working with NRTC, Pakistan.

Ranjan Bala Jain is graduate in Electronics and Telecommunication and post graduate in optical fiber communication from SGSITS Indore. She obtained her doctrate degree from IIT Bombay. Her topic of thesis is "Capacity analysis in Cellular based wireless System". She worked in the capacity of lecturer, Assistant professor, and Associate professor in various institutes. She presented various around 42 papers in national, international conferences and journals. She got best paper award for analysis of intercell interference in relay based OFDMA system in international conference ICIP in Banglore 2011. She is member of ISTE, IETE, IWSA, UACEE etc. Her area of interest includes Resource allocation and performance modeling in Wireless networks, OFDMA system, Wireless Sensor Networks, Cognitive Radio. Presently she is working as professor in Dept. of EXTC in Vivekanand Education Society's Institute of Technology Chembur.

Sanjay Jamwal received his MCA degree in 1998 from Jammu University, MPhil degree in 2006 from Annamalai University and Ph.D degree in 2011 from Jammu University, India. He is currently working as Senior Assistant Professor in the Department of Computer Sciences with 20 years of teaching experience. He has published more than 30 papers in various International and peer reviewed journals. Presented more than 45 papers in various International and National conferences. He has attended more than 30 workshops on Computer science topics. He is also a lifetime member of Computer Society of India, Institution of Electronics and Telecommunication Engineers and Indian Science Congress. His research interest includes wireless networks, Sensor networks, Congestion control and e-governance.

Hakima Khelifi is currently pursuing a Ph.D. degree from Beijing Institute of Technology, Beijing, China. She received a B.Sc. degree in Computer Science (2013) from Constantine 2 University, Constantine, Algeria, and M.Sc. degree (2015) from Mohamed El Bachir El Ibrahimi University, Bordj Bou-Arreridj, Algeria. Her current research interest includes Next-Generation Networking and Internet, Internet of Things, Vehicular ad hoc networks, Information-Centric Networking and Named Data Networking. She is a Student Member of IEEE and ACM.

Senlin Luo received the B.E. and M.E. degrees from the college of Electrical and Electronic Engineering, Harbin University of Science and Technology, Harbin, China, in 1992 and 1995 respectively, and the Ph.D. degree from the School of Information and Electronics, Beijing Institute of Technology, Beijing, China, in 1998. He is currently a Deputy Director, Laboratory Director, and Professor of Information System and Security Countermeasures Experimental Center, Beijing Institute of Technology. His current research interests include Machine Learning, Medical Data Mining, and Information Security.

Daniel Minoli, Principal Consultant, DVI Communications, New York, has published 60 technical books, 300 papers and made 90 conference presentations. He has many years of technical and managerial experience in planning, designing, deploying, and operating secure IP/IPv6-, VoIP, telecom-, wireless-, satellite- and video networks for carriers and financial companies. He has published and lectured in the area of M2M/ IoT, network security, satellite systems, wireless networks, IP/IPv6/Metro Ethernet and has taught at New York University, Stevens Institute of Technology, and Rutgers University.

Tabasum Mirza has received a Master's Degree in Computer Applications from University of Kashmir in 2008. She is presently working as Lecturer in the Department of Computer Science, School Education, Government of Jammu and Kashmir, India. She has a 6.5 years experience of working in JK Bank Pvt. Ltd. Her specialization is software Engineering, Java Programming, Data Mining, and e-learning. She has published more about15 research articles/papers in the reputed national/international journals.

Sushruta Mishra has completed his Ph.D in Computer Science and Engineering from KIIT Deemed University. His area of Interests are Machine Learning and Data Science.

Hassine Moungla is Associate Professor at the University of Paris Descartes and a member of the Paris Descartes Computer Science Laboratory (LIPADE) since October 2008. He was a researcher at INRIA until 2008 and research fellow at CNRS-LIPN laboratory of the Paris Nord University until 2007. His research interests lie in the field of Wireless Area Body Networking (WBAN) for medical and health applications, Wireless Sensor Networking, QoS in WSN, Middleware for 5G Mobile and Sensor Networks. He participated and still participates to several national and international research projects. He is on the technical program committee of different ACM and IEEE conferences, including Globecom, ICC, WCNC, PIMRC, IWCMC and chaired some of their sessions. He is also reviewer on a regular basis for major international journals. Dr. Moungla is a member of IEEE and IEEE Communication Society.

Boubakr Nour is a student at Beijing Institute of Technology, Beijing, China. Previously, he received both M.Sc. and B.Sc. degrees with distinction in Computer Science from Djillali Liabes University of Sidi Bel Abbes, Algeria, in 2016, and 2014 respectively. His research interests include next-generation networking and Internet, information-centric networking, internet of things, edge computing, and wireless sensor networks. He served as a reviewer and TPC member on different journals, conferences, and workshops, including IEEE Communications Magazine, Internet of Things Journal, Transactions on Services Computing, VTC, ISNCC. He is a student member of the IEEE, and ACM.

Benedict Occhiogrosso is a Co-Founder of DVI Communications, New York. He is a graduate of New York University Polytechnic School of Engineering. His experience encompasses technical and managerial disciplines including sales, marketing, business development, systems development program management, procurement and contract administration. He also served as a testifying expert witness in cases encompassing patent infringement, and other legal matters.

Utkarsha S. Pacharaney received her M.E and B.E degree in Electronics and Telecommunication from the University of Mumbai in the year 2011 and 1999 respectively. She is currently working towards the PhD degree from the University of Mumbai. Her research interest includes wireless communication and sensor networks. She has a teaching experience of over 16 years and currently working with Datta Meghe College of Engineering, Airoli.

Lata L. Ragha has received her Ph. D. degree from Jadavpur University, Kolkata in 2011. She received her B. E. and M.Tech degree in Computer Science and Engineering from Karnatak University in 1987 and Vishveshwarai Technological University, Karnataka, in 2000 respectively. She is currently working as Professor and Head, Department of Computer Engineering at Fr. C. Rodrigues Institute of Technology, Vashi, Navi-Mumbai. Her research interests include Networking, Security, Internet Routing, and Data Mining. She has more than 100 research publications in International Journals and conferences.

Vaishali Sarbhukan has received B. E. Degree in Computer Science and Engineering and M. E. Degree in Computer Science from SSGMCE Shegaon, Amravati University in 2002 and Mumbai University in 2013 respectively. She is currently working towards Ph. D. Degree with Department of Information Technology, Terna Engineering College, Nerul, Mumbai University. She has published 28 papers in National and International conferences and journals. Also she has published one patent. She has 12 years teaching experience and one-year industry experience.

Archana Sharma has completed PhD in Computer Science. She has over 24 years of experience spanning the IT industry and academia in different capacities and she has published 33 research papers of which 12 are in international journals. She has also authored one text book for MCA and B.Tech. students. She has organised and attended various conferences, Faculty Development Programmes, workshops and seminars during her stint in different organisations and has been credited with awards and commendations. Her major areas of competencies include Advanced Database, DBMS, Distributed systems, Mobile Commerce, Operating systems and C++.

Jhum Swain is currently working as Assistant Professor in the Department of Computer Science and Engineering at Centurion University of Technology and Management, India. Her research interests are Mobile ad-hoc networks and computer security.

Meena T., VIT AP University, India, is currently pursuing PhD in Computer Science and Engineering from VIT AP University, India.

Index

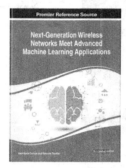
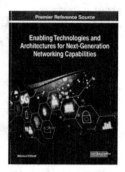

Ensure Quality Research is Introduced to the Academic Community

Become an IGI Global Reviewer for Authored Book Projects

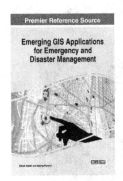
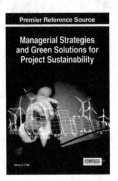

The overall success of an authored book project is dependent on quality and timely reviews.

In this competitive age of scholarly publishing, constructive and timely feedback significantly expedites the turnaround time of manuscripts from submission to acceptance, allowing the publication and discovery of forward-thinking research at a much more expeditious rate. Several IGI Global authored book projects are currently seeking highly-qualified experts in the field to fill vacancies on their respective editorial review boards:

Applications and Inquiries may be sent to:
development@igi-global.com

Applicants must have a doctorate (or an equivalent degree) as well as publishing and reviewing experience. Reviewers are asked to complete the open-ended evaluation questions with as much detail as possible in a timely, collegial, and constructive manner. All reviewers' tenures run for one-year terms on the editorial review boards and are expected to complete at least three reviews per term. Upon successful completion of this term, reviewers can be considered for an additional term.

If you have a colleague that may be interested in this opportunity,
we encourage you to share this information with them.

IGI Global Proudly Partners With eContent Pro International

Receive a 25% Discount on all Editorial Services

Editorial Services

IGI Global expects all final manuscripts submitted for publication to be in their final form. This means they must be reviewed, revised, and professionally copy edited prior to their final submission. Not only does this support with accelerating the publication process, but it also ensures that the highest quality scholarly work can be disseminated.

English Language Copy Editing

Let eContent Pro International's expert copy editors perform edits on your manuscript to resolve spelling, punctuaion, grammar, syntax, flow, formatting issues and more.

Scientific and Scholarly Editing

Allow colleagues in your research area to examine the content of your manuscript and provide you with valuable feedback and suggestions before submission.

Figure, Table, Chart & Equation Conversions

Do you have poor quality figures? Do you need visual elements in your manuscript created or converted? A design expert can help!

Translation

Need your documjent translated into English? eContent Pro International's expert translators are fluent in English and more than 40 different languages.

Email: customerservice@econtentpro.com   **www.igi-global.com/editorial-service-partners**

Printed in the United States
By Bookmasters